Lecture Notes in Artificial Intelligence   6877

Subseries of Lecture Notes in Computer Science

LNAI Series Editors

Randy Goebel
University of Alberta, Edmonton, Canada
Yuzuru Tanaka
Hokkaido University, Sapporo, Japan
Wolfgang Wahlster
DFKI and Saarland University, Saarbrücken, Germany

LNAI Founding Series Editor

Joerg Siekmann
DFKI and Saarland University, Saarbrücken, Germany

W0193329

Pascal Schreck   Julien Narboux
Jürgen Richter-Gebert (Eds.)

Automated Deduction in Geometry

8th International Workshop, ADG 2010
Munich, Germany, July 22-24, 2010
Revised Selected Papers

 Springer

Series Editors

Randy Goebel, University of Alberta, Edmonton, Canada
Jörg Siekmann, University of Saarland, Saarbrücken, Germany
Wolfgang Wahlster, DFKI and University of Saarland, Saarbrücken, Germany

Volume Editors

Pascal Schreck
LSIIT, Pôle API
Boulevard Sébastien Brant, BP 10413
67412 Illkirch Cédex, France
E-mail: schreck@unistra.fr

Julien Narboux
LSIIT, Pôle API
Boulevard Sébastien Brant, BP 10413
67412 Illkirch Cédex, France
E-mail: narboux@unistra.fr

Jürgen Richter-Gebert
Technische Universität München, Zentrum Mathematik (M10)
Lehrstuhl für Geometrie und Visualisierung
Boltzmannstraße 3
85748 Garching, Germany
E-mail: richter@ma.tum.de

ISSN 0302-9743   e-ISSN 1611-3349
ISBN 978-3-642-25069-9   ISBN 978-3-642-25070-5 (eBook)
DOI 10.1007/978-3-642-25070-5
Springer Heidelberg Dordrecht London New York

Library of Congress Control Number: 2011940216

CR Subject Classification (1998): I.2.3, I.3.5, F.4.1, F.3, G.2-3, D.2.4

LNCS Sublibrary: SL 7 – Artificial Intelligence

Typesetting: Camera-ready by author, data conversion by Scientific Publishing Services, Chennai, India

Printed on acid-free paper

Springer is part of Springer Science+Business Media (www.springer.com)

Preface

From July 22 to July 24, 2010, the Technische Universität München, Germany, hosted the eighth edition of the now well-established ADG workshop dedicated to Automatic Deduction in Geometry. From the first edition, which was held in Toulouse in 1996, to ADG 2010, a slow mutation has taken place. The workshop that was formerly centered around computer algebra became a larger forum where several communities could exchange new ideas coming from various domains, such as computer algebra, logic, computer-assisted proof, combinatorial geometry or even software development, but all focused on proof in geometry.

ADG 2010 was a fruitful meeting where 19 papers, from 22 submissions, were selected for presentation after a review process involving at least two reviewers per article. The set of presentations was completed by an invited talk given by Robert Joan-Arinyo from the Universitat Politècnica de Catalunya, Spain. ADG 2010 was also an enjoyable meeting thanks to the rigorous and flawless organization of the Munich team (see the Organizing Committee list).

After the meeting, a new call for papers was launched, accepting contributions not necessarily related to a presentation at ADG 2010.

The present volume of the LNAI series is the result of this selection process, which includes a new review process and discussions within the Program Committee. It is composed of 13 papers which present original research reflecting the current state of the art in this field. The following categorization proposes a key to understanding the papers. But, obviously as with all categorizations, it is rather arbitrary and it should not be taken strictly. Most papers can indeed also be considered from a radically different point of view.

Three papers deal with incidence geometry using some kind of combinatoric argument. Susanne Apel and Jürgen Richter-Gerbert explore two ways to automatically prove a geometric theorem by discovering cancellation patterns. Dominique Michelucci studies incidence geometry leading to two papers: one deals with an abstract notion of line and the other concerns human readable proofs in geometry.

Three papers fall in the domain of computer algebra. Daniel Lichtblau studies a problem related to the locus of the midpoint of a triangle in a corner, which is a variant of the "penny in a corner" problem, by using numeric, formal and graphical tools. Pavel Pech exposes a method to automatically prove theorems related to inequalities in geometry. Yu Zou and Jingzhong Zhang propose a way to generate readable proofs using the so-called Mass Point Method involving barycentric calculations with real or complex masses.

Four papers are more related to software implementation. Michael Gerhäuser and Alfred Wassermann present a Web-integrated software for dynamic geometry which includes a Gröbner-based tool able to compute plane loci. Fadoua Ghourabi, Tetsuo Ida and Asem Kasem expose methods to produce readable proofs of theorem within the Origami problematics. Pedro Quaresma describes TGTP—a library of problems for automated theorem proving in geometry. Phil Scott and Jacques Fleuriot present the concurrent implementation of a forward chaining algorithm in the Isabelle/HOL framework.

Last but not least, logic and proof assistants are the subject of three papers of this book. Following his own work on non-standard analysis, Jacques Fleuriot explores the foundations of discrete geometry in Isabelle/HOL. Laurent Fuchs and Laurent Théry represent here both the Coq and the geometric algebra communities by presenting the formalization in Coq of Grassmann Caley Algebra and its application to automatize the production of proofs in projective geometry. Sana Stojanović, Vesna Pavlović and Predrag Janičić expose a framework where coherent logic is used to implement a geometric prover able to deliver readable proofs.

Our gratitude goes to the Chairs of the previous editions of ADG. We thank them for their guidance and for having made ADG what it is now. We would also to thank the Program Committee and the numerous referees who did a lot of work to improve the quality of the workshop and of this book.

July 2011   Pascal Schreck

Organization

Organizing Committee

Jürgen Richter-Gebert (Germany), Chair
Jutta Niebauer (Germany)

Program Committee

Pascal Schreck (France), Chair
Hirokazu Anai (Japan)
Francisco Botana (Spain)
Jacques Fleuriot (UK)
Xiao-Shan Gao (China)
Predrag Janičić (Serbia)
Deepak Kapur (USA)
Ulrich Kortenkamp (Germany)
Montserrat Manubens (Spain)
Dominique Michelucci (France)
Bernard Mourrain (France)

Julien Narboux (France)
Pavel Pech (Czech Republic)
Tomás Recio (Spain)
Georg Regensburger (Austria)
Jürgen Richter-Gebert (Germany)
Meera Sitharam (USA)
Thomas Sturm (Spain)
Dongming Wang (France)
Bican Xia (China)

Invited Speaker

Robert Joan-Arinyo (Spain)

External Reviewers

Xiaoyu Chen
Oliver Labs
Filip Maric
Vesna Pavlović
Phil Scott

Sana Stojanović
John Sullivan
Hitoshi Yanami
Lu Yang
Christoph Zengler

Table of Contents

Cancellation Patterns
in Automatic Geometric Theorem Proving

Susanne Apel* and Jürgen Richter-Gebert

Technical University of Munich, Department of Mathematics,
Boltzmannstr. 3, 85748 Garching, Germany

Abstract. This article is about the equivalence of two seemingly different methods for proving incidence theorems in projective geometry. The
first proving method is essentially an algebraic certificate for the non-
existence of a counterexample—via biquadratic final polynomials [13].
For the second method the theorems of Ceva and Menelaus are elementary building blocks and are used as faces of an oriented topological
2-cycle, with their geometric structure on the edges identified appropriately. The fact that the cycle finally closes up translates into the proof
of the theorem. We start by formalizing both methods. After this we
present a bijective translation process that establishes the equivalence
of the two methods. The proving methods and the translation process
will be illustrated by a (quite well-natured) example. Using our methods
one gains additional structural insight in the purely algebraic proofs (biquadratic final polynomials) by reconstructing an underlying topological
structure of the proof.

1 Introduction

A quite general strategy for automatic proving in geometry can be paraphrased
as follows: "Translate the hypotheses and the conclusion of a theorem into
polynomials—search for an algebraic dependence which shows that the conclusion can be derived from the hypotheses". Various theorem provers follow this
general strategy. Depending on the concrete setup the main emphasis here is
either on the algebraic translation process (like determinant or exterior algebra
based approaches [13,17,20] or the "Area method" [8,9,10,11,7]) or on the process of finding dependencies (like in Groebner bases based approaches or Ritt's
characteristic set method), or on both. Despite the fact that the latter class
of methods is more general, it suffers from the effect that these proofs are often only checkable by a computer, and there may be no explicit control over
non-degeneracy conditions.

In this article we will deal with two proving techniques for which the main
emphasis is on finding an algebraic translation of a geometric theorem in a way
that a theorem can be proved by comparably simple cancellation arguments.

* The author gratefully acknowledge the support of the TUM's Thematic Graduate
Center TopMath at Technische Universität München.

P. Schreck, J. Narboux, and J. Richter-Gebert (Eds.): ADG 2010, LNAI 6877, pp. 1–33, 2011.
© Springer-Verlag Berlin Heidelberg 2011

Both methods are capable of generating human readable proofs that often provide additional insight in the structure of the theorem under consideration. The first method we will deal with is known as the *binomial proving technique* (compare [13,3]). It has its origins in the automatic generation of non-realizability proofs for oriented matroids [2] and has proven to be a powerful tool in this context. The second method is known as *Ceva-Menelaus-technique* [16] and is based on cyclic, manifold like structures build from triangles that are induced by the incidence structure of the geometric theorem. These cycles shed additional light on the structure of the theorems.

In fact it turns out that both methods are essentially equivalent (under the mild assumption that one is allowed to add auxiliary points in generic position, to impose natural consistency and generic non-degeneracy assumptions on the theorem and to express the conclusion by an equivalent one). In fact, the equivalence of both methods is surprising. To see this one has to know that the binomial proving method creates many equations of the form *"products of determinants = products of determinants"* and then—more or less blindly—searches for a dependence among these expressions by solving a rather big system of associated linear equations. Compared to this a Ceva-Menelaus proof is much more structured and in a sense synthetic. It consists of a concrete combinatorial manifold composed from triangles whose vertices are associated with certain vertices of the configuration (for details see below). Each triangle of the manifold is associated to either a Ceva or a Menelaus configuration. The fact that two triangles share an edge will be interpreted on two different levels. In the geometry of the theorem it corresponds to the fact that a certain coincidence holds, on an algebraic level it translates into the cancellation of a certain term. By this incidences become associated with possible algebraic cancellations. The cycle structure of the oriented manifold then translates to the existence of a global cancellation pattern that proves the theorem. Compared to the binomial proving technique this method is by far more structured and at first sight seems to be much more restrictive. Nevertheless it turns out that whenever a binomial proof is found the cancellation pattern (in a non-obvious way) translates into the existence of a manifold on which a Ceva-Menelaus proof can be based and vice versa.

In the following sections we will demonstrate the two proving techniques and the translation method, along with an instructive (though still well-natured) example. We will also introduce the concept of a base graph (see [4]) that plays a crucial role in the translation process.

2   Definitions: Theorems and Proving Techniques

We will exemplify some concepts of this article in the case of a simple running example. This is the well known theorem of Pappos in its real projective version as illustrated in Figure 1. Whenever we use it we will refer to the labeling of this picture. In a very precise sense Pappos's theorem is the smallest theorem that is only based on incidence relations between points and lines. For a more complicated example see [16]. Pappos's theorem states that if we have two triples

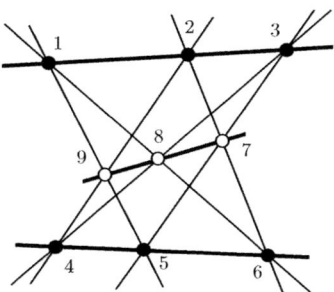

Fig. 1. The claim of Pappos's theorem is that the white points are collinear

$\{1,2,3\}$ and $\{4,5,6\}$ of collinear points then the points 7, 8 and 9 as constructed in Figure 1 are collinear as well (provided that in the construction process no degeneracy like intersecting two identical lines arises). Similarly one can say that as long as no two lines in the theorem coincide the collinearity of the eight triples

$$\mathbf{H} = \big\{\{1,2,3\}, \{1,5,9\}, \{1,6,8\}, \{2,4,9\}, \{4,5,6\}, \{3,4,8\}, \{2,6,7\}, \{3,5,7\}\big\}$$

implies the collinearity of $C = \{7,8,9\}$. In our modeling of an incidence assertion $\mathcal{T} = (\mathbf{H}, \mathbf{B}, C)$ we will express the hypotheses by a set $\mathbf{H}$ of triples of points, indicating which points *should* be collinear, along with a set $\mathbf{B}$ of triples (the non-degeneracy conditions) that *should not* be collinear. The conclusion will be expressed by a single triple C. One should keep in mind that formulating an assertion does not claim anything about its *validity*. We call the list of points $P = (p_1, \ldots, p_n)$ with concrete coordinates an *instance* of the assertion if it satisfies the collinearities $\mathbf{H}$ and the non-collinearities $\mathbf{B}$. The assertion $\mathcal{T}$ is *valid* (and by this becomes a theorem), if for every instance of $\mathcal{T}$ also the triple indicated by the conclusion C is collinear.

In the geometry literature quite frequently geometric theorems are stated without specific information on non-degeneracy conditions. So in a very strict sense it often happens that theorems are simply stated falsely, since some implicitly assumed and sometimes obvious non-degeneracy conditions may be missing. In our setup we will impose certain *generic* non-degeneracy assumptions that are derived directly from $\mathbf{H}$ and C. These generic non-degeneracies will mainly assert that different points on a line of the theorem are also geometrically distinct and that two lines through a theorem point are also distinct. Only in very rare cases it will be necessary to add additional non-degeneracy assumptions beyond these generic ones. Details on the representation of incidence theorems will be given later in Section 2.2.

2.1   The Binomial Proving Method

Let us now explain how a binomial proof works. Our entire considerations will be valid for incidence theorems in projective planes over an arbitrary field $\mathbb{K}$. However, to be more illustrative we here focus on the real projective plane $\mathbb{RP}^2$.

We will assume that we have an index set $E_n := \{1, \ldots, n\}$ whose members serve as the labels of corresponding point configuration $P = (p_1, \ldots, p_n) \in \mathbb{R}^{3 \cdot n}$ (here each point is represented by its homogeneous coordinates). We use $[a, b, c]_P := \det(p_a, p_b, p_c)$ as a shortcut for the determinant of the homogeneous coordinates of three points. Later on we will use $[a, b, c]$, without a subscript as a purely formal symbol, that models the behavior of an abstract determinant. In the real projective plane we have that the triple of points labeled by $\{a, b, c\}$ is collinear if and only if $[a, b, c]_P = 0$. Now for each collinearity $\{a, b, c\} \in \mathbf{H}$ in the hypotheses of a theorem, the binomial proving method considers equations of the form $[a, b, x]_P[a, c, y]_P = [a, b, y]_P[a, c, x]_P$, the *biquadratic equations*. Usually we will assume that the points x and y are not incident with the line abc, but for the moment this is not yet important. We can interpret biquadratic equations as fragments of Grassmann-Plücker relations in the following way. Since for arbitrary points in $\mathbb{RP}^2$ the Grassmann-Plücker relation

$$[a, b, c]_P[a, x, y]_P - [a, b, x]_P[a, c, y]_P + [a, b, y]_P[a, c, x]_P = 0$$

always holds (see for instance [15]), we get the equivalence

$$[a, b, x]_P[a, c, y]_P = [a, b, y]_P[a, c, x]_P \iff \Big(\{a, b, c\} \text{ or } \{a, x, y\} \text{ are collinear} \Big).$$

Observe that the above statement is invariant under a simultaneous exchange of $b \leftrightarrow x$ and $c \leftrightarrow y$.

The binomial proving method (see references [14,13,3]) now creates (usually by an algorithm) binomial expressions from the collinearity hypotheses $\mathbf{H}$ and from the conclusion C in a way that "obviously" the conclusion can be expressed by a combination of the hypotheses. A proof of Pappos's theorem along these lines may look as follows. From the set $\mathbf{H}$ of collinearities we can conclude the following eight biquadratic equations.

$$\{1, 2, 3\} \text{ collinear} \implies [1, 2, 4]_P[1, 3, 7]_P = [1, 2, 7]_P[1, 3, 4]_P,$$
$$\{1, 5, 9\} \text{ collinear} \implies [1, 5, 4]_P[1, 9, 7]_P = [1, 5, 7]_P[1, 9, 4]_P,$$
$$\{1, 6, 8\} \text{ collinear} \implies [1, 8, 4]_P[1, 6, 7]_P = [1, 8, 7]_P[1, 6, 4]_P,$$
$$\{2, 4, 9\} \text{ collinear} \implies [4, 2, 7]_P[4, 9, 1]_P = [4, 2, 1]_P[4, 9, 7]_P,$$
$$\{4, 5, 6\} \text{ collinear} \implies [4, 5, 7]_P[4, 6, 1]_P = [4, 5, 1]_P[4, 6, 7]_P,$$
$$\{3, 4, 8\} \text{ collinear} \implies [4, 8, 7]_P[4, 3, 1]_P = [4, 8, 1]_P[4, 3, 7]_P,$$
$$\{2, 6, 7\} \text{ collinear} \implies [7, 2, 1]_P[7, 6, 4]_P = [7, 2, 4]_P[7, 6, 1]_P,$$
$$\{3, 5, 7\} \text{ collinear} \implies [7, 5, 1]_P[7, 3, 4]_P = [7, 5, 4]_P[7, 3, 1]_P.$$

The derived equations are automatically valid for every instance of Pappos's theorem. We now assert (as additional non-degeneracy assumptions) that all determinants that occur in these expressions are non-zero. This is not a strong restriction. As one can easily check this only implies that the lines in Figure 1 do not coincide. Now, multiplying all left sides and all right sides of these equations and canceling determinants that occur on both sides (we can do this since we assumed that they are non-zero) we are left with

$$[7, 8, 4]_P[7, 9, 1]_P = [7, 8, 1]_P[7, 9, 4]_P$$

This in turn is again a biquadratic equation and so it is equivalent to the fact that either $\{7, 8, 9\}$ is collinear or that $\{7, 1, 4\}$ is collinear. The first one is the desired conclusion. The second possibility has to be excluded by another non-degeneracy condition that $\{1, 4, 7\}$ is non-collinear. All in all this argument shows that under the assumptions—which are encoded by the eight Pappos hypothesis collinearities $\mathbf{H}$ and a non-degeneracy list $\mathbf{B}$ (which contains all the triples corresponding to determinants in the above equations together with $\{1, 4, 7\}$)— in every instance also the collinearity of the conclusion $C = \{7, 8, 9\}$ is established.

We are now going to formalize this proving pattern for general incidence theorems. For this we start by formalizing the exact notion of a theorem.

Definition 1. *A real projective incidence assertion $\mathcal{T}$ on n points is a triple $(\mathbf{H}, \mathbf{B}, C)$ such that:*

- $\mathbf{H}, \mathbf{B} \subset \{\{i, j, k\} \mid 1 \leq i < j < k \leq n\}$
- $C = \{a, b, c\}$ *with* $1 \leq a < b < c \leq n$ *and* $C \notin \mathbf{H}$.

A point configuration $P = (p_1, \ldots, p_n)$ in $\mathbb{RP}^2$ is called an instance *of $\mathcal{T}$ if $[i, j, k]_P = 0$ for all $\{i, j, k\} \in \mathbf{H}$ and $[i, j, k]_P \neq 0$ for all $\{i, j, k\} \in \mathbf{B}$. If in addition for every instance $P = (p_1, \ldots, p_n)$ of $\mathcal{T}$ also $[a, b, c]_P = 0$ holds, then $\mathcal{T}$ is called a* valid assertion *or a* theorem.

Later on we will consider only theorems satisfying additional consistency assumptions (we will call them *theorems in canonic saturated form*). However, for the time being it is sufficient to just stick to the above more general definition. Formalizing the notion of a biquadratic proof we have to face an additional technical problem. In the proof the determinants play merely the role of formal symbols. It is never necessary to actually evaluate a determinant since the cancellation process entirely takes place on the level of the determinants. To deal with this we will introduce formal symbols (called *brackets*) $[a, b, c]$ (without subscript) that will play the role of formal determinants and inherit from real determinants only the properties that are relevant for our proofs. In particular these symbols should satisfy the alternating determinant rules:

$$[a, b, c] = [b, c, a] = [c, a, b] = -[b, a, c] = -[c, b, a] = -[a, c, b].$$

Furthermore for the signature we must impose the canonical rule "$- \cdot - = +$". There are several approaches to deal with formal determinant-like expressions as we need them for our purposes. One approach was introduced by Neil White [20] and is known as the *bracket ring*. There one defines a ring over all formal determinant symbols and "mods out" all natural relations known to hold (permutation rules, Grassmann-Plücker relations, collinearities.) The other one is due to Andreas Dress and Walter Wenzel [4,5,6,18,19]. They define several variations of what is known as the *Tutte group of a matroid*. There all calculations take place in a group (not a ring). A symbol ϵ with the property $\epsilon^2 = 1$ is introduced playing the role of a formal "-1". Equality of product terms is expressed as certain fractions in this group being 1. While in the bracket ring approach it is easy to deal with formal Grassmann-Plücker relations, in the Tutte group approach there is a natural notion of division (fractions will turn out to be important later in this article). Our approach will be close to the one of Dress and Wenzel, however restricted to the necessities of our treatment.

We will introduce a multiplicative group in which identifications are made that model the above rules. The (formal) brackets that are allowed to occur in this group are those that correspond to the non-degeneracy assumptions in **B**, since these brackets posses a natural multiplicative inverse. An additional special element ϵ will play the role of a formal -1. We define a set of formal group elements (brackets):

$$Br(\mathbf{B}) := \{[a, b, c] \mid \{a, b, c\} \in \mathbf{B}\} \cup \{\epsilon\}.$$

We consider the free Abelian (multiplicatively written) group $\mathcal{F}_\mathbf{B} = (Br(\mathbf{B}), \cdot)$ over these elements and define a subgroup that models the determinant identities:

$$\mathcal{A}_\mathbf{B} := \left\langle \left\{ \frac{[a, b, c]}{[b, c, a]} \mid \{a, b, c\} \in \mathbf{B} \right\} \cup \left\{ \epsilon \cdot \frac{[a, b, c]}{[b, a, c]} \mid \{a, b, c\} \in \mathbf{B} \right\} \cup \{\epsilon^2\} \right\rangle.$$

Here $\langle \ldots \rangle$ denotes the group generated by elements in $\mathcal{F}_\mathbf{B}$. For each instance P (over the reals) for which all the brackets in **B** are non-zero there is a natural homomorphism Φ_P from the factor group $\mathcal{B}_\mathbf{B} := \mathcal{F}_\mathbf{B}/\mathcal{A}_\mathbf{B}$ to $(\mathbb{R} \setminus \{0\}, \cdot)$ induced by the relations $\Phi_P([a, b, c]) = [a, b, c]_P$ and $\Phi_P(\epsilon) = -1$.

The collinearities stated in the hypotheses **H** imply (via the formal equivalent of Grassmann-Plücker relations) additional relations that we can impose on $\mathcal{B}_\mathbf{B}$. We collect such elements of $\mathcal{B}_\mathbf{B}$ that must be 1 due to collinearity conditions **H** in a set $\mathcal{H}_{\mathbf{H},\mathbf{B}}$.

$$\mathcal{H}_{\mathbf{H},\mathbf{B}} := \left\{ \frac{[a, b, x][a, c, y]}{[a, b, y][a, c, x]} \in \mathcal{B}_\mathbf{B} \ \middle| \ \{a, b, c\} \in \mathbf{H}; x, y \in E_n \right\}$$

The fractions occurring in this set are derived from the equations $[\ldots][\ldots] = [\ldots][\ldots]$ we considered so far. Under our canonical homomorphism Φ_P the images of elements of $\mathcal{H}_{\mathbf{H},\mathbf{B}}$ will automatically turn out to be 1 for every instance of $\mathcal{T} = (\mathbf{H}, \mathbf{B}, C)$. As an immediate consequence of the definitions we get:

Lemma 1. *Let P be an instance of $\mathcal{T} = (\mathbf{H}, \mathbf{B}, C)$ and Φ_P be the corresponding homomorphism, then for every element $\alpha \in \mathcal{H}_{\mathbf{H},\mathbf{B}}$ we have $\Phi_P(\alpha) = 1$.*

Now we define a similar set of fractions $\mathcal{C}_{C,\mathbf{B}}$ that express the conclusion:

$$\mathcal{C}_{C,\mathbf{B}} := \left\{ \frac{[a, b, x][a, c, y]}{[a, b, y][a, c, x]} \in \mathcal{B}_\mathbf{B} \ \middle| \ \{a, b, c\} = C; \{a, x, y\} \in \mathbf{B} \right\}$$

Again we get an immediate consequence of our definitions.

Lemma 2. *Let P be an instance of $\mathcal{T} = (\mathbf{H}, \mathbf{B}, C)$ and Φ_P be the corresponding homomorphism. If we there is $\alpha \in \mathcal{C}_{C,\mathbf{B}}$ with $\Phi_P(\alpha) = 1$ then the triple corresponding to C must be collinear in P.*

Now a binomial proof of $\mathcal{T} = (\mathbf{H}, \mathbf{B}, C)$ corresponds to identifying an element of $\mathcal{C}_{C,\mathbf{B}}$ as an element in the subgroup $\langle \mathcal{H}_{\mathbf{H},\mathbf{B}} \rangle$.

Definition 2. *A binomial proof for an incidence assertion $\mathcal{T} = (\mathbf{H}, \mathbf{B}, C)$ consists of $\alpha \in \mathcal{H}_{\mathbf{H},\mathbf{B}}$ and $b_i \in \mathcal{H}_{\mathbf{H},\mathbf{B}}$ pairwise distinct (for $1 \leq i \leq k$, $k \in \mathbb{N}_0$) and an exponent vector $(e_1, \ldots, e_k) \in \mathbb{N}_0^k$ which witness $\mathcal{C}_{C,\mathbf{B}} \cap \langle \mathcal{H}_{\mathbf{H},\mathbf{B}} \rangle \neq \emptyset$, i.e.*

$$\alpha = b_1^{e_1} \cdot b_2^{e_2} \cdots b_k^{e_k}$$

In addition the exponent vector $(e_1, \ldots, e_k) \in \mathbb{N}_0^k$ is assumed to be be minimal with this property and with respect to the partial order induced by "$<$" componentwise. The elements of $\mathcal{H}_{\mathbf{H},\mathbf{B}} \cup \mathcal{C}_{C,\mathbf{B}}$ are called biquadratic fractions.

As a direct consequence of Lemma 1 and Lemma 2 we get:

Theorem 1. *If a projective incidence assertion $\mathcal{T} = (\mathbf{H}, \mathbf{B}, C)$ has a binomial proof, then it is a theorem.*

Let us spend a word on the automated finding of a proof with the above structure. If we consider a collinearity of three points (a, b, c) and consider all binomial expressions that arise from considering three term Grassmann-Plücker relations involving these three points and two other configuration points there are altogether $3 \cdot \binom{n-3}{2}$ biquadratic equations that are consequences of this hypothesis. For Pappos's theorem this makes altogether $8 \cdot 3 \cdot \binom{6}{2} = 360$ relations coming from all hypotheses. Then one has to search for a way to express a binomial equation representing the conclusion as a combination of these hypotheses. This problem can in principle be attacked by solving systems of linear equations (for this consider the exponent vectors of the bracket equations involved). In [14] a detailed complexity analysis of this scenario is given that also includes a few tricks to cut down the hypothesis space.

Nevertheless finding a cancellation pattern falls back to an algorithmic process on an in general quite huge (though still polynomial-sized) space of binomials describing the hypotheses. It seems kind of unreasonable to consider every possible cancellation pattern between these equations. A big question is whether one can take advantage of the incidence structure in advance and by this cut down the algorithmic complexity of the search process.

2.2   How to Represent a Theorem?

As we already mentioned, non-degeneracy conditions like non-coinciding points, or non-coinciding lines are quite often implicitly assumed without stating them explicitly. We here will introduce a set of *generic non-degeneracies* that are directly generated from the set $\mathbf{H}$ and the conclusion C, which will model exactly these assumptions. In fact, if we use these generic non-degeneracy assumptions they will ensure that sufficiently many brackets in the equations of potential binomial proof will not vanish (see [14]). We will introduce a notion of *saturation* for the collinearities $\mathbf{H}$, which models that trivial conclusions are already taken.

In addition we impose consistency conditions on the non-degeneracy assumptions **B**. We start by explaining the saturation process. Under the assumption that all points on a line are distinct one can conclude from $\{a, b, c\}, \{b, c, d\} \in \mathbf{H}$ that p_a, p_b, p_c, p_d lie on a common line in each instance $P = (p_1, \ldots, p_n)$. So we could w.l.o.g. assume that $\{a, b, d\} \in \mathbf{H}$, too. The collinearities **H** are *saturated*, if

$$\{a, b, c\} \in \mathbf{H} \text{ and } \{b, c, d\} \in \mathbf{H} \implies \{a, b, d\} \in \mathbf{H}.$$

Note that we still require $C \notin \mathbf{H}$, even after saturation. We now can express the generic non-degeneracy assumptions. For this we first consider a set $\mathbf{A} = \mathbf{A}(\mathbf{H}, C)$ of triples, which is made saturated by the same rule but with groundset $\mathbf{H} \cup \{C\}$.

We define the **A**-flat (i.e. the maximal sets of dependent elements) supported by a triple $\{a, b, c\} \in \mathbf{A}$ by

$$f(\{a, b, c\}) := \{i \mid \{a, b, i\} \in \mathbf{A}\}.$$

f is well-defined and does not depend on the order of a, b, c since **A** is saturated. We call these **A**-flats *derived lines* and collect them in a set $\mathcal{G}$

$$\mathcal{G} := \big\{ f(\{a, b, c\}) \mid \{a, b, c\} \in \mathbf{A} \big\}.$$

The derived lines correspond to the lines visible in the drawing of an instance of the theorem. Within this setting we also require a consistency for the set **B** of non-degeneracy conditions. The idea is as follows: we can paraphrase $\{a, b, c\} \in \mathbf{B}$ as p_c not lying on the line spanned by p_a, p_b. So if there is another (different) point p_i $(1 \leq i \leq n)$ on this line, $[a, i, c]_P$ will not evaluate to zero either. We will call **B** *consistent* if

$$\{a, b, c\} \in \mathbf{B} \text{ and } \{a, b, i\} \in \mathbf{A} \Rightarrow \{a, i, c\} \in \mathbf{B}$$

Now we are able to describe the generic non-degeneracy assumptions $\mathbf{B}(\mathbf{A})$. They ensure that no two (derived) lines collapse if they meet in a theorem point. This is only the dual version of the assumption that no two points on a line indicated by **H** should coincide. So $\mathbf{B}(\mathbf{A})$ consists of all triples which would indicate a collapse of two lines (see also Figure 2).

$$\mathbf{B}(\mathbf{A}) := \Big\{ \{a, b, c\} \mid \exists G, H \in \mathcal{G} \text{ with } |G \cap H| = 1$$
$$\text{and } a \in G \setminus H, \ b \in H \setminus G, \ c \in H \cup G \Big\}$$

Observe that if for a configuration the generic non-degeneray assumptions $\mathbf{B}(\mathbf{A})$ are satisfied and a derived line G intersects at least one other derived line H, then automatically all points on G are distinct. Isolated lines that are not incident with another derived line cannot be of any relevance for the theorem, and we can w.l.o.g. neglect them completely. Furthermore the non-degeneracies $\mathbf{B}(\mathbf{A})$ are by definition automatically consistent. From now on, we will consider only theorems in canonic saturated form as defined as follows.

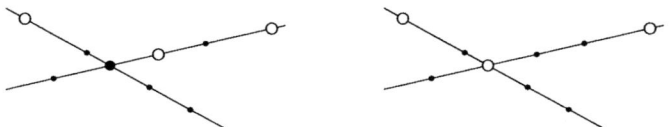

Fig. 2. Two non-degeneracy triples (white points)

Definition 3. *Let $\mathcal{T} = (\mathbf{H}, \mathbf{B}, C)$ be a theorem and let $\mathbf{A} = \mathbf{A}(\mathbf{H}, C)$ be defined as above. We say that it is in* canonic saturated form, *if*

- *the set $\mathbf{H}$ is saturated,*
- *with $\mathbf{B}(\mathbf{A})$ defined as above $\mathbf{B}(\mathbf{A}) \subseteq \mathbf{B}$,*
- *$\mathbf{B}$ is consistent,*
- *there is no derived line which is disjoint from any other derived line,*
- *$\mathbf{B}$ and $\mathbf{A}$ are disjoint.*

Remark 1. If one prefers not to deal with generically generated non-degeneracies one may weaken the assumptions on $\mathbf{B}$ and not require that $\mathbf{B}(\mathbf{A}) \subset \mathbf{B}$ but only the consistency of $\mathbf{B}$. In this case, one has to additionally ensure that no two points on common derived lines can coincide in any instance. However, if one wants to derive a binomial proof in this wider setup, one has to explicitly ensure every non-degeneracy of brackets involved in biquadratic fractions—in contrast to the situation with $\mathbf{B}(\mathbf{A})$.

It might be the case that the conclusion line contains more than three points. In this case there are several triples of points which can play the role of the conclusion. It might also be the case that we can conclude other collinearities from C and $\mathbf{H}$ by applying the rules that saturate sets of triples about collinearities. Theorems with C changed in this way are considered as equivalent:

Definition 4. *Let $\mathcal{T} = (\mathbf{H}, \mathbf{B}, C)$ be a real projective theorem. Then a theorem $\mathcal{T}' = (\mathbf{H}, \mathbf{B}, C')$ is* equivalent *to $\mathcal{T}$ if $C' \in \mathbf{A}(\mathbf{H}, C)$ and $C \in \mathbf{A}(\mathbf{H}, C')$.*

2.3   The Ceva-Menelaus Proving Method

The Idea: We now discuss another proving method that by definition considers only cancellation patterns of points that are in a certain sense "close" in the incidence graph of the theorem. For this proving method consider the well known (affine) theorems of Ceva and Menelaus:

- Ceva's theorem states that if in a triangle the sides are cut by three concurrent lines that pass through the corresponding opposite vertex, then the product of the three (oriented) length ratios along each side equals 1.
- Menelaus's theorem states that this product is -1 if the cuts along the sides come from a single line.

Both situations are illustrated in Figure 3.

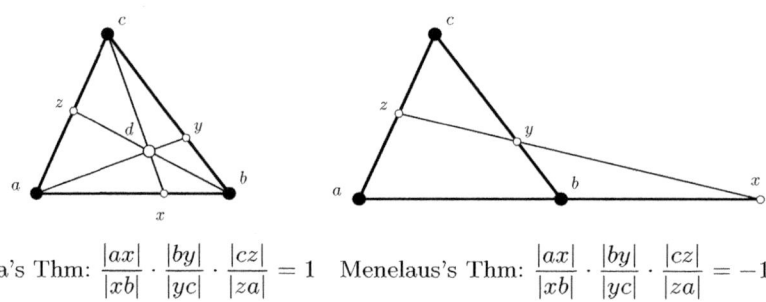

Ceva's Thm: $\dfrac{|ax|}{|xb|} \cdot \dfrac{|by|}{|yc|} \cdot \dfrac{|cz|}{|za|} = 1$ Menelaus's Thm: $\dfrac{|ax|}{|xb|} \cdot \dfrac{|by|}{|yc|} \cdot \dfrac{|cz|}{|za|} = -1$

Fig. 3. Theorems of Ceva and Menelaus

For the moment we assume that in any instance P of a Ceva or Menelaus configuration each point p is finite and represented by special homogeneous coordinates of the form $(p_x, p_y, 1)$. Then the concrete determinant $[a, b, c]_P$ for the instance P equals $2 \cdot \text{area}(a, b, c)$ and we can express the ratios of oriented lengths on a line also by ratios of suitable determinants.

$$\frac{|ax|}{|xb|} = \frac{[a, \, u, v]_P}{[b, \, v, u]_P} \qquad \text{where } u \text{ and } v \text{ span a line which intersects } \overline{ab} \text{ in } x \qquad (1)$$

(compare this also with the "area principle" of [7,8,9,10]). Using this identity we can rewrite the expressions in the theorems of Ceva and Menelaus as quotients of products of determinants. By this we obtain almost trivial proofs of these two theorems. (Observe that the numerator and denominator of the expressions obviously cancel up to the corresponding sign—for the labeling see again Figure 3):

$$\frac{[a, d, c]_P}{[b, c, d]_P} \cdot \frac{[b, d, a]_P}{[c, a, d]_P} \cdot \frac{[c, d, b]_P}{[a, b, d]_P} = +1 \quad \Longleftrightarrow \quad \frac{[a, \, c, d]_P}{[b, \, c, d]_P} \cdot \frac{[b, \, a, d]_P}{[c, \, a, d]_P} \cdot \frac{[c, \, b, d]_P}{[a, \, b, d]_P} = -1 \tag{2}$$

in the Ceva case and

$$\frac{[a, d, e]_P}{[b, e, d]_P} \cdot \frac{[b, d, e]_P}{[c, e, d]_P} \cdot \frac{[c, d, e]_P}{[a, e, d]_P} = -1 \quad \Longleftrightarrow \quad \frac{[a, \, d, e]_P}{[b, \, d, e]_P} \cdot \frac{[b, \, d, e]_P}{[c, \, d, e]_P} \cdot \frac{[c, \, d, e]_P}{[a, \, d, e]_P} = +1 \tag{3}$$

in the Menelaus case, where d and e are two arbitrary distinct points on the line x, y, z. Observe that all these equations can again be modeled in the language of the multiplicative group $\mathcal{B}_{\mathbf{B}}$ with $\mathbf{B}$ appropriately chosen.

We will call the terms on the right in (2) resp. (3) *Ceva* or *Menelaus expressions*, respectively. It is important to notice that in these expressions each point occurs as often in the numerator as in the denominator. By this the expression becomes projectively invariant (i.e. it is stable under projective transformations and rescaling of the homogeneous coordinates). This follows from standard arguments for bracket invariants [15]. So we can drop the assumption that the points were represented in the standard embedding of the form $(p_x, p_y, 1)$. Also the expressions are invariant under various obvious permutations of the points involved (cyclic rotation of a, b, c, transposition of d, e in the Menelaus case).

By introducing determinants we also eliminated the explicit occurrence of the *edge points* x, y and z from the representations of the algebraic expression. They are represented implicitly by the corresponding *cutting lines*. (We will use this terminology of edge points and cutting lines later on.)

We will now describe the fundamental idea of a Ceva-Menelaus proof. For this idea the edge points will play a crucial role. We start by explaining the argument on the level of oriented lengths. Later on we will have to switch to the bracket representation in order to make the connection to binomial proofs. As a binomial proof, a Ceva-Menelaus proof does also work with specified cancellation patterns. However, here the primary objects are *ratios of determinants*. The construction technique presented below will produce incidence theorems whose proof is already implicitly given by their construction. The following process illustrates the argument:

Start with any triangulated oriented combinatorial 2-manifold. The structure of this manifold serves as a kind of *frame* for the construction of an incidence theorem. Consider this manifold as being realized by flat triangles (it does not matter if these triangles intersect, coincide or are coplanar as long as they represent the combinatorial structure of the manifold, i.e. triangles that share an edge in the manifold must also share an edge in the geometric realization). Since we here deal with theorems in the plane, we embed these triangles in $\mathbb{R}^2$. Let us be concrete and take the projection of a tetrahedron (a, b, c, d) to $\mathbb{R}^2$. Now we chose one additional point on each of the edges of the realized manifold. So in our example we choose six points u, v, w, x, y, z, one on each of the edges of the projected tetrahedron. Assume that for three of the faces these points satisfy Ceva's condition. Then they automatically satisfy Ceva's condition also for the last face—an incidence theorem.

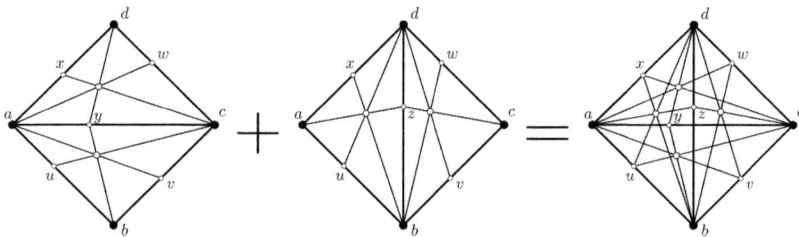

The proof of this theorem is almost obvious from the algebraic characterization of Ceva's condition. Consider the following formula that arises by multiplying the four Ceva conditions.

$$\left(\frac{|au|}{|ub|} \cdot \frac{|bv|}{|vc|} \cdot \frac{|cy|}{|ya|}\right) \cdot \left(\frac{|cw|}{|wd|} \cdot \frac{|dx|}{|xa|} \cdot \frac{|ay|}{|yc|}\right) \cdot$$

$$\left(\frac{|ax|}{|xd|} \cdot \frac{|dz|}{|zb|} \cdot \frac{|bu|}{|ua|}\right) \cdot \left(\frac{|bz|}{|zd|} \cdot \frac{|dw|}{|wc|} \cdot \frac{|cv|}{|vb|}\right) = 1. \qquad (4)$$

This formula obviously already holds on a symbolic level, since each of the oriented lengths occurs once in the numerator and once in the denominator. On the other hand each of the four factors being 1 states the Ceva condition for one of the faces. Thus three of these conditions imply the last one. The essential fact about the structure of this proof is that whenever two faces meet in an edge the two corresponding ratios cancel. In general we obtain:

> For any triangulated oriented 2-manifold we may choose a point on each edge such that for every except one face either a Ceva or a Menelaus condition is generated. Then the edge points of the last triangle auto-matically support a Ceva or Menelaus configuration such that the total number of Menelaus configurations will be even.

Notice: for stating equations like (4) we need the orientability of the manifold: it is necessary to equip each triangle with an orientation in such a way that triangles traverse a common edge in different directions. Many details on this proving pattern can be found in [16].

Instead of "filling the last triangle" there is also a different way of looking at this proving method. The essence of the argument is that if triangles sup-porting Ceva or Menelaus configurations are glued along their edges such that corresponding edge points along edges are identified, then the corresponding edge ratios along the edges cancel. So finally consider such a structure having a *boundary*. The product of the corresponding ratios of this of the boudary must be either $+1$ or -1 depending on whether the number of Menelaus configura-tions involved is even or odd. The fact that the final triangle can be consistently equipped with a Ceva or Menelaus configuration comes from the fact that the product along a 3-cycle of boundary edges is $+1$ or -1. The same argument can be applied to arbitrary cycle lengths along a boundary. For a 2-cycle we obtain the following version of the same argument:

> Assume that in a triangulated oriented 2-manifold each triangle is equipped with a Ceva or a Menelaus configuration such that for all but one of the edges the corresponding edge points of adjacent triangles are identified. Then, if the number of Menelaus configurations is even, the two edge points on the final edge coincide automatically.

We will see that this version of the argument is already very close to the con-struction of binomial proofs.

Ceva-Menelaus-theorems: We will call an incidence structure constructed in the way we just described a *CM-theorem* (independent from our previous notion of a theorem from Section 2.2). We briefly summarize all the important points we need for such a theorem. First we collect the necessities of the framing manifold like structure.[1]

(i) An index set $V_{\mathcal{M}} = \{v_1, \ldots, v_n\}$ that plays the role of the vertices of the underlying manifold like structure.

---

[1] There is also an alternative way to formulate these by triangulated CW complexes, but we will avoid these rather subtle topological approach here.

(ii) A list of abstract triangles $T = (t_1, \ldots, t_k)$; $t_i \in V^3_{\mathcal{M}}$; $|t_i| = 3$.
   A triangle is explicitly allowed to occur multiple times in this list.

(iii) The oriented edges of a triangle $t = (i, j, k)$ are $\partial(t) := \{(i, j), (j, k), (k, i)\}$.

(iv) In the family of all edges of triangles there should exist a (further on fixed) matching such that each edge (a, b) is matched with an edge with same endpoints but in opposite direction (b, a) (this models the gluing process).

(v) The chosen matching induces a graph structure with nodes in T, where two triangles are connected if they are matched by at least one edge. This graph must be connected.

(vi) For each triangle in $t \in T$ we must specify whether it is of type $\mathfrak{C}$ (Ceva) or of type $\mathfrak{M}$ (Menelaus).

(vii) The number of Menelaus triangles must be even.

So far these properties just encode the underlying triangle structure. They do not carry incidence information yet. We now add incidence information by introducing edge points and Ceva or Menelaus configurations for each triangle. For this we also have to specify what should be considered as an instance of the so far purely combinatoric description of this framework. For the frame described above this is just an assignment of positions in $\mathbb{R}^2$ for every vertex in V (triangles may even collapse completely). We assume that there are γ triangles of type $\mathfrak{C}$ and μ triangels of type $\mathfrak{M}$. We now need:

(viii) An index set $E_{\mathcal{M}} = \{e_1, \ldots, e_k\}$ with one index for each of the pairs of matched edges.

(ix) An index set $C_{\mathcal{M}} = \{c_1, \ldots, c_\gamma\}$ one for each triangle of type $\mathfrak{C}$.

(x) An index set $M_{\mathcal{M}} = \{m_1, \ldots, m_{2\mu}\}$ two for each triangle of type $\mathfrak{M}$.

An instance of the overall structure is an assignment of coordinates in $\mathbb{R}^2$ to the points of $V_{\mathcal{M}}$, $E_{\mathcal{M}}$, $C_{\mathcal{M}}$ and $M_{\mathcal{M}}$ such that

- For each edge point e_i the corresponding representation is collinear with the two endpoints of the corresponding edge. (Note that this also covers the case in which the two endpoints of the edge coincide).
- For type $\mathfrak{C}$ triangles the three lines joining edge points to the opposite triangle vertex meet in the representation of the corresponding Ceva point c_i.
- The three edge points of any triangle of type $\mathfrak{M}$ are collinear. A line through these three points is spanned by the two corresponding points in $M_{\mathcal{M}}$.
- Each edge point is not allowed to coincide with either of the endpoints of the corresponding edge. This ensures suitable non-degeneray of the instance. In particular this implies that Ceva points are not allowed to be on the edges of their triangle.

Our earlier considerations on cancellation patterns show that for finding an instance of a CM-theorem we do not have to care about the very last incidence. It will be satisfied automatically. The reason for this is the previously described cancellation pattern on the oriented edge ratios. The non-degeneracy assumptions ensure that all lengths that occur and all brackets involved in our cancellation argument are indeed non-zero. For details see [1,16].

2.4   Modeling Theorems by Ceva-Menelaus Constructions

Assume that a geometric theorem is given that was generated by the process described in the preceding paragraphs (so its proof is implicit by the construction). We now want to discuss the question when such a theorem corresponds to a real projective incidence theorem described in *canonic saturated form* $\mathcal{T} = (\mathbf{H}, \mathbf{B}, C)$ as introduced in Section 2.2. It is clear that affine proving methods for incidence theorems translate into projective ones because we can w.l.o.g. assume that all points are finite. To prove a theorem given by $\mathcal{T} = (\mathbf{H}, \mathbf{B}, C)$ by a CM-theorem we will associate the points of $\mathcal{T}$ to the points of $\mathcal{M}$ in such a way that the hypotheses of $\mathcal{T}$—i.e. $\mathbf{B}$ and $\mathbf{H}$—will certify that every instance of $\mathcal{T}$ is automatically an instance of $\mathcal{M}$. The conclusion C of $\mathcal{T}$ must be such that it can be interpreted as a closing condition of the CM-theorem. By construction a CM-theorem comes along with four sets of points ($V_{\mathcal{M}}$, $E_{\mathcal{M}}$, $C_{\mathcal{M}}$ and $M_{\mathcal{M}}$), and with certain collinearity conditions involving them. The points of $\mathcal{T}$ are taken from an index set E_n. We will see that the vertices $V_{\mathcal{M}}$ of the frame of the CM-theorem, the Ceva points $C_{\mathcal{M}}$ and the Menelaus points $M_{\mathcal{M}}$ will be taken from this index set E_n. It may happen that some edge points in $E_{\mathcal{M}}$ of a CM-theorem are implicitly present by the intersection of two lines and need not be associated to points in E_n. Details will be explained in the next few paragraphs.

Let $\mathcal{M}$ be a CM-theorem with vertex set $V_{\mathcal{M}}$, Ceva/Menelaus points $C_{\mathcal{M}}$, $M_{\mathcal{M}}$ and edge points $E_{\mathcal{M}}$. Furthermore let $\mathcal{T} = (\mathbf{H}, \mathbf{B}, C)$ be a theorem in saturated canonic form. Since $\mathcal{T}$ is saturated we have a natural notion of flats of collinear points (each triple of points of a $\mathbf{H}$-flat will correspond to a triple in the saturated $\mathbf{H}$). A subset $S \subseteq E_n$ is called $\mathbf{H}$-*collinear* if all three-element subsets of S are in the collinearity hypotheses $\mathbf{H}$. Note that by this definition also sets of cardinality less than three are automatically $\mathbf{H}$-collinear. Each element in $V_{\mathcal{M}} \cup C_{\mathcal{M}} \cup M_{\mathcal{M}}$ and some points in $E_{\mathcal{M}}$ will correspond to an element of E_n. Thereby the same point in E_n may play multiple of these roles. We will model this correspondence by a mapping $f \colon V_{\mathcal{M}} \cup C_{\mathcal{M}} \cup M_{\mathcal{M}} \to E_n$. This mapping will be extended later on to some of the edge points. We define this mapping in a way that instances of the theorem $\mathcal{T}$ will automatically be instances of the CM-theorem $\mathcal{M}$. For better readability we will set $\overline{x} := f(x)$. So if x is a point label in $\mathcal{M}$ then $\overline{x}$ is the corresponding label in the theorem $\mathcal{T}$.

The definition of an instance of an CM-theorem requires certain points to be collinear (the edge point with the two corresponding endpoints of the edge and the collinearities coming from the Ceva or Menelaus constructions). These collinearities should be induced by corresponding collinearities from $\mathcal{T}$. For this we have to see how the triples in $\mathbf{H}$ interact with the incidence structure of $\mathcal{M}$. We have to examine the situation that occurs around the edge along which two triangles are glued. We first single out one matched pair of edges in $\mathcal{M}$ that will later on play the role of the conclusion of the theorem. For all other edges we now consider the situation of the two triangles adjacent to this edge. Assume that the two triangles are (a, b, c) and (a, c, d) in $\mathcal{M}$ (compare Figure 4). They have the edge (a, c) in common which they traverse in opposite directions. On this edge there is an edge point z.

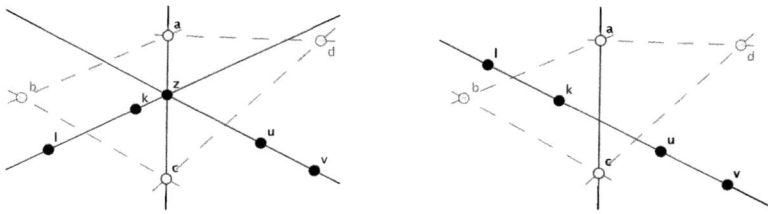

Fig. 4. Combinatoric possibilities that the subconfigurations of two glued triangles fit together

For each of the two triangles we have a corresponding *cutting line* that cuts (a, c) in the point $z \in E_\mathcal{M}$. We can specify two specific points spanning each line (in the drawing l, k and u, v—they may coincide with other points of the configuration). They come from the corresponding lines in the Ceva or Menelaus configurations and will be called *canonical cutting points*—a term we will use frequently later on. In the case of a Ceva triangle these two points are the central Ceva point (taken from $C_\mathcal{M}$) and the point opposite to the edge (taken from $V_\mathcal{M}$). In the case of a Menelaus triangle these are two points spanning the Menelaus line (both taken from $M_\mathcal{M}$). The required non-degeneracies on instances of $\mathcal{M}$ are implied by:[2]

- The triples $\{\overline{k}, \overline{l}, \overline{a}\}$ and $\{\overline{k}, \overline{l}, \overline{c}\}$ as well as $\{\overline{u}, \overline{v}, \overline{a}\}$ and $\{\overline{u}, \overline{v}, \overline{c}\}$ lie in **B**.

Now two fundamentally different situations may arise depending on whether the two cutting lines already coincide as a consequence of the collinearities in **H** or not. These situations correspond to the two situations shown in Figure 4. We say that the situation around the edge is *properly represented* by $\mathcal{T}$ if one of the following two cases arises:

1. The map f is extended to map $f(z) = \overline{z}$ and the sets $\{\overline{a}, \overline{c}, \overline{z}\}, \{\overline{k}, \overline{l}, \overline{z}\}$, $\{\overline{u}, \overline{v}, \overline{z}\}$, are all **H**-collinear in $\mathcal{T}$. (This is the situation in which $\overline{z}$ is a part of the theorem.)
2. The set $\{\overline{k}, \overline{l}, \overline{u}, \overline{v}\}$ is **H**-collinear in $\mathcal{T}$. (This is the situation in which $\overline{z}$ is implicitly present as intersection of two lines.)

If all but one edge of $\mathcal{M}$ are properly presented by the mapping f and the theorem $\mathcal{T}$ and these non-degeneracy requirements are met then automatically every instance of $\mathcal{T}$ will induce a corresponding instance of $\mathcal{M}$. The construction of the CM-theorem $\mathcal{M}$ implies that the last coincidence for the so far excluded edge is satisfied automatically.

Now for the conclusion of $\mathcal{T}$. We have to associate this conclusion with the situation around the so far excluded edge of $\mathcal{M}$. Again there are different combinatorial situations in which the conclusion is represented by this final coincidence.

---

[2] Observe that these requirements ensure, that all brackets in *all* Ceva resp. Menelaus expressions involved do not vanish.

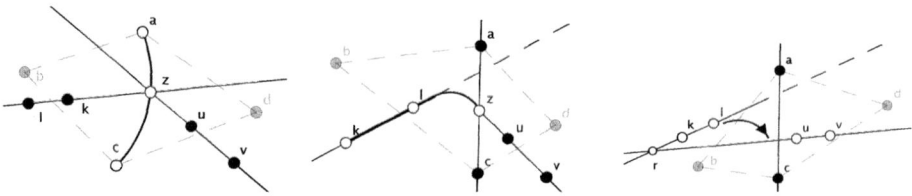

Fig. 5. Combinatoric possibilities to encode the conclusion (indicated by bent lines)

The three possible situations are shown in Figure 5. The construction of the CM-theorem $\mathcal{M}$ tells us that the edge and the two cutting lines coming from the two triangles adjacent to it must have a point in common—this is the conclusion in $\mathcal{M}$. If there is a corresponding proper edge point in which the three lines meet, we can model this easily in terms of $\mathcal{T} = (\mathbf{H}, \mathbf{B}, C)$—see cases 1. and 2. below. If not, by the construction of $\mathcal{M}$ we only know that both cutting lines have a point with the edge in common in every instance. A way to conclude a non-trivial collinearity of points is to assume, that both lines have another different point r in common. So both lines collaps in any instance—modelled by case 3. below. The triangles are taken to be (a, b, c) and (a, c, d) with corresponding edge point $z \in E_{\mathcal{M}}$ and the cutting lines are spanned by l, k and u, v. As before these points may partially coincide. We say that the conclusion is properly presented if one of the following cases arises.

1. The map f is extended to map $f(z) = \overline{z}$, $C = \{\overline{a}, \overline{c}, \overline{z}\}$, each of the sets $\{\overline{l}, \overline{k}, \overline{z}\}$ and $\{\overline{u}, \overline{v}, \overline{z}\}$ is $\mathbf{H}$-collinear, and $\{\overline{k}, \overline{z}, \overline{u}\} \in \mathbf{B}$.
2. The map f is extended to map $f(z) = \overline{z}$, $C = \{\overline{k}, \overline{l}, \overline{z}\}$, each of the sets $\{\overline{a}, \overline{c}, \overline{z}\}$ and $\{\overline{u}, \overline{v}, \overline{z}\}$ is $\mathbf{H}$-collinear, and $\{\overline{a}, \overline{c}, \overline{k}\} \in \mathbf{B}$.
3. There is an $r \in E_n$ such that each of the $\{r, \overline{u}, \overline{v}\}$ and $\{r, \overline{k}, \overline{l}\}$ is $\mathbf{H}$-collinear, C indicates the collapse of both $\mathbf{H}$-flats, and $\{\overline{a}, \overline{c}, r\} \in \mathbf{B}$.[3]

3   Fractions, Groups and Graphs

We now aim at showing the equivalence of the two methods. For the entire section let $\mathcal{T} = (\mathbf{H}, \mathbf{B}, C)$ be a theorem in canonic saturated form. We will relate the both proving techniques on the level of underlying group calculations as well as on the level of a so called *base graph*—a diagrammatic tool that helps to visualize the cancellation patterns. For this we define a *set* $\mathcal{Q}_{\mathbf{B}}$ of formal fractions in $\mathcal{B}_{\mathbf{B}}$:

$$\mathcal{Q}_{\mathbf{B}} := \left\{ \frac{[a, b, c]}{[d, e, f]} \in \mathcal{B}_{\mathbf{B}} \;\middle|\; |\{a, b, c\} \cap \{d, e, f\}| = 2 \right\}.$$

These fractions may be interpreted as ratios of oriented lengths (compare Section 2.3). There are formal equations of elements of $\mathcal{Q}_{\mathbf{B}}$ that obviously evaluate

---

[3] Recall the definition for a set being $\mathbf{H}$-collinear. It is a typical case which will also arises later on in a proof for Pappos's theorem, that $\overline{k} = r = \overline{u}$.

to 1 or ϵ in $\mathcal{B}_{\mathbf{B}}$. For instance we have $\frac{[x,y,a]}{[x,y,b]} \cdot \frac{[x,y,b]}{[x,y,c]} \cdot \frac{[x,y,c]}{[x,y,a]} = 1$. Recall that this is the expression that was related to Menelaus's theorem. Similarly the equation $\frac{[d,c,a]}{[d,c,b]} \cdot \frac{[d,a,b]}{[d,a,c]} \cdot \frac{[d,b,c]}{[d,b,a]} = \epsilon$ resembles Ceva's expression. On the other hand the biquadratic fractions, i.e. the elements of $\mathcal{H}_{\mathbf{H},\mathbf{B}} \cup \mathcal{C}_{C,\mathbf{B}}$, can be expressed as the product of two elements in $\mathcal{Q}_{\mathbf{B}}$. So the set $\mathcal{Q}_{\mathbf{B}}$ is our starting point to relate the two worlds of proving methods.

3.1   The Base Graph Γ

We introduce a diagrammatic way to deal with products of such fractions. The understanding of both proving methods on the level of the base graph $\Gamma(\mathcal{T})$ of a given theorem $\mathcal{T} = (\mathbf{H}, \mathbf{B}, C)$ will be the key to see the equivalence between the proving methods. The idea behind $\Gamma(\mathcal{T})$ is that this graph shall have the elements of $\mathbf{B}$ as nodes. Two nodes are joined by an edge if the triples differ by exactly one element. This approach is closely related to the work of Dress and Wenzel [4,5,6,19]. In contrast to us they used the bases of the matroid of a concrete instance P as vertices of a graph $\Gamma(P)$. In this approach it can be shown that $\Gamma(P)$ has the helpful property of being a Maurer Graph as defined there. This property was motivated by [12] and revived in [19]. Unfortunately, we here have only the partial information encoded in a theorem $\mathcal{T} = (\mathbf{H}, \mathbf{B}, C)$. So, if P is an instance of $\mathcal{T}$ then $\Gamma(\mathcal{T})$ is a subgraph of $\Gamma(P)$. In general $\Gamma(\mathcal{T})$ will not be a Maurer Graph. We consider $\Gamma(\mathcal{T})$ to be directed in the way that each edge is contained in Γ in both directions.

$$\Gamma(\mathcal{T}) := \Big(\mathbf{B}, \big\{(A, B) \,\big|\, A, B \in \mathbf{B}, |A \cap B| = 2\big\}\Big)$$

In what follows we will consider various subgraphs in $\Gamma(\mathcal{T})$. Each edge of $\Gamma(\mathcal{T})$ may be considered as a graphical equivalent of an element in $\mathcal{Q}_{\mathbf{B}}$. We will consider the directed edge $(\{a, b, x\}, \{a, b, y\})$ as representing $\frac{[a,b,x]}{[a,b,y]}$. Here the order inside the brackets (and thereby the permutation rules and the sign) are left aside since they have no resemblance on the level of $\Gamma(\mathcal{T})$. Roughly speaking, the edge is directed from the numerator to the denominator. Conversely, we can also interpret any collection of edges in Γ as product of ratios of brackets.

Later on, we will consider two edges canceled if they are traversed in different directions. This models the fact that multiplying a bracket ratio by its inverse will provide a cancellation. In order to visualize also products where some ratios occur multiply, we also have to consider weighted subgraphs S_δ of $\Gamma(\mathcal{T})$. Here δ is a weight function giving each edge in S a value in $\mathbb{N}$. This allows us to draw several copies of the same edge. We will make heavily use of the (affine) interpretation for the objects given in (1) in the next sections. This is no restriction since by a suitable projective transformation we can always assume that all points involved are finite. Next we illustrate the basic building blocks of both proving methods within the base graph $\Gamma(\mathcal{T})$. We will see later on that the building blocks of both proving methods complement each other in the base graph.

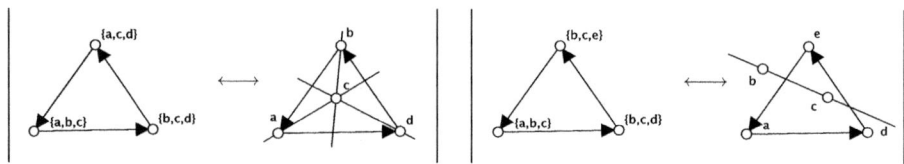

Fig. 6. The two kinds of triangles that can arise in Γ and how they are interpreted using (1)

3.2   Triangles in Ceva-Menelaus Proofs Are Triangles in Γ

As mentioned before, the Ceva and Menelaus expressions (2) and (3) correspond to triangles in $\Gamma(\mathcal{T})$ due to the brackets involved in these expressions (see Figure 6). However, the combinatorial type of these triangles is different for Ceva and Menelaus configurations—and corresponds to the two possible combinatorics for a 3-cycle in Γ. In a Ceva triangle four points are involved and each triangle contains the Ceva center point. In a Menelaus triangle five points are involved and each bracket contains the two spanning points of the Menelaus line. Observe that this interpretation associates an edge in the graph triangle with an edge of the Ceva or Menelaus triangle by using again the interpretation given in (1). So the points spanning the cutting line of this edge correspond to the two points that are shared by the bases joined by the corresponding edge in $\Gamma(\mathcal{T})$. This property is another characterisation for the *canonical cutting points* for an edge as introduced earlier.

3.3   Biquadratic Fractions Are Quadrangles in Γ

Consider a biquadratic fraction $\alpha = \frac{[b,a,x][b,c,y]}{[b,a,y][b,c,x]}$, i.e. an element in $\mathcal{H}_{\mathbf{H},\mathbf{B}} \cup \mathcal{C}_{C,\mathbf{B}}$. By definition, all brackets involved belong to $\mathbf{B}$. In Γ, the index triples corresponding to the brackets in this fraction are the vertices of a quadrangle in $\Gamma(\mathcal{T})$. If we want to write α as products of elements in $\mathcal{Q}_{\mathbf{B}}$, we have to split α. There are two different ways for this each one corresponds to singling out a specific pair of opposite edges in this quadrangle. We get the two possibilities:

$$\frac{[b,a,x]}{[b,c,x]} \cdot \frac{[b,c,y]}{[b,a,y]} \quad \text{or} \quad \frac{[b,a,x]}{[b,a,y]} \cdot \frac{[b,c,y]}{[b,c,x]}. \tag{5}$$

We investigate the left expression a bit closer (the other one can be treated analogously). The corresponding picture in Γ is given in Figure 7. Vertices occurring in the numerators are drawn in white, the other ones in black. There is more information about the expression, which is not visualized in this quadrangle: there must be a collinear triple by which the expression is contained in $\mathcal{H}_{\mathbf{H},\mathbf{B}} \cup \mathcal{C}_{C,\mathbf{B}}$. (Hence, not every quadrangle in Γ corresponds to a biquadratic fraction). In the style of Lemma 2 and using Grassmann-Plücker relations we get:

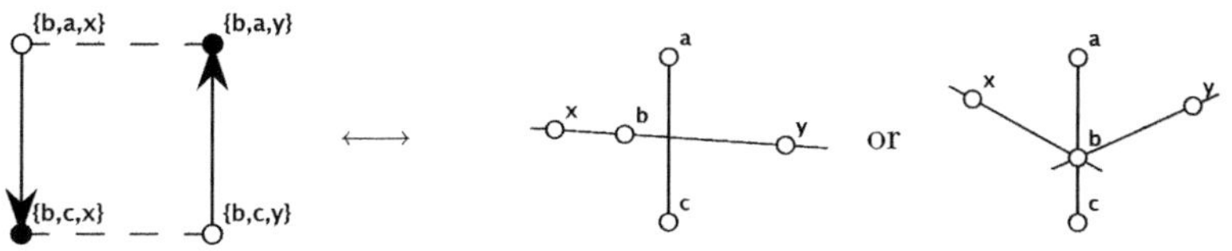

Fig. 7. The split biquadratic fraction $\frac{[b,a,x]}{[b,c,x]}\frac{[b,c,y]}{[b,a,y]}$ visualized in Γ and realized by points

Fact 1. *For* $\alpha = \frac{[b,a,x][b,c,y]}{[b,a,y][b,c,x]} \in \mathcal{B}_\mathbf{B}$ *and an instance* $P = (p_1, \ldots, p_n)$ *holds:*

$$\Phi_P(\alpha) = 1 \ \Leftrightarrow \ \{p_a, p_b, p_c\} \ or \ \{p_b, p_x, p_y\} \ are \ collinear \ \Leftrightarrow \ \frac{[b,a,x]_P}{[b,c,x]_P} = \frac{[b,a,y]_P}{[b,c,y]_P}$$

We now want to find an affine interpretation by applying (1) to both hand-sides of the right equation. Doing this we see, that a biquadratic fraction indicates an equality of oriented length ratios both along the edge ac for every instance satisfying the equation. Thus, the points b and x resp. b and y play the role of points spanning the cutting line for both parts of a biquadratic fraction. So a biquadratic fraction exchanges one of the points spanning the cutting line, without changing the length ratio. The different situations are visualized in Figure 7 and correspond directly to Fact 1. We see that this closely resembles the situations in Figure 4 and Figure 5. While a single biquadratic fraction is able to exchange *one* point on a cutting line the situation in those figures deals with the more general equality of oriented length ratios, with two potentially disjoint pairs (k, l) and (u, v) spanning the cutting lines.

3.4   Chains of Quadrangles

Fortunately, we can model the more general case of exchanging arbitrary cutting points by considering several biquadratic fractions which successively change the points spanning the cutting line. They form a picture like $\uparrow\!\vdots\!\Uparrow\!\vdots\!\Uparrow\!\vdots\!\downarrow$ in the base graph. This corresponds to a product of biquadratic fractions $\alpha_1, \ldots, \alpha_k$. Here each $\alpha_i = \tau_i \rho_i$ is split in in two fractions $\tau_i, \rho_i \in \mathcal{Q}_\mathbf{B}$ with the additional property $\rho_i = (\tau_{i+1})^{-1}$ for $i \in \{1, \ldots, k-1\}$. Multiplying all these biquadratic fractions we end up with the more general bracket ratio that expresses the exchange of two arbitrary pairs of cutting points:

$$\underbrace{(\tau_1 \cdot \overbrace{\rho_1) \cdot (\tau_2}^{1} \cdot \overbrace{\rho_2) \cdot (\tau_3}^{1} \cdot \rho_3) \cdots (\tau_k \cdot \rho_k)}_{\alpha_1 \quad\quad \alpha_2 \quad\quad \alpha_3 \quad\quad\quad \alpha_k} = \tau_1 \cdot \rho_k.$$

Such a sequence $\alpha_1, \ldots, \alpha_k$ of biquadratic fractions with these properties will further on be called a *chain*. We define $\mathcal{C}$ forming a *closed chain* in the same way by in addition requiring $\rho_k = (\tau_1)^{-1}$.

Lemma 3. *In every instance P in which all $\Phi_P(\alpha_1) = \ldots = \Phi_P(\alpha_k) = 1$ we must have $\Phi_P(\tau_1) = \ldots = \Phi_P(\tau_k)$ and in particular $\Phi_P(\tau_1) = \Phi_P(\rho_k)^{-1}$.*

Proof. Use the rightmost equivalence in Fact 1 for each $\Phi_P(\alpha_i) = 1$ $(1 \leq i \leq k)$.

In each of the fractions τ_i there is exactly one letter (say a) in which the numerator differs from the denominator (let the new letter there be c). The structure of the binomial fractions implies that in each of the τ_i these two differing letters are the same. Thus each τ_i expresses an oriented length ratio along the edge ac. The lemma above shows that under Φ_P all length ratios expressed by the τ_i are identical. To emphasise the relation to this edge we say $\alpha_1, \ldots, \alpha_k$ *forms a chain along the edge ac.*

Any such chain of split biquadratic fractions can be interpreted as a successive exchange of one of the points, which span the cutting line. Each step corresponds to one of the situations in Figure 7: We either exchange a point on the same cutting line or we pivot to another cutting line. So in total we pass from two cutting points to two other ones. With this terminology we can state two technical lemmata about the structure of binomial proofs. We will need them also as technical tools later on.

Lemma 4. *Let $\alpha = b_1^{e_1} \cdot b_2^{e_2} \cdots b_k^{e_k}$ be a binomial proof for a theorem $\mathcal{T} = (\mathbf{H}, \mathbf{B}, C)$ in canonic saturated form. Let $\mathcal{C}$ be a sequence of biquadratic fractions taken from the set $\{\alpha^{-1}, b_1, \ldots, b_k\}$ which forms a chain along the edge ac. If $\{a, c, z\} \in \mathbf{A}$ and $\{a, c, z'\} \in \mathbf{A}$ are collinear triples belonging to fractions in $\mathcal{C}$, then $z = z'$.*

Proof. Let $\mathcal{C} = (\alpha_1, \ldots, \alpha_m)$ be the chain. Let $\alpha_i = \frac{[a, x_i, z_i][c, y_i, z_i]}{[c, x_i, z_i][a, y_i, z_i]}$. Then the collinear triple of α_i is either $\{a, c, z_i\}$ (second case in Figure 7) or $\{x_i, y_i, z_i\}$ (first case in Figure 7). W.l.o.g. we may assume that the collinear triple of α_1 is $\{a, c, z\}$, the collinear triple of α_m is $\{a, c, z'\}$, and furthermore that the collinear triples of α_i are of the form $\{x_i, y_i, z_i\}$ for $i = 2, \ldots, m - 1$ (if this were not the case we may just consider some suitable subchain of $\mathcal{C}$). Since we have a chain the collinear triples of successive α_i, α_{i+1} for $i = 2, \ldots, m - 2$ must have two points in common. By our saturation assumption this implies that *all* points x_i, y_i, z_i for $i = 2, \ldots, m - 1$ must lie on a common line (a cutting line of ac). Futhermore we must have $z_1 = z$ and $z_m = z'$. So either $z = z'$ or these two points span the cutting line. However the latter would be a contradiction to the collinearities $\{a, c, z\}$, $\{a, c, z'\}$ and the non-degeneracy assumptions $\mathbf{B}$ that prevent ac to be identical to the cutting line in any biquadratic fraction.

Lemma 5. *Let $\mathcal{T} = (\mathbf{H}, \mathbf{B}, C)$ be a theorem in canonic saturated form. There can be no binomial proof $\alpha = b_1^{e_1} \cdot b_2^{e_2} \cdots b_k^{e_k}$ such that there is a closed chain using elements from the set $\{\alpha^{-1}, b_1, \ldots, b_k\}$.*

Proof. Assume on the contrary that we have such a chain and that this chain is along the edge ac. Furthermore let $\frac{[b, a, x]}{[b, c, x]} \cdot \frac{[b, c, y]}{[b, a, y]}$ be the way one has to rewrite α in order to draw the chain in Γ. So either $\{x, b, y\} \in \mathbf{B}$, $C = \{a, b, c\}$ or

$\{a, b, c\} \in \mathbf{B}$, $C = \{x, b, y\}$. The rest of the proof by contradiction can be done by a case distinction on the indicated shape of C, and similar arguments as in the last lemma and an additional case distinction. We omit the technical details.

4 Equivalence!

4.1 How to Derive a Ceva-Menelaus Proof from a Binomial Proof

The main question of this section is: *Given a theorem $\mathcal{T} = (\mathbf{H}, \mathbf{B}, C)$ in canonic saturated form along with a binomial proof, how to construct a Ceva-Menelaus proof for an equivalent theorem.* Most of our considerations will be performed on the level of the base graph Γ. The main strategy will be to take the different quadrangles corresponding to biquadratic fractions and—like a combinatoric puzzle—by gluing them create cycles that can be translated into Ceva-Menelaus structures. If all these cycles are triangles they directly correspond to Ceva's or Menelaus's theorem and this translation process will be comparably easy. Unfortunately there are situations in which these cycles can become considerably longer. There will be two principle ways of how to deal with such situations:

- We can add two generic points to the theorem and thereby also enlarge $\Gamma(\mathcal{T})$ in a trivial way. In this case we can find a Ceva-Menelaus proof by further decomposition of the long cycles.

- Alternatively, we can treat these long cycles as first class citizens and see that they themselves encode theorems. We will call them Γ-cycle theorems. We can consider them as additional building blocks in our manifold proofs.

Here we will mainly consider the first option and make some brief remarks on the second option later on. So now we model the augmentation of $\mathcal{T}$ with two generic points. This leads to a new theorem $\mathcal{T}'$ with enlarged groundset E_{n+2}. For reasons of better readability we write $\mathbf{g}$ and $\mathbf{h}$ for the generic points $n+1$ and $n+2$. These points can be interpreted as not lying on any derived line of the theorem. So we define the set $\mathbf{B}'$ to be the set $\mathbf{B}$ extended by all triples of the form $\{a, b, \mathbf{g}\}, \{a, b, \mathbf{h}\}, \{a, \mathbf{g}, \mathbf{h}\}$ for each element $\{a, b, c\}$ of $\mathbf{B}$ and each order of a, b and c.

Observe that $\mathbf{B}'$ is still in saturated form.[4] We now want to derive a Ceva-Menelaus proof from the given biquadratic proof. We allow ourselves to prove a theorem equivalent to $\mathcal{T}' = (\mathbf{H}, \mathbf{B}', C)$ with the two generic points added. We have seen that both types of proofs can be considered as calculations with elements from the set of fractions $\mathcal{Q}_{\mathbf{B}}$. The equivalence proof will be mainly a matter of regrouping these fractions.

To start with, we will illustrate the approach in the case of our running example, Pappos's theorem, and its biquadratic proof presented in Section 2.1.

---

[4] The set $\mathbf{A}$ remains unchanged. When checking the consistency constraints for the additional triples in $\mathbf{B}'$ it is helpful to remember that $\mathbf{B}$ separates any two points on a common derived line.

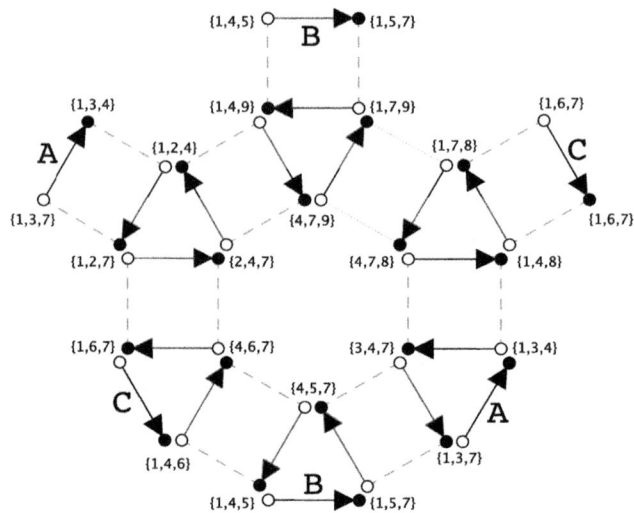

Fig. 8. Visualization of the cancellation patterns in the biquadratic proof of Pappos's theorem in $\Gamma(\mathcal{T})$. Edges labeled with the same capital letter are identified. The biquadratic fraction with dotted lines encodes the conclusion.

This proof can be considered as the product of eight biquadratic fractions (expressing the hypotheses) being equal to another one (expressing the conclusion). Equivalently we may write it as:

$$\left(\frac{[124]\,[137]}{[127]\,[134]}\right) \cdot \left(\frac{[154]\,[197]}{[157]\,[194]}\right) \cdot \left(\frac{[184]\,[167]}{[187]\,[164]}\right) \cdot \left(\frac{[427]\,[491]}{[421]\,[497]}\right) \cdot \left(\frac{[457]\,[461]}{[451]\,[467]}\right) \cdot$$

$$\left(\frac{[487]\,[431]}{[481]\,[437]}\right) \cdot \left(\frac{[721]\,[764]}{[724]\,[761]}\right) \cdot \left(\frac{[751]\,[734]}{[754]\,[731]}\right) \cdot \left(\frac{[781]\,[794]}{[784]\,[791]}\right) = 1$$

The fact that we have a biquadratic proof implies that each bracket occurs in the numerator as often as in the denominator. From every binomial proof $\alpha = b_1^{e_1} \cdot b_2^{e_2} \cdots b_k^{e_k}$ a similar equation $1 = b_1^{e_1} \cdot b_2^{e_2} \cdots b_k^{e_k} \cdot \alpha^{-1}$ can be derived. We name this equation $(*)$. As a product of split biquadratic fractions the different factors induce a (weighted) subgraph in Γ. See Figure 8 for the case of Pappos's theorem. Just like in Figure 7 we also indicated which ratios of brackets belong together via a biquadratic fraction. The fraction encoding the conclusion is emphasized by using dotted lines.

Regrouping the fractions: In this (very special!) case of Pappos's theorem, equation $(*)$ and the splitting indicated in Figure 8 imply cancellation cycles in Γ which are already triangles. By our considerations of Section 3.2, they can directly be interpreted as Ceva or Menelaus triangles. We get exactly six Ceva configurations. The biquadratic fractions indicate which triangles should be glued

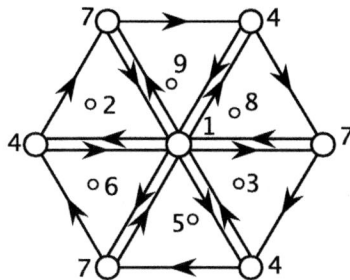

Fig. 9. The six big equilateral triangles in the picture correspond to the six Ceva configurations that arise in the proof together with the central Ceva point. Opposite sides of the hexagon have to be identified, such that the edges are traversed in opposite directions. So the overall topology of the proof is a torus.

together. They model the exchange of a single point on a cutting line. The result is shown in Figure 9. This structure should be considered as the frame of a CM-theorem. Each of the six triangles thereby is of type $\mathfrak{C}$ (Ceva). The structure of the underlying manifold is a combinatorial torus composed of six triangles; opposite edges of the hexagon have to be identified. The presence of five of the Ceva configurations implies the presence of the last one (which can be considered as the final incidence of Pappos's theorem).

In fact, one has to look a bit closer to rediscover Pappos's theorem from this proof, since many edge points used in the CM-theorem do not play a role in Pappos's theorem. To see this consider Figure 10. The first picture visualizes the framing manifold of the CM-theorem. Opposite sides of the hexagon have to be identified. The second picture shows the situation after the identification. A careful investigation of the biquadratic fractions involved shows that we are always in the left case in Figure 7, in which the edge points are not part of the theorem and can be removed from the drawing. Therefore none of the edge points in the CM-theorem is mapped to a point in Pappos's theorem. Something similar holds for the lines supporting the triangle: they only contain two theorem points and are of no relevance for the theorem. Therefore they can be omitted in the picture, too. Now we are left with a drawing of Pappos's theorem (third picture) and one can easily check that we really found a Ceva-Menelaus proof for it.

Back to the general case: Also in the general case, the equation corresponding to (∗) indicates a weighted subgraph S_δ of Γ consisting of cycles. However, the above conversion is special in three different ways. Firstly, in general the cycles that occur may be larger than triangles. Secondly, the biquadratic fractions matching the edges of two cycles may in general be replaced by a whole chain of quadrangles with each quadrangle representing a biquadratic fraction (compare Section 3.3). Thirdly, we have to deal with multiple edges in S_δ. Multiple edges correspond to fractions that are used more than once in the cancellation pattern. As before δ is the weight function that expresses these multiplicities.

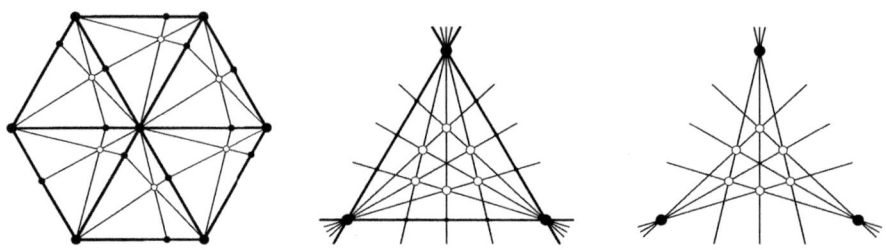

Fig. 10. Extracting Pappos's theorem from the Ceva-Menelaus proof

We first deal with the problem of large cycles and consider S_δ. As mentioned before, we now consider edges canceled if they are traversed in different directions. So any 2-cycle vanishes. At this point, a vanishing 2-cycle stands for two biquadratic fractions which are adjacent in a chain (and the inner edges cancel). So we are left with cycles of length ≥ 3. If possible, we decompose cycles of length ≥ 4 into smaller cycles. This means modifying the subgraph S_δ of $\Gamma(\mathcal{T})$ together with its weight function δ by introducing new pairs of opposite edges. We may introduce such a 2-cycle, between two non-adjacent edges in a cycle in S_δ, whenever the corresponding vertices have two labels in common (thus are connected in $\Gamma(\mathcal{T})$). So we are left with *irreducible* cycles of unknown length. Here irreducible means that the cycle cannot be decomposed in $\Gamma(\mathcal{T})$ into smaller cycles. So there can be no multiple edge in irreducible cycle, since they could be decomposed otherwise. Hence the irreducible cycles are (ordinary) subgraphs of $\Gamma(\mathcal{T})$. For ordinary cycles being irreducible implies that no two non-adjacent vertices of a cycle are connected in $\Gamma(\mathcal{T})$. If we were in a base graph $\Gamma(P)$ of an instance P, by results from [4] and [12] and some relatively simple considerations on the fact that the matroid underlying the instance is (simple and) realizable all irreducible cycles were of length 3. However, we are dealing with $\Gamma(\mathcal{T})$ and many of the edges present in $\Gamma(P)$ may be missing. So we are forced to deal with the remaining irreducible cycles (of length ≥ 4) by other means, since the partial information of **B** may not suffice to decompose $\Gamma(\mathcal{T})$. As mentioned before there are two strategies to dealing with them.

Two Additional Generic Points: In this approach we will find a Ceva-Menelaus proof of a theorem equivalent to $\mathcal{T}' = (\mathbf{H}, \mathbf{B}', C)$. Of course the collection of irreducible cycles found above is also a subgraph of the even bigger graph $\Gamma(\mathcal{T}')$ in which two generic points have been added. In fact these two points in general position added before introduce enough new bases and edges in $\Gamma(\mathcal{T}') = \Gamma((\mathbf{H}, \mathbf{B}', C))$ that arbitrarily large cycles become decomposable, if they are already a subgraph of $\Gamma(\mathcal{T})$. A very general example for this method is given in Figure 11. To the large cycle of length m (drawn with thick lines) we first add a corona of bases only involving the generic point **g**. This introduces a ring of m Menelaus triangles adjacent to the irreducible cycle followed by a ring of m Ceva triangles adjacent to them. The vertices of the non-matched edges of these Ceva triangles all share point **g**. The two vertices of such an edge have one

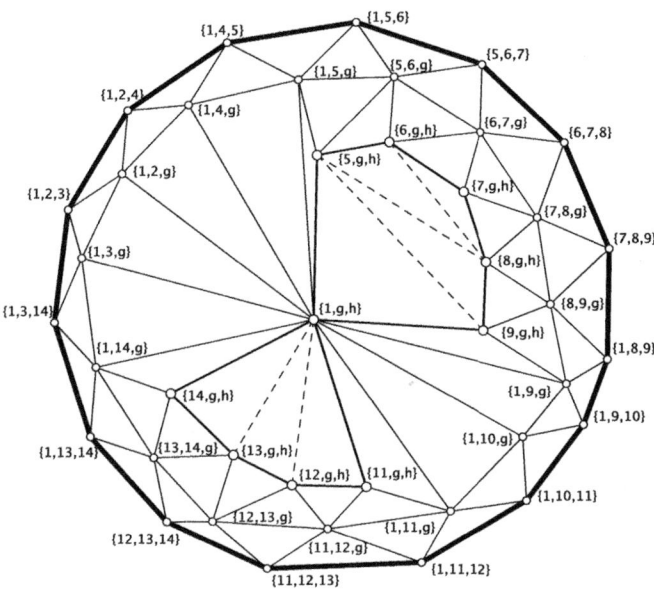

Fig. 11. An example for decomposing an irreducible cycle

more point in common. Finally considering bases of the form $(x, \mathbf{g}, \mathbf{h})$ helps to fill the remaining hole of the structure by Ceva and Menelaus triangles.

It is not hard to verify that this pattern works for *arbitrary* irreducible cycles in $\Gamma(\mathcal{T})$. So we managed to decompose all cycles in (∗) with length ≥ 4 into triangles. Just as in the previous example these triangles will all indicate Ceva and Menelaus triangles. We will now start to check the formal requirements for a Ceva-Menelaus proof as stated in Section 2.3. By construction we already meet some requirements for the framing manifold like structure. The collection of graph triangles just found gives us: a list of triangles (one for each graph triangle), but without explicit matching. It also gives an orientation and a type ($\mathfrak{C}$ or $\mathfrak{M}$) induced by the interpretation given in Section 3.2. Therefore we can determine sets $C_{\mathcal{M}}$ and $M_{\mathcal{M}}$. As soon as we can find a matching, we also can explicitly describe $V_{\mathcal{M}}$. The interpretation of $V_{\mathcal{M}}$ as vertices of graph triangles as described in Section 3.2 will induce a function $f \colon V_{\mathcal{M}} \cup C_{\mathcal{M}} \cup M_{\mathcal{M}} \to E_{n+2}$. Furthermore, the vertices of the graph triangles lie by definition in **B′**. On the other hand, this gives us exactly the required non-degeneracies needed in Ceva-Menelaus proofs—they encoded that the vertices of an edge are not incident with the cutting line.

The remaining requirements are mostly concerned with the way triangles are glued. We fix a binomial proof $1 = b_1^{e_1} \cdot b_2^{e_2} \cdots b_k^{e_k} \cdot \alpha^{-1}$ for the following considerations. So if we take any edge in one of the graph triangles just found we want to find an edge in another graph triangle to glue to it (i.e. we search for the matching of the edges). To do so we track how the edge got into the cycle in Γ. If it originates from the decomposition of a bigger cycle, it has to cancel with another edge and we found its counterpart. Otherwise it will be part of a (split) biquadratic fraction in (∗). So the other part of the fraction (which is also an edge) has to be part of the cancellation pattern, too. This edge in turn is either an edge of an triangle or it vanishes by the decomposition rules. If it vanishes we can restart the game with the edge responsible for this cancellation process. Thus by induction, we discover a (maximal) chain of biquadratic fractions in the cancellation pattern. It has to end somewhere[5] and the last edge is the counterpart of the edge we started with. If we iterate this process for every edge in the irreducible cycles, we can find a matching. It induces $V_{\mathcal{M}}$. Observe that the induced orientations on the matched edges have opposite directions by the way we split biquadratic fractions. Furthermore, the graph structure with a node for each triangle induced by the matching as described in Section 2.3 is connected. Assuming the contrary, one connected component would induce a proper subfamily $\mathcal{S}$ of the binomial proof that gives a complete cancellation pattern, in conflict to the definition of a biquadratic proof (the exponent vectors there were minimal).

Now consider two matched edges (a, c) and (c, a). Say that k, l resp. u, v are the canonical cutting points induced by the triangles. We want to show that the situation around the edge is properly represented (see Section 2.3 and Figure 4). If the graph edges corresponding to (a, c) and (c, a) form 2-cycles, then $\{\overline{k}, \overline{l}\} = \{\overline{u}, \overline{v}\}$ and $\{\overline{k}, \overline{l}, \overline{u}, \overline{v}\}$ is $\mathbf{H}$-collinear. So now we can assume that the graph edges are connected via biquadratic fractions in (∗). Due to Lemma 5, there will be exactly one pair of matched triangles where this chain contains α^{-1}, the fraction expressing the conclusion. This pair will play the role of the conclusion in the CM-proof. All remaining pairs of edges are glued with chains of fractions contained in $\{b_1, \ldots, b_k\}$. As discussed in Section 3.3, such a chain induces a situation which resembles one of the pictures in Figure 4. The fact that biquadratic fractions in $\{b_1, \ldots, b_k\}$ come along with a collinear triple in $\mathbf{H}$ and the fact that $\mathbf{H}$ is saturated ensures that also the formal requirements for the proper representation are met. (The details are similar, but also easier as in the following considerations.)

Now we want to analyse the conclusion edges. Lemma 5 ensures that the edges really exist and do not vanish in Γ. Say

$$\alpha^{-1} = \frac{[b, \overline{a}, x]}{[b, \overline{c}, x]} \cdot \frac{[b, \overline{c}, y]}{[b, \overline{a}, y]}, \tag{6}$$

split in the same way as it occurs in (∗) and expressed in terms of the existing preimages of f. So (a, c) is the conclusion edge. By investigating the conclusion chain—name it $\mathcal{C}$—in detail we will see that it is exactly an equivalent formulation of the conclusion being properly represented (corresponding to Figure 5). So it should be no surprise, that we will have to do a case distinction. We will speak of left and right ends of the chain to tell them apart. Let l, k resp. u, v be the canonical cutting points at the left resp. right end of the chain, which are the edges directly corresponding to the conclusion edges in triangles on our manifold like structure.

---

[5] Since we started with a non-vanishing edge. Vanishing edges always vanish in pairs.

Case (1): $C = \{\overline{a}, b, \overline{c}\}$ and $\{x, b, y\} \in \mathbf{B}$. With $z \in E_{\mathcal{M}}$ the corresponding edge point and $\overline{z} := b$ this case will match with Figure 5 (left) and resembles the first possibility for the conclusion in Ceva-Menelaus proofs. The details are: In no fraction between α^{-1} and an edge at the end of the chain $\mathcal{C}$, the cutting line can be changed due to Lemma 4. Since each fraction comes along with a triple in $\mathbf{H}$ and since $\mathbf{H}$ is saturated we can conclude that $\{\overline{k}, \overline{l}, \overline{z} = b, x\}$ and $\{\overline{u}, \overline{v}, \overline{z}, y\}$ are $\mathbf{H}$-collinear. W.l.o.g. $\overline{k} \neq \overline{z}$ and $\overline{u} \neq \overline{z}$.[6] The following scheme indicates how we can exploit $\mathbf{B}$ being consistent, triples on the top of the arrows indicate $\mathbf{H}$-collinear sets: $\{x, \overline{z}, y\} \xrightarrow{\{\overline{k}, \overline{z}, x\}} \{\overline{k}, \overline{z}, y\} \xrightarrow{\{\overline{u}, \overline{z}, y\}} \{\overline{k}, \overline{z}, \overline{u}\}$. In total we meet the requirements for the first possibility of the conclusion being properly represented.

Case (2): $C = \{x, b, y\}$ and $\{\overline{a}, b, \overline{c}\} \in \mathbf{B}$.

Case (2a): No biquadratic fraction in $\mathcal{C}$ has $\{\overline{a}, \overline{c}, w\}$ as collinear triple (for any $w \in E_n$). With this we can conclude, that (w.l.o.g) $\{b, \overline{k}, \overline{l}, x\}$ and $\{b, \overline{u}, \overline{v}, y\}$ are $\mathbf{H}$-collinear. So C indicates a collaps of both lines. With $r := b$ this is exactly the third case in the definition of a Ceva-Menelaus proof.

Case (2b): There is a biquadratic fraction in $\mathcal{C}$ which $\{\overline{a}, \overline{c}, w\} \in \mathbf{H}$ as collinear triple (for any $w \in E_n$). The fraction α^{-1} cannot be located between two fractions in $\mathcal{C}$ with collinear triple $\{\overline{a}, \overline{c}, w\}$.[7] So w.l.o.g. $\{\overline{a}, \overline{c}, w\}$ is a collinear triple in $\mathcal{C}$ only on the right side of α^{-1}. So on this side $\{w, \overline{u}, \overline{v}\}$ is $\mathbf{H}$-collinear. From the fractions between α^{-1} and where $\{\overline{a}, \overline{c}, w\}$ is first (seen from the left side) collinear triple and by Lemma 4 we can conclude that $\{b, y, w\}$ is $\mathbf{H}$-collinear. From the left side we conclude that $\{\overline{k}, \overline{l}, x, b\}$ is $\mathbf{H}$-collinear.

The collinearity of $C' := \{\overline{k}, \overline{l}, w\}$ can be deduced from this information together with $C = \{x, b, y\}$ ad vice versa. So $(\mathbf{H}, \mathbf{B}', C')$ is an equivalent formulation of the theorem $(\mathbf{H}, \mathbf{B}', C)$. In addition we can deduce $\{\overline{a}, \overline{c}, \overline{k}\} \in \mathbf{B}$ (w.l.o.g. $\overline{k} \neq w$) and fulfill the requirements in the second possibility of a Ceva-Menelaus proof. The deduction is again indicated by the scheme

$$\{\overline{a}, b, \overline{c}\} \xrightarrow{\{\overline{a}, \overline{c}, w\}} \{\overline{a}, w, b\} \xrightarrow{\{w, b, \overline{k}\}} \{\overline{a}, w, \overline{k}\} \xrightarrow{\{\overline{a}, w, \overline{c}\}} \{\overline{a}, \overline{c}, \overline{k}\}.$$

The only thing left to show is that *the number of Menelaus triangles is even*. The total number of triangles is even, since otherwise, there could not exist a matching of edges. So it is sufficient to show that the number of Ceva triangels is even. So we can reformulate our claim: Form the Ceva or Menelaus expression (as indicated in (2) or (3) and with this order of indices inside the brackets) and multiply them. All ratios in (2) or (3) have the shape $\frac{[a, c, d]}{[b, c, d]}$ (for some elements $a, b, c, d \in E_n$). The number of Ceva triangles is even if and only if this products equals 1.

To show this, we undo all we did so far without permuting indices inside the brackets and by this simplify the expression: consider only the edge ratios of

---

[6] Since: $\{\overline{a}, \overline{k}, \overline{l}\} \in \mathbf{B} \Longrightarrow \overline{k} \neq \overline{l} \Longrightarrow$ w.l.o.g $\overline{k} \neq \overline{z}$. Analogously for $\overline{u} \neq \overline{z}$.

[7] Otherwise, we could construct a closed chain contradicting Lemma 5.

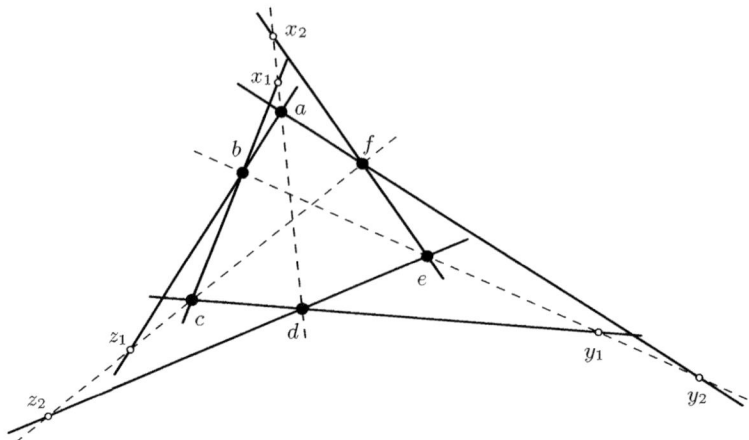

Fig. 12. A Γ-cycle theorem that corresponds to a certain 6-cycle of bases. It can be used to derive another proof of Pappos's theorem.

those triangles added by decomposing an irreducible cycle. We know by construction that in the product a lot of ratios cancel and we are left with the edge ratios of the irreducible cycle. If we proceed in this way of thinking we will end up just with the cancelling cycles S_δ we started with. They were—up to a permutation of indices—the left side of ($*$). But by considering the ratios in a split biquadratic fraction as in (5) and the shape of ratios present, we see that we can permute the indices to obtain the left side of ($*$) *without changing the sign*. And the right side of the equation ($*$) is 1.

Considering Irreducible Cycles in Γ as Theorems: We now want to leave the path just seen which uses the generic points and indicate another way to treat the irreducible cycles in Γ. Another, in a sense more compact approach is to consider those cycles themselves as fundamental objects and evaluate their role in the assembling of our manifold like structures. They encode theorems similar to Ceva's or Menelaus's theorems, we call them Γ-*cycle theorems*. They can be proved by our previous decomposition argument by glueing Ceva and Menelaus triangles. The structure of these Γ-cycle theorems is simple: Take any (irreducible) cycle in Γ, form the corresponding expression of ratios of brackets. It forms a product which equals 1.

Again we may interpret this equation on the level of ratios of lengths. The bracket expression gives another almost trivial proof for the Γ-cycle theorem. Although the proofs of such theorems are simple and straight forward they give rise to surprising theorems on the level of length ratios. A natural task is to enumerate such Γ-cycle theorems (up to isomorphism and only irreducible cycles). If one does so one finds that there are two such theorems for 3-cycles, one for 4-cycles, two for 5-cycles, five for 6-cycles, ten for 7-cycles and 23 for 8-cycles.[8]

---

[8] The enumeration together with a geometric analysis of the theorems will be completed in a forthcoming paper.

One example of particular beauty is shown in Figure 12. It corresponds to the 6-cycle

$$\{a, b, c\} \rightarrow \{b, c, d\} \rightarrow \{c, d, e\} \rightarrow \{d, e, f\} \rightarrow \{e, f, a\} \rightarrow \{f, a, b\} \rightarrow \{a, b, c\}.$$

Expressed in terms of length ratios this implies for the incidence configuration of Figure 12 that

$$\frac{|a, x_1|}{|x_1, d|} \cdot \frac{|d, x_2|}{|x_2, a|} \cdot \frac{|b, y_1|}{|y_1, e|} \cdot \frac{|e, y_2|}{|y_2, b|} \cdot \frac{|c, z_1|}{|z_1, f|} \cdot \frac{|f, z_2|}{|z_2, c|} = 1$$

This implies that if $x_1 = x_2$ and $y_1 = y_2$ we must also have $z_1 = z_2$. This is another proof of Pappos's theorem. In Γ, the coincidence of points corresponds to gluing opposite sides of the hexagon representing the above 6-cycle: Again the structure of a torus!

4.2   From a Ceva-Menelaus Proof to a Binomial Proof

So far we explained how a binomial proof can be transferred into a Ceva-Menelaus proof. The opposite direction is comparably simple. Let $T = (\mathbf{H}, \mathbf{B}, C)$ be a real projective incidence theorem in canonical saturated form with a Ceva-Menelaus proof given. So we are given sets $V_\mathcal{M}$, $C_\mathcal{M}$, $M_\mathcal{M}$ and $E_\mathcal{M}$, a list of triangles T together with indicated types ($\mathfrak{C}$ or $\mathfrak{M}$), a matching on the edges, a location for the conclusion singled out and a function f that specifies the relation of E_n and the Ceva-Menelaus construction. They meet all the requirements of Section 2.3. By Section 3.2 the triangles in T correspond to triangles in Γ via the corresponding Ceva or Menelaus expressions. For each pair of edges glued together in the Ceva-Menelaus proof, we want to connect the corresponding edges in Γ with biquadratic fractions. Roughly speaking, collecting all these fractions will give us a binomial proof for a theorem $T' = (\mathbf{H}, \mathbf{B}, C')$ equivalent to T.

Now consider two triangles glued together along the edge (a, c) and with edge point $z \in E_\mathcal{M}$ (and (a, c) not the conclusion edge). Let k, l resp. u, v be the canonical cutting points in the triangles corresponding to (a, c). So $(\{\overline{a}, \overline{k}, \overline{l}\}, \{\overline{c}, \overline{k}, \overline{l}\})$ and $(\{\overline{c}, \overline{u}, \overline{v}\}, \{\overline{a}, \overline{u}, \overline{v}\})$ are the graph edges corresponding to the edge (a, c) in both triangles. We want to express $\frac{[\overline{a}, \overline{k}, \overline{l}]}{[\overline{c}, \overline{k}, \overline{l}]} \frac{[\overline{c}, \overline{u}, \overline{v}]}{[\overline{a}, \overline{u}, \overline{v}]}$ as a product of biquadratic fractions. By the assumptions on the matching in Ceva-Menelaus proofs and our considerations of Section 3.2 we can successively exchange the points spanning the cutting line via biquadratic fractions. Together with the fact that $\mathbf{B}$ is consistent we can conclude, that in each intermediate step, the triples occurring lie in $\mathbf{B}$.

We exemplify this in the case where f is extended to z and $\{\overline{l}, \overline{k}, \overline{z}\}$ and $\{\overline{u}, \overline{v}, \overline{z}\}$ and $\{\overline{a}, \overline{c}, \overline{z}\}$ are $\mathbf{H}$-collinear (left picture in Figure 4—the other case can be done similarly). W.l.o.g $\overline{k} \neq \overline{z}$ and $\overline{u} \neq \overline{z}$. Consider the biquadratic fractions:

$$\frac{[\overline{a},\overline{k},\overline{l}]\,[\overline{c},\overline{k},\overline{z}]}{[\overline{c},\overline{k},\overline{l}]\,[\overline{a},\overline{k},\overline{z}]}, \quad \frac{[\overline{a},\overline{k},\overline{z}]\,[\overline{c},\overline{u},\overline{z}]}{[\overline{c},\overline{k},\overline{z}]\,[\overline{a},\overline{u},\overline{z}]}, \quad \frac{[\overline{a},\overline{u},\overline{z}]\,[\overline{c},\overline{u},\overline{v}]}{[\overline{c},\overline{u},\overline{z}]\,[\overline{a},\overline{u},\overline{v}]} \tag{7}$$

Since $\{\overline{a},\overline{k},\overline{l}\}, \{\overline{c},\overline{k},\overline{l}\}, \{\overline{a},\overline{u},\overline{v}\}, \{\overline{c},\overline{u},\overline{v}\} \in \mathbf{B}$ and since $\mathbf{B}$ is consistent, the fractions are indeed elements in $\mathcal{B}_{\mathbf{B}}$. Bearing in mind the definition for $\{\overline{l},\overline{k},\overline{z}\}$ resp. $\{\overline{u},\overline{v},\overline{z}\}$ resp. $\{\overline{a},\overline{z},\overline{c}\}$ being $\mathbf{H}$-collinear, the fractions either evaluate to 1 or are elements of $\mathcal{H}_{\mathbf{H},\mathbf{B}}$ due to $\{\overline{l},\overline{k},\overline{z}\} \in \mathbf{H}$ resp. $\{\overline{u},\overline{v},\overline{z}\} \in \mathbf{H}$ resp. $\{\overline{a},\overline{z},\overline{c}\} \in \mathbf{H}$. Multiplying the fractions gives the desired expression.

We collect all those biquadratic fractions for any two glued edges in our manifold like structure which do not model the conclusion. If we multiply all these fractions (with the order of indices inside the brackets as just stated) the chains just produced cancel and we are exactly left with fractions corresponding to the triangles in the graph we started with (compare (2) and (3) and Figure 6). So in $\mathcal{B}_{\mathbf{B}}$ these brackets cancel in 3-cycles and leave a factor ϵ for each Ceva configuration. Since the overall number of Ceva configurations is even, in total we are left with a 2-cycle along the conclusion edge. More precisely: if we call the conclusion edge (a,c) and let k, l resp. u, v be the canonical cutting points in the associated triangles, then due to cancellation this overall product must have the form:

$$\frac{[\overline{c},\overline{k},\overline{l}]\,[\overline{a},\overline{u},\overline{v}]}{[\overline{a},\overline{k},\overline{l}]\,[\overline{c},\overline{u},\overline{v}]} \tag{8}$$

If we do the same translation process as for the normally glued edges as far as possible (with the given hypotheses), we will be left with a biquadratic fraction encoding an equivalent conclusion. So we have to differentiate between the different possibilities to encode a conclusion in a Ceva-Menelaus proof (compare also Figure 5). Therefore let $z \in E_{\mathcal{M}}$ be the edge point corresponding to the edge (a,c).

Case (1): f is extended to map $f(z) = \overline{z}$, $C = \{\overline{a},\overline{c},\overline{z}\}$, $\{\overline{l},\overline{k},\overline{z}\}$ and $\{\overline{u},\overline{v},\overline{z}\}$ are $\mathbf{H}$-collinear, $\{\overline{k},\overline{z},\overline{p}\} \in \mathbf{B}$. With the same considerations as before we can again conclude that the rightmost and the leftmost fractions in (7) are in $\mathcal{H}_{\mathbf{H},\mathbf{B}}$ or evaluate to 1. Multiplying them with (8), we are left with the biquadratic fraction in the middle of (7) which lies in $\mathcal{C}_{C,\mathbf{B}}$ since $C = \{\overline{a},\overline{c},\overline{z}\}$ and $\{\overline{k},\overline{z},\overline{u}\} \in \mathbf{B}$. If we collect all biquadratic fractions used we found the desired certificate for $\mathcal{C}_{C,\mathbf{B}} \cap \langle \mathcal{H}_{\mathbf{H},\mathbf{B}}\rangle \neq \emptyset$.

Case (2): f is extended to map $f(z) = \overline{z}$, $C = \{\overline{k},\overline{l},\overline{z}\}$, $\{\overline{a},\overline{c},\overline{z}\}$ and $\{\overline{u},\overline{v},\overline{z}\}$ are $\mathbf{H}$-collinear, $\{\overline{a},\overline{c},\overline{k}\} \in \mathbf{B}$. Again assume w.l.o.g. $\overline{k} \neq \overline{z}$ and $\overline{u} \neq \overline{z}$. Multiply expression (8) by the elements

$$\frac{[\overline{c},\overline{u},\overline{z}]\,[\overline{a},\overline{k},\overline{z}]}{[\overline{a},\overline{u},\overline{z}]\,[\overline{c},\overline{k},\overline{z}]}, \quad \frac{[\overline{c},\overline{u},\overline{v}]\,[\overline{a},\overline{u},\overline{z}]}{[\overline{a},\overline{u},\overline{v}]\,[\overline{c},\overline{u},\overline{z}]} \text{ of } \mathcal{H}_{\mathbf{H},\mathbf{B}} \text{ gives } \frac{[\overline{c},\overline{k},\overline{l}]\,[\overline{a},\overline{k},\overline{z}]}{[\overline{a},\overline{k},\overline{l}]\,[\overline{c},\overline{k},\overline{z}]}$$

which is an element of $\mathcal{C}_{C,\mathbf{B}}$ due to $\{\overline{a},\overline{c},\overline{k}\} \in \mathbf{B}$. In total we found a certificate for $\mathcal{C}_{C,\mathbf{B}} \cap \langle \mathcal{H}_{\mathbf{H},\mathbf{B}}\rangle \neq \emptyset$. For the fractions lying in $\mathcal{H}_{\mathbf{H},\mathbf{B}}$ we need also $\mathbf{B}$ being saturated.

Case (3): C indicates the collaps of the two **H**-flats given by the **H**-collinear sets $\{r, \overline{u}, \overline{v}\}$ and $\{r, \overline{k}, \overline{l}\}$ (with an $r \in E_n$). Furthermore holds $\{\overline{a}, \overline{c}, r\} \in \mathbf{B}$. Multiplying expression (8) with the elements

$$\frac{[\overline{c}, \overline{k}, r]\,[\overline{a}, \overline{k}, \overline{l}]}{[\overline{a}, \overline{k}, r]\,[\overline{c}, \overline{k}, \overline{l}]}, \quad \frac{[\overline{c}, \overline{u}, \overline{v}]\,[\overline{a}, \overline{u}, r]}{[\overline{a}, \overline{u}, \overline{v}]\,[\overline{c}, \overline{u}, r]} \; \text{ of } \mathcal{H}_{\mathbf{H},\mathbf{B}} \; \text{ gives } \; \frac{[\overline{c}, \overline{k}, r]\,[\overline{a}, \overline{u}, r]}{[\overline{a}, \overline{k}, r]\,[\overline{c}, \overline{u}, r]},$$

an element in $\mathcal{C}_{C',\mathbf{B}}$ with $C' := \{\overline{k}, r, \overline{u}\}$. This gives a certificate for $\mathcal{C}_{C',\mathbf{B}} \cap \langle \mathcal{H}_{\mathbf{H},\mathbf{B}} \rangle \neq \emptyset$ and proofs the equivalent theorem $\mathcal{T}' := (\mathbf{H}, \mathbf{B}, C')$.

It is clear that we can also modify the certificates found to obtain a (minimal) binomial proof.

5   Conclusion

We will close this paper with a few remarks and questions that we consider relevant and worth a further investigation.

1. On glue and matter: There is a nice duality between the binomial and the Ceva-Menelaus proving techniques: In the Ceva-Menelaus construction we saw that by multiplying biquadratic fractions we can pass from one pair of (canonical) spanning points to the others. So one could say that the (chains of) biquadratic fractions are the *glue* in the Ceva-Menelaus world. On the other hand, if we transfer a binomial proof to a Ceva-Menelaus proof, we can consider the cancellation patterns to consist of triangles (which correspond to Ceva or Menelaus triangles)—at least in the approach with **g** and **h**. So here the Ceva-Menelaus triangles play the role of the glue.

2. Finding Ceva-Menelaus proofs: Our translation process is capable to translate a binomial proof (which itself can be found algorithmically) to a Ceva-Menelaus proof. Nevertheless so far we know of no good algorithm that is capable of producing a Ceva-Menelaus proof directly from $\mathcal{T} = (\mathbf{H}, \mathbf{B}, C)$. The ultimate goal here would be to find an algorithm that uses the incidence structure of the configuration directly to extract a very small search tree for unveiling the underlying manifold proof.

3. Some proofs are nicer than others: Our considerations show that an algorithm that is capable of finding a Ceva-Menelaus proof cannot be more powerful than the known algorithms for finding binomial proofs (see [13]). But we could try to find an (efficient) algorithm which transfers a binomial proof into a Ceva-Menelaus proof. Ceva-Menelaus proofs are in a sense more visual. They provide a manifold structure that serves as a framework for the proof of the theorem. In a way looking at a Ceva-Menelaus proof one directly "sees" that a theorem is true. Since there are several degrees of freedom in the proof, it could be an interesting task to find an algorithm that produces the "most beautiful" Ceva-Menelaus proof based on a binomial proof.

4. What are interesing Γ-cycle theorems? Γ-cycle theorems form a natural class of geometric theorems with almost trivial proofs. Classify them!

5. Is there a genus of an incidence theorem?   The manifold proofs come along with a natural topology. In particular we have seen two proofs of Pappos's theorem and both beared the structure of a torus. Is the topological type of the proof an invariant of the theorem? So far we were not able to find Ceva-Menelaus proofs for Pappos's theorem that had a different topological type (as long as we exclude the possibility of adding additional generic points).

6. Where is the complexity?   All our algorithms can find proofs in polynomial time. However proving of incidence theorems is provably hard, since one can in principle encode NP-hard problems. So if $P \neq NP$ there must be theorems that are not provable by these two methods. Is there still a way to add auxiliary points to those configurations such that manifold proofs become available? The complexity of finding the proof would then be hidden in finding the right auxiliary constructions.

References

1. Apel, S.: A comparison of Binomial proofs and Ceva-/Menelaus proofs for real projective incidence theorems, Bachelor Thesis, TU Munich (2009)
2. Bokowski, J., Richter, J.: On the finding of final polynomials. Europ. J. Combinatorics 11, 21–34 (1990)
3. Crapo, H., Richter-Gebert, J.: Automatic proving of geometric theorems. In: White, N. (ed.) Invariant Methods in Discrete and Computational Geometry, pp. 107–139. Kluwer Academic Publishers, Dordrecht (1995)
4. Dress, A.W.M., Wenzel, W.: Endliche Matroide mit Koeffizienten. Bayreuth. Math. Schr. 24, 94–123 (1978)
5. Dress, A.W.M., Wenzel, W.: Geometric Algebra for Combinatorial Geometries. Adv. in Math. 77, 1–36 (1989)
6. Dress, A.W.M., Wenzel, W.: Grassmann-Plücker Relations and Matroids with Coefficients. Adv. in Math. 86, 68–110 (1991)
7. Fearnley-Sander, D.: Plane Euclidean Reasoning. In: Wang, D., Yang, L., Gao, X.-S. (eds.) ADG 1998. LNCS (LNAI), vol. 1669, pp. 86–110. Springer, Heidelberg (1999)
8. Grünbaum, B., Shephard, G.C.: Ceva, Menelaus, and the Area Principle. Mathematics Magazine 68, 254–268 (1995)
9. Grünbaum, B., Shephard, G.C.: A new Ceva-type theorem. Math. Gazette 80, 492–500 (1996)
10. Grünbaum, B., Shephard, G.C.: Ceva, Menelaus, and Selftransversality. Geometriae Dedicata 65, 179–192 (1997)
11. Grünbaum, B., Shephard, G.C.: Some New Transversality Properties. Geometriae Dedicata 71, 179–208 (1998)
12. Maurer, S.B.: Matroid basis graphs I. J. Combin. Theory B 14, 216–240 (1973)
13. Richter-Gebert, J.: Mechanical theorem proving in projective geometry. Annals of Mathematics and Artificial Intelligence 13, 139–172 (1995)
14. Richter-Gebert, J.: On the Realizability Problem of Combinatorial Geometries – Decision Methods, Doktoral Thesis, TU Darmstadt (1992)
15. Richter-Gebert, J.: Perspective on Projective Geometry, p. 580. Springer, Heidelberg (2011)

16. Richter-Gebert, J.: Meditations on Ceva's Theorem. In: Davis, C., Ellers, E. (eds.) The Coxeter Legacy: Reflections and Projections, pp. 227–254. American Mathematical Society, Fields Institute (2006)
17. Sturmfels, B.: Algorithms in Invariant Theory. Springer, Wien (1993)
18. Wenzel, W.: A Group-Theoretic Interpretation of Tutte's Homotopy Theory. Adv. in Math. 77, 27–75 (1989)
19. Wenzel, W.: Maurer's Homotopy Theory fo Even Δ-Matroids and Related Combinatorial Geometries. J. Combin. Theory A 71, 19–59 (1995)
20. White, N.: The Bracket Ring of Combinatorial Geometry I. Transactions AMS 202, 79–95 (1975)

Exploring the Foundations of Discrete Analytical Geometry in Isabelle/HOL

Jacques Fleuriot

Centre for Intelligent Systems and their Applications
School of Informatics, University of Edinburgh,
EH8 9AB, United Kingdom
jdf@inf.ed.ac.uk

Abstract. This paper gives an overview of the formalization of the Harthong-Reeb integer number system (HR_ω) in the proof-assistant Isabelle. The work builds on an existing mechanization of nonstandard analysis and describes how the basic notions underlying HR_ω can be recovered and shown to have their expected properties, without the need to introduce any axioms. We also look at the formalization of the well-known Euler method over the new integers and formally prove that the algorithmic approximation produced can be made to be infinitely-close to its continuous counterpart. This enables the discretization of continuous functions and of geometric concepts such as the straight line and ellipse and acts as the starting point for the field of discrete analytical geometry.

Keywords: discrete geometry, nonstandard analysis, Harthong-Reeb numbers, Euler method, mechanical theorem proving, Isabelle.

1 Introduction

The Harthong-Reeb number system HR_ω is a well-established discrete model of the continuum [10] stemming from nonstandard analysis (NSA) [15]. It can provide a systematic framework in which a continuous function, e.g. an ellipse, can be discretized and the results then shown to be equivalent to the original continuous function through the use of rigorous arguments based on infinitesimals and infinitely large numbers.

In this work, we carry out a rigorous formalization of HR_ω in the proof assistant Isabelle [13]. This is both an exercise in formalized mathematics and a means of extending the number systems available for reasoning about geometric problems in the theorem prover. In some past work [7,4], we already combined NSA with geometry to formally explore the properties of novel infinitesimal and infinite geometric notions. Although the current work is in the same spirit, we do *not* claim to have established any of the mathematical foundations but merely to be verifying some of the interesting results that a team of people have been producing recently in discrete geometry [1,2,9].

As a result, our article deliberately follows in structure much of the exposition given by Chollet *et al.* [2], which purposefully revisits the original NSA approach

P. Schreck, J. Narboux, and J. Richter-Gebert (Eds.): ADG 2010, LNAI 6877, pp. 34–50, 2011.
© Springer-Verlag Berlin Heidelberg 2011

to discrete analytical geometry (DAG). We examine their approach and show that their results can be rigorously mechanized in Isabelle/HOL as a conservative extension of our existing theories i.e. without introducing any new axioms.

2   On Nonstandard Analysis in Isabelle

The development of nonstandard analysis in Isabelle [8] is based on the extensional approach, first introduced by Robinson [15]. The system contains new types of numbers which are nonstandard extensions of the usual number systems. Thus, the hypernaturals $\mathbb{N}^*$ and hyperreals $\mathbb{R}^*$ are the extensions of the natural $\mathbb{N}$ and real numbers $\mathbb{R}$, respectively, and contain new, well-defined notions such as infinitely large numbers and infinitesimals. In Isabelle, all the nonstandard number systems are obtained through the so-called ultrapower construction [8].

While we shall not delve into the details of the development of NSA in Isabelle (the interested reader may consult a number of papers on this [8,6]), we wish to note that this approach differs from the one that is usually used in the presentation of the Harthong-Reeb numbers, where a minimal, axiomatic form of nonstandard analysis, related to Nelson's Internal Set Theory (IST) [12], is preferred [1]. We note also that the aim of the current work is not to advocate our version of NSA as an alternative to the one that is usually used but to show that the Harthong-Reeb number system, its properties, and use can be formalized as conservative extensions of our existing mechanization.

2.1   Nonstandard Numbers

In the axiomatic version of NSA, a new predicate $\lim(x)$ is introduced to indicate that a number x is limited (the predicate standard(x) is also often used e.g. in IST). Informally, this enables the theory to distinguish between what the extensional version of NSA would classify as finite and infinite nonstandard numbers. If we consider the hypernaturals, for instance, then the limited numbers are just the familiar natural numbers — denoted by Nats in Isabelle — while the non-limited numbers correspond to the infinitely large hypernaturals, denoted by HNatInfinite in Isabelle. Using this idea, we can straightforwardly define the set of *limited hyperreal* numbers in Isabelle/HOL:[1]

$$\text{Limited} = \{x :: \text{hypreal}. \, \exists n \in \text{Nats}. \, |x| < \text{hypreal\_of\_hypnat } n\}$$

where hypreal_of_hypnat (n) is the function that embeds a limited hypernatural number (cf. a finite natural number) in the hyperreals. We note here the need for such embedding functions that enable one type of number to be mapped into another type. These are pervasive to our formalization as the simply-typed system of Isabelle does not allow subtyping. For the rest of this paper, though, unless essential, we shall omit these functions from our descriptions.

---

[1] In Isabelle, $x :: \tau$ means that x is of type τ.

Now, our existing theories of NSA in Isabelle already define the notion of finite hyperreal numbers:

$$\mathsf{HFinite} = \{x :: \mathsf{hypreal}.\, \exists r \in \mathbb{R}.\, |x| < r\}$$

and we easily prove using the Archimedean properties of the reals that the sets HFinite and Limited are in fact equal. This means that we can use all our existing theorems about finite hyperreals when dealing with limited numbers in Isabelle.

In a similar fashion, we define various other sets of numbers [16] and in each case prove that they are simply variants of already mechanized definitions:

- InfiniteLarge $= \{x.\, \forall n \in \mathsf{Nats}.\, n < |x|\} = \{x.\, \forall r \in \mathbb{R}.\, r < |x|\}$.
- InfiniteSmall $= \{x.\, \forall n.\, |x| < \mathsf{inverse}\,(\mathsf{Suc}\,n)\} = \{x.\, \forall r \in \mathbb{R}.\, 0 < r \to |x| < r\}$.
- Appreciable $= -(\mathsf{InfiniteSmall} \cup \mathsf{InfiniteLarge})$, $-$ denoting set complement.

where InfiniteSmall denotes the set of numbers smaller than any real numbers i.e. infinitesimals.

2.2   Two New Relations on the Hyperreals

Aside from the usual relations, such as equality and ordering, and the usual operations, such as addition, multiplication, and inverse, it is possible to define new relations on the hyperreals. In particular, a crucial one that arises from the existence of infinitely small numbers is the *infinitely close* relation:

$$\mathsf{inf\_close} :: [\mathsf{hypreal}, \mathsf{hypreal}] \Rightarrow \mathsf{bool}\ (\mathsf{infixl} \approx 50)$$
$$(x \approx y) \equiv (x - y) \in \mathsf{InfiniteSmall}$$

This relation, which is easily shown to be an equivalence relation, has numerous properties with respect to the usual relations and algebraic operations on numbers that are already formalized in Isabelle [8]. We can also use it to relax the partial ordering that $\leq$ imposes on the hyperreals and introduce a new relation:

$$\mathsf{less\_inf\_close} :: [\mathsf{hypreal}, \mathsf{hypreal}] \Rightarrow \mathsf{bool}\ (\mathsf{infixl} \lesssim 50)$$
$$(x \lesssim y) \equiv (x < y) \vee (x \approx y)$$

With these definitions mechanized, the next step is to formalize the actual Harthong-Reeb numbers (HR_ω) and show that, within our current interpretation, the operations and relations on HR_ω have the expected properties.

3   The Harthong-Reeb System

The construction of HR_ω requires as unit an infinitely large natural number ω (i.e. a positive infinite integer), whose existence could be asserted via an axiom [1], but which in our case can be picked from our existing set of infinite hypernaturals

NatInfinite. As we wish to pick a fixed but arbitrary infinite hypernatural number as our ω, we use the Hilbert epsilon operator to do so:

$$\omega \equiv (\varepsilon n.\, n \in \mathsf{HNatInfinite})$$

This ensures that none of our definitions or proofs involving ω actually depend on its value but merely on its existence. Once this is done, we capture directly the notion of limited integers at the scale ω [1]:

$$\mathsf{HR}_\omega :: \mathsf{hypint\ set}$$
$$\mathsf{HR}_\omega = \{x.\, \exists n \in \mathsf{Nats}.\, |x| \leq n\omega\}$$

where the type hypint denotes the type of hyperintegers $\mathbb{Z}^*$. Thus, HR_ω is just a subset of $\mathbb{Z}^*$ in our current formalization (although we could go one step further and define it as a new Isabelle type in its own right).

Next, we define various relations and algebraic operations that can be used on HR_ω. These include equality and ordering at the scale ω, addition, multiplication, inverse, as well as zero 0_ω and the unit 1_ω:

$$\mathsf{eq\_mega} \quad :: [\mathsf{hypint}, \mathsf{hypint}] \Rightarrow \mathsf{bool}\ (\mathsf{infixl} =_\omega 50)$$
$$(x =_\omega y) \equiv (\forall n \in \mathbb{N}.\, n\,|x - y| \leq \omega)$$
$$\mathsf{le\_omega} \quad :: [\mathsf{hypint}, \mathsf{hypint}] \Rightarrow \mathsf{bool}\ (\mathsf{infixl} \leq_\omega 50)$$
$$(x \leq_\omega y) \equiv (x \leq y \vee x =_\omega y)$$
$$\mathsf{less\_omega} :: [\mathsf{hypint}, \mathsf{hypint}] \Rightarrow \mathsf{bool}\ (\mathsf{infixl} <_\omega 50)$$
$$(x <_\omega y) \equiv (\exists n \in \mathbb{N}.\, \omega < n(y - x))$$

$$\mathsf{omegazero} :: \mathsf{hypint}$$
$$0_\omega \equiv 0$$
$$\mathsf{omegaone} :: \mathsf{hypint}$$
$$1_\omega \equiv \omega$$
$$\mathsf{uminus\_omega} :: \mathsf{hypint} \Rightarrow \mathsf{hypint}\ (-_\omega\ \_[81]80)$$
$$-_\omega\, x \equiv -x$$
$$\mathsf{inverse\_omega} :: \mathsf{hypint} \Rightarrow \mathsf{hypint}\ (\mathsf{inverse}_\omega\ \_)\ \mathsf{where}$$
$$\mathsf{inverse}_\omega\, x \equiv \left\lfloor \frac{\omega^2}{x} \right\rfloor$$
$$\mathsf{add\_omega} :: [\mathsf{hypint}, \mathsf{hypint}] \Rightarrow \mathsf{hypint}\ (\mathsf{infixl} +_\omega 65)$$
$$x +_\omega y \equiv x + y$$
$$\mathsf{mult\_omega} :: [\mathsf{hypint}, \mathsf{hypint}] \Rightarrow \mathsf{hypint}\ (\mathsf{infixl} \star_\omega 70)$$
$$x \times_\omega y \equiv \left\lfloor \frac{xy}{\omega} \right\rfloor$$

Note that our definitions are actually over the hyperintegers rather than the subset HR_ω. This is not a problem, and in fact, we can formally prove that many of the properties hold independently of HR_ω. For instance, aside from proving that $=_\omega$ is an equivalence relation and that $\leq_\omega$ is reflexive, anti-symmetric, and transitive, we also mechanize the following (expected) theorems directly over $\mathbb{Z}^{*}$:[2]

- $\neg\, x <_\omega x$
- $x \leq_\omega y \leftrightarrow x <_\omega y \vee x =_\omega y$
- $x <_\omega y \leftrightarrow x <_\omega y \vee x =_\omega y$
- $[\![x =_\omega x'; y =_\omega y'; x \leq_\omega y]\!] \implies x' \leq_\omega y'$ which, though "easy to see" [16], required some effort to prove formally.

We also show that the algebraic operations are well-behaved over HR_ω by deriving all the expected closure rules:

- $x \in \mathsf{HR}_\omega \implies -_\omega x \in \mathsf{HR}_\omega$
- $[\![x \in \mathsf{HR}_\omega; y \in \mathsf{HR}_\omega]\!] \implies x + y \in \mathsf{HR}_\omega$
- $x \in \mathsf{HR}_\omega \Rightarrow \mathsf{inverse}_\omega\, x \in \mathsf{HR}_\omega$
- $[\![x \in \mathsf{HR}_\omega; y \in \mathsf{HR}_\omega]\!] \implies x \times_\omega y \in \mathsf{HR}_\omega$

While the first two rules are trivially proved, the last two require somewhat more work as they involve case-splits on the variables involved.

With this done, we then mechanize all the additional properties — the commutativity and associativity of addition and multiplication, the existence of additive and multiplicative identities (0_ω and 1_ω respectively,) and inverses ($-_\omega$ and $\mathsf{inverse}_\omega$ respectively), and the distributivity of multiplication over addition — required to demonstrate that HR_ω form a field. With the exception of the proofs involving multiplication and the inverse operation, the properties are all straightforward to prove. For these two operations, though, we need to decompose the hyperreals into their integral and fractional parts using the floor operation. In particular, our NSA theory defines the fractional part of a hyperreal as follows:

$$\mathsf{hpart} :: \mathsf{hypreal} \Rightarrow \mathsf{hypreal}$$
$$\{x\} \equiv x - \mathsf{hypreal\_of\_hypint}\, \lfloor x \rfloor \tag{1}$$

As an illustration, we can consider the proof of the following theorem as mechanized in Isabelle:

$$[\![\neg(x =_\omega 0_\omega); x \in \mathsf{HR}_\omega]\!] \implies x \times_\omega (\mathsf{inverse}_\omega\, x) =_\omega 1_\omega$$

Proof. From the definition of 1_ω, multiplication, inverse, and $=_\omega$, the conclusion becomes:

$$n \left| \left| \frac{x \left\lfloor \frac{\omega^2}{x} \right\rfloor}{\omega} \right| - \omega \right| \leq \omega$$

---

[2] Note that in Isabelle the notation $[\![\alpha_1, \ldots, \alpha_n]\!] \implies \beta$ can be informally read as "if $\alpha_1 \wedge \ldots \wedge \alpha_n$ then β".

where n is an arbitrary limited (finite) natural number. Using definition (1), we can replace the floor function once to get:

$$n \left| \frac{x \left\lfloor \frac{\omega^2}{x} \right\rfloor}{\omega} - \left\{ \frac{x \left\lfloor \frac{\omega^2}{x} \right\rfloor}{\omega} \right\} - \omega \right| \leq \omega$$

which simplifies to:

$$n \left| x \left\lfloor \frac{\omega^2}{x} \right\rfloor - \omega \left\{ \frac{x \left\lfloor \frac{\omega^2}{x} \right\rfloor}{\omega} \right\} - \omega^2 \right| \leq \omega^2$$

and using (1) again (where appropriate) and simplifying, we have:

$$n \left| -x \left\{ \frac{\omega^2}{x} \right\} - \omega \left\{ \frac{x \left\lfloor \frac{\omega^2}{x} \right\rfloor}{\omega} \right\} \right| \leq \omega^2$$

and using the triangle equality (and some arithmetic), the goal becomes:

$$n \left| x \left\{ \frac{\omega^2}{x} \right\} \right| + n \left| \omega \left\{ \frac{x \left\lfloor \frac{\omega^2}{x} \right\rfloor}{\omega} \right\} \right| \leq \frac{\omega^2}{2} + \frac{\omega^2}{2}$$

which means having to prove: $n \left| x \left\{ \frac{\omega^2}{x} \right\} \right| \leq \frac{\omega^2}{2}$ and $n \left| \left\{ \frac{x \left\lfloor \frac{\omega^2}{x} \right\rfloor}{\omega} \right\} \right| \leq \frac{\omega^2}{2}$.

Since for any x, we have $0 \leq \{x\} < 1$, and $n < \omega$ for all limited n, the second subgoal is easily proved, while the first one becomes $n |x| \leq \frac{\omega^2}{2}$. Now, since $x \in \mathsf{HR}_\omega$, this means that $|x| \leq m\omega$ for some limited m. Thus, $n |x| \leq nm\omega \leq \frac{\omega^2}{2}$, since both n and m are limited. □

A number of other theorems, some of them given by Wallet [16], are also mechanized:

- $[\![x =_\omega x'; y =_\omega y'; x \leq_\omega y]\!] \implies x +_\omega x' +_\omega y'$
- $[\![u \in \mathsf{HR}_\omega; x =_\omega y]\!] \implies u \times_\omega x =_\omega u \times_\omega y$
- $[\![x \in \mathsf{HR}_\omega; y \in \mathsf{HR}_\omega; x' \in \mathsf{HR}_\omega; y' \in \mathsf{HR}_\omega; x =_\omega x'; y =_\omega y'; x \leq_\omega y]\!]$
  $\implies x +_\omega x' \leq_\omega y +_\omega y'$
- $[\![x \in \mathsf{HR}_\omega; x' \in \mathsf{HR}_\omega; \neg(x =_\omega 0_\omega); \neg(x' =_\omega 0_\omega); x =_\omega x']\!]$
  $\implies \mathsf{inverse}_\omega\, x =_\omega \mathsf{inverse}_\omega\, x'$

We note the need for the assumption $\neg(x' =_\omega 0_\omega)$ in the last theorem, which seems to have been omitted in the paper by Wallet [16].

3.1   From HR_ω to the (Limited) Hyperreals and Back

The next step in our formalization is to relate HR_ω to our existing limited hyperreals. This can be done through the formalization of two maps [1]:

$$\mathsf{Limited\_of\_HR} \;::\; \mathsf{hypint} \Rightarrow \mathsf{hypreal}$$
$$\varphi_\omega(x) \equiv \frac{x}{\omega}$$
$$\mathsf{HR\_of\_Limited} \;::\; \mathsf{hypreal} \Rightarrow \mathsf{hypint}$$
$$\psi_\omega(x) \equiv \lfloor \omega x \rfloor$$

We note here that the two functions are defined over the hyperintegers and the hyperreals respectively. However, the following easily mechanized lemmas show that the maps behave as expected:

- $z \in \mathsf{HR}_\omega \implies \varphi_\omega(z) \in \mathsf{Limited}$
- $x \in \mathsf{Limited} \implies \psi_\omega(x) \in \mathsf{HR}_\omega$

We then verify the claim [1] that φ_ω is an isomorphism from the system $(\mathsf{HR}_\omega, =_\omega, \leq_\omega, +_\omega, \times_\omega)$ to $(\mathsf{Limited}, \approx, \lesssim, +, \times)$ and that ψ_ω is the inverse isomorphism by mechanizing all of the following theorems:

- $\varphi_\omega(0_\omega) \approx 0$, $\varphi_\omega(1_\omega) \approx 1$ and $\psi_\omega(0) =_\omega 0_\omega$, $\psi_\omega(1) =_\omega 1_\omega$
- $x =_\omega y \leftrightarrow \varphi_\omega(x) \approx \varphi_\omega(y)$
- $x \leq_\omega y \leftrightarrow \varphi_\omega(x) \lesssim \varphi_\omega(y)$
- $\varphi_\omega(\psi_\omega(x)) \approx x$ and $\psi_\omega(\varphi_\omega(z)) =_\omega z$
- $\varphi_\omega(x +_\omega y) \approx \varphi_\omega(x) + \varphi_\omega(y)$ and $\psi_\omega(x + y) =_\omega \psi_\omega(x) +_\omega \psi_\omega(y)$
- $\varphi_\omega(x \times_\omega y) \approx \varphi_\omega(x) \times \varphi_\omega(y)$ and
  $[\![x \in \mathsf{Limited}; y \in \mathsf{Limited}]\!] \implies \psi_\omega(x \times y) =_\omega \psi_\omega(x) \times_\omega \psi_\omega(y)$ $\hspace{2em}(\star)$
- $x \in \mathsf{Limited} \Rightarrow \exists z \in \mathsf{HR}_\omega. \varphi_\omega(z) \approx x$ and $\hspace{2em}(\dagger)$
  $y \in \mathsf{HR}_\omega \implies \exists x \in \mathsf{Limited}. \psi_\omega(x) =_\omega y$

Most of these properties are fairly easy to formalize, with the exception of $(\dagger)$, which required some more effort since it involves unfolding the various definitions and working with the floor function. We note also that theorem $(\star)$ is not provable without explicitly stating that all the numbers involved are limited. Finally, it may be worth remarking on the similarity between theorem $(\dagger)$ and a well-know theorem of NSA known as the Standard Part theorem:

$$x \in \mathsf{Limited} \implies \exists r \in \mathbb{R}. x \approx r$$

although $(\dagger)$ is much stronger since for any limited hyperreal, it tells us that we can find a Harthong-Reeb integer infinitely close to it.

4   Arithmetizing and Mechanizing Euler's Method

Consider the problem of approximating a continuous function $y = f(x)$ on $x \geq a$ which satisfies the differential equation

$$y' = F(x, y)$$

and the initial condition

$$y(a) = b$$

in which b is a given constant. There is a well-known geometrical method developed by Euler [3] to prove that the initial value problem given by these equations has a solution. This may given by the following set of recursive equations (we follow a presentation similar to our main source [1] here):

$$
\begin{cases}
x_0 = a \\
x_{n+1} = x_n + h \\
\\
y_0 = b \\
y_{n+1} = y_n + hF(x_n, y_n)
\end{cases}
\tag{2}
$$

Thus, geometrically, beginning from the starting point y_0, the algorithm takes a (small) step h along the tangent at x_0 to give the new point y_1. This reasoning is then repeated for y_1 and so on, thereby computing the approximation $y_0, y_1, y_2, y_3, \ldots$ to the original function $y = f(x)$. It can be shown that, if $y = f(x)$ is C^2 and f is Lipschitz then, as h decreases to zero, the error $|y(x_n) - y_n|$ also decreases to zero i.e. the Euler method converges.

The next step in our formalization is to capture (2) above in our Isabelle framework. We do so directly using Isabelle's primitive recursive package. We first specify a step function:

$$
\begin{aligned}
&\text{hEulerStep} &&:: \ \text{hypreal} \Rightarrow \text{hypreal} \Rightarrow \text{nat} \Rightarrow \text{hypreal} \\
&\text{hEulerStep } h \, a \, 0 &&= a \\
&| \ \text{hEulerStep } h \, a \, (\text{Suc } n) &&= \text{hEulerStep } h \, a \, n + h
\end{aligned}
\tag{3}
$$

and then the actual Euler method approximation (or Euler scheme):[3]

$$
\begin{aligned}
&\text{hEulerScheme} &&:: \ (\text{hypreal} \Rightarrow \text{hypreal} \Rightarrow \text{hypreal}) \\
& && \Rightarrow \text{hypreal} \Rightarrow \text{hypreal} \Rightarrow \text{hypreal} \Rightarrow \text{nat} \Rightarrow \text{hypreal} \\
&\text{hEulerScheme } F \, h \, a \, b \, 0 &&= b \\
&| \ \text{hEulerScheme } F \, h \, a \, b \, (\text{Suc } n) &&= \text{hEulerScheme } F \, h \, a \, b \, n \\
& && + hF(\text{hEulerStep } h \, a \, n) \, (\text{hEulerScheme } F \, h \, a \, b \, n)
\end{aligned}
\tag{4}
$$

Note that we define (3) and (4) over the hyperreals as we wish to consider the behaviour of (2) with infinitely small steps h and, intuitively, the hyperreals of our extensional NSA theory can be viewed as the reals obtained using the internal version of NSA (cf. Section 2.1).

4.1 Formally Verifying an Arithmetization at the Scale ω

With (3) and (4) formalized, the next step is to provide an alternative, discrete version of the Euler scheme over HR_ω that can be shown to be infinitely close to (4) under the mapping φ_ω. For this, we can use the definition provided by Chollet et al. as a candidate discrete scheme [1], namely:

---

[3] Note that our Isabelle functions are curried so instead of $F(x, y)$, we write $F \, x \, y$.

$$\begin{cases} X_0 = A \\ X_{n+1} = X_n + \alpha \\ \\ Y_0 = B \\ Y_{n+1} = Y_n + F_\omega(X_n, Y_n) \div \beta \end{cases} \tag{5}$$

where $\div$ denotes Euclidean division, $A \equiv \lfloor \omega a \rfloor = \psi_\omega(a)$, $B \equiv \lfloor \omega b \rfloor = \psi_\omega(b)$, and $F_\omega(X_n, Y_n) \equiv \lfloor \omega F(\frac{X_n}{\omega}, \frac{Y_n}{\omega}) \rfloor = \psi_\omega(F(\varphi_\omega(X_n), \varphi_\omega(Y_n)))$ is an *arithmetization* or discretization of $F(X_n, Y_n)$. The interested reader should consult the article [1] from which (5) is extracted for some further motivation. These functions are formalized recursively as follows:

$$
\begin{aligned}
&\mathsf{HR\_EulerStep} &&:: \ \mathsf{hypint} \Rightarrow \mathsf{hypreal} \Rightarrow \mathsf{nat} \Rightarrow \mathsf{hypint} \\
&\mathsf{HR\_EulerStep}\, \alpha\, a\, 0 &&= \psi_\omega(a) \\
&|\ \mathsf{HR\_EulerStep}\, \alpha\, a\, (\mathsf{Suc}\, n) &&= \mathsf{HR\_EulerStep}\, \alpha\, a\, n + \alpha
\end{aligned}
\tag{6}
$$

and

$$
\begin{aligned}
&\mathsf{HR\_EulerScheme} \quad :: (\mathsf{hypreal} \Rightarrow \mathsf{hypreal} \Rightarrow \mathsf{hypreal}) \Rightarrow \mathsf{hypint} \\
&\qquad\qquad\qquad\qquad \Rightarrow \mathsf{hypint} \Rightarrow \mathsf{hypreal} \Rightarrow \mathsf{hypreal} \Rightarrow \mathsf{nat} \Rightarrow \mathsf{hypint} \\
&\mathsf{HR\_EulerScheme}\, F\, \alpha\, \beta\, a\, b\, 0 = \psi_\omega(b) \\
&|\ \mathsf{HR\_EulerScheme}\, F\, \alpha\, \beta\, a\, b\, (\mathsf{Suc}\, n) = \mathsf{HR\_EulerScheme}\, F\, \alpha\, \beta\, a\, b\, n \\
&\quad + \psi_\omega(F(\varphi_\omega(\mathsf{HR\_EulerStep}\, \alpha\, a\, n))\, (\varphi_\omega(\mathsf{HR\_EulerScheme}\, F\, \alpha\, \beta\, a\, b\, n)))\, \mathsf{div}\, \beta
\end{aligned}
\tag{7}
$$

We note that the Isabelle definitions (6) and (7) capture scheme (5) exactly. Now, to formally prove that (5) is an arithmetization of (2), we need to show that:

- $\varphi_\omega(A) \approx a$,
- $\varphi_\omega(B) \approx b$, and
- $\varphi_\omega(Y_n) \approx y_n$, which boils down to proving $\varphi_\omega(F_\omega(X_n, Y_n)) \approx F(x_n, y_n)$ when doing a proof by induction on n

given that $x_n \equiv \frac{X_n}{\omega} = \varphi_\omega(X_n)$, $y_n \equiv \frac{Y_n}{\omega} = \varphi_\omega(Y_n)$, and $\omega = \alpha\beta$, where β is an infinitely large divisor of ω. The first two parts of the arithmetization proof are trivially mechanized as follows:

- $\varphi_\omega(\mathsf{HR\_EulerStep}\, \alpha\, a\, 0) \approx \mathsf{hEulerStep}\, h\, a\, 0$
- $\varphi_\omega(\mathsf{HR\_EulerScheme}\, F\, \alpha\, \beta\, a\, b\, 0) \approx \mathsf{hEulerScheme}\, F\, h\, a\, b\, 0$, a simple consequence of the fact that $\varphi_\omega(\psi_\omega(b)) \approx a$.

The third property is trickier and its mechanization will be discussed next. Formulated in Isabelle, we wish to mechanically prove, given that $\beta \in \mathsf{HNatInfinite}$ and $\omega = \alpha\beta$ that:

$$\varphi_\omega(\mathsf{HR\_EulerScheme}\, f\, \alpha\, \beta\, a\, b\, n) \approx \mathsf{hEulerScheme}\, f\, \frac{1}{\beta}\, a\, b\, n \tag{8}$$

We proceed by induction on the (standard) natural n. The base case is trivially discharged (as expected). Now for the step case, given the induction hypothesis:

$$\varphi_\omega(\mathsf{HR\_EulerScheme}\, f\, \alpha\, \beta\, a\, b\, n) \approx \mathsf{hEulerScheme}\, f\, \frac{1}{\beta}\, a\, b\, n$$

we need to prove:

$$\varphi_\omega(\mathsf{HR\_EulerScheme}\, f\, \alpha\, \beta\, a\, b\, n +$$
$$\psi_\omega(F(\varphi_\omega(\mathsf{HR\_EulerStep}\, \alpha\, a\, n))\,(\varphi_\omega(\mathsf{HR\_EulerScheme}\, F\, \alpha\, \beta\, a\, b\, n)))\,\mathrm{div}\,\beta)$$
$$\approx \mathsf{hEulerScheme}\, f\, \frac{1}{\beta}\, a\, b\, n + \frac{1}{\beta} F(\mathsf{hEulerStep}\, \frac{1}{\beta}\, a\, n)\,(\mathsf{hEulerScheme}\, F\, \frac{1}{\beta}\, a\, b\, n)$$

Since φ_ω is additive and $[\![x \approx y; x' \approx y']\!] \implies x + x' \approx y + y'$, the goal reduces to:

$$\varphi_\omega(\psi_\omega(F(\varphi_\omega(\mathsf{HR\_EulerStep}\, \alpha\, a\, n))\,(\varphi_\omega(\mathsf{HR\_EulerScheme}\, F\, \alpha\, \beta\, a\, b\, n)))\,\mathrm{div}\,\beta)$$
$$\approx \frac{1}{\beta} F(\mathsf{hEulerStep}\, \frac{1}{\beta}\, a\, n)\,(\mathsf{hEulerScheme}\, F\, \frac{1}{\beta}\, a\, b\, n)$$

By unfolding the definitions of φ_ω and of ψ_ω and using the fact that $x\,\mathrm{div}\,y = \left\lfloor \frac{x}{y} \right\rfloor = \left\lfloor \frac{1}{y} x \right\rfloor$ for $x, y \in \mathbb{Z}^*$, this can be rewritten to:

$$\frac{1}{\omega} \left(\left\lfloor \frac{1}{\beta} \left\lfloor \omega F\left(\frac{\mathsf{HR\_EulerStep}\, \alpha\, a\, n}{\omega} \right) \left(\frac{\mathsf{HR\_EulerScheme}\, F\, \alpha\, \beta\, a\, b\, n}{\omega} \right) \right\rfloor \right\rfloor \right)$$
$$\approx \frac{1}{\beta} F\left(\mathsf{hEulerStep}\, \frac{1}{\beta}\, a\, n \right)\left(\mathsf{hEulerScheme}\, F\, \frac{1}{\beta}\, a\, b\, n \right)$$

and using $\lfloor x \rfloor = x - \{x\}$ from (1), this becomes:

$$\frac{1}{\omega} \frac{1}{\beta} \left\lfloor \omega F\left(\frac{\mathsf{HR\_EulerStep}\, \alpha\, a\, n}{\omega} \right) \left(\frac{\mathsf{HR\_EulerScheme}\, F\, \alpha\, \beta\, a\, b\, n}{\omega} \right) \right\rfloor$$
$$-\frac{1}{\omega} \left\{ \frac{1}{\beta} \left\lfloor \omega F\left(\frac{\mathsf{HR\_EulerStep}\, \alpha\, a\, n}{\omega} \right) \left(\frac{\mathsf{HR\_EulerScheme}\, F\, \alpha\, \beta\, a\, b\, n}{\omega} \right) \right\rfloor \right\}$$
$$\approx \frac{1}{\beta} F\left(\mathsf{hEulerStep}\, \frac{1}{\beta}\, a\, n \right)\left(\mathsf{hEulerScheme}\, F\, \frac{1}{\beta}\, a\, b\, n \right)$$

Since $0 \leq \{x\} < 1$ for any x, and $\frac{1}{\omega}$ is infinitely large, this reduces to:

$$\frac{1}{\beta\omega} \left\lfloor \omega F\left(\frac{\mathsf{HR\_EulerStep}\, \alpha\, a\, n}{\omega} \right) \left(\frac{\mathsf{HR\_EulerScheme}\, F\, \alpha\, \beta\, a\, b\, n}{\omega} \right) \right\rfloor$$
$$\approx \frac{1}{\beta} F\left(\mathsf{hEulerStep}\, \frac{1}{\beta}\, a\, n \right)\left(\mathsf{hEulerScheme}\, F\, \frac{1}{\beta}\, a\, b\, n \right)$$

and using (1) again:

$$\frac{1}{\beta\omega} \left(\begin{array}{c} \omega F\left(\frac{\mathsf{HR\_EulerStep}\, \alpha\, a\, n}{\omega} \right) \left(\frac{\mathsf{HR\_EulerScheme}\, F\, \alpha\, \beta\, a\, b\, n}{\omega} \right) \\ - \left\{ \omega F\left(\frac{\mathsf{HR\_EulerStep}\, \alpha\, a\, n}{\omega} \right) \left(\frac{\mathsf{HR\_EulerScheme}\, F\, \alpha\, \beta\, a\, b\, n}{\omega} \right) \right\} \end{array} \right)$$
$$\approx \frac{1}{\beta} F\left(\mathsf{hEulerStep}\, \frac{1}{\beta}\, a\, n \right)\left(\mathsf{hEulerScheme}\, F\, \frac{1}{\beta}\, a\, b\, n \right)$$

and simplifying:

$$F\left(\frac{\text{HR\_EulerStep } \alpha\, a\, n}{\omega}\right)\left(\frac{\text{HR\_EulerScheme } F\, \alpha\, \beta\, a\, b\, n}{\omega}\right)$$
$$\approx F\left(\text{hEulerStep } \tfrac{1}{\beta}\, a\, n\right)\left(\text{hEulerScheme } F\, \tfrac{1}{\beta}\, a\, b\, n\right)$$

At this point in our mechanical proof though, we faced a difficulty as we could not proceed further with our existing hypotheses. The problem was a missing assumption about the continuity of F (which is assumed but does not seem to be used explicitly in the pen-and-paper proof [1]). This can be captured by having:

$$\forall xyx'y'.\, x \approx x' \wedge y \approx y' \rightarrow f\, x\, y \approx f\, x'\, y'$$

as an extra fact. Our goal now becomes the following 2 subgoals:

$$\tfrac{1}{\omega}\text{HR\_EulerStep } \alpha\, a\, n \approx \text{hEulerStep } \tfrac{1}{\beta}\, a\, n$$

$$\tfrac{1}{\omega}\text{HR\_EulerScheme } F\, \alpha\, \beta\, a\, b\, n \approx \text{hEulerScheme } F\, \tfrac{1}{\beta}\, a\, b\, n$$

which are easily proved since the second one is just the induction hypothesis while the first one becomes:

$$\frac{1}{\omega}(\lfloor \omega a \rfloor + n\alpha) \approx a + \frac{n}{\beta}$$

This last subgoal can be discharged since $\frac{1}{\omega}\lfloor \omega a \rfloor \approx a$ and $\frac{1}{\omega}n\alpha = \frac{n}{\beta}$ from our assumption that $\omega = \alpha\beta$. The theorem once mechanized looks thus:

$$[\![\beta \in \text{HNatInfinite}; \omega = \alpha\beta; \forall xyx'y'.\, x \approx x' \wedge y \approx y' \rightarrow f\, x\, y \approx f\, x'\, y']\!]$$
$$\implies \varphi_\omega(\text{HR\_EulerScheme } f\, \alpha\, \beta\, a\, b\, n) \approx \text{hEulerScheme } f\, \tfrac{1}{\beta}\, a\, b\, n$$

This result formally proves that, under the given assumptions, the Euler scheme (5) over HR_ω is an arithmetization at the scale ω of the initial scheme (2) over $\mathbb{R}^*$. We also note that our result differs slightly from the one by Chollet *et al.* [1] in that we did not require α to be an infinitely large hypernatural (in fact, this does not seem to be required for the proof if we assume that β is an infinitely large divisor of ω).

4.2   Interpreting the Arithmetization at the Scale β

As discussed by Chollet *et al.* [1], in the arithmetized Euler scheme (5), if α is taken to be infinitely large then the solution (X_n, Y_n) returned is a sequence of infinitely distant points since $X_{n+1} = X_n + \alpha$. In order to obtain points that are closer together, the authors perform a scaling that allows (5) to be interpreted at an intermediary scale. This is achieved by assuming that $\alpha = \beta$, thereby making $\omega = \beta^2$, and moving from the scale ω to the scale β by mapping every element

$x \in \mathsf{HR}_\omega$ to an element $x \div \beta$ since moving from HR_ω to HR_β can be achieved by taking the map $\psi_\beta(\varphi_\omega(x)) = \lfloor \beta \frac{x}{\omega} \rfloor = \lfloor \frac{x}{\beta} \rfloor = x \div \beta$.

This leads to a new discrete Euler scheme, which involves the quotient and remainder of integers in HR_ω under Euclidean division. Thus, we need to decompose every $x \in \mathsf{HR}_\omega$ as follows:

$$z = \widetilde{z}\beta + \widehat{z}$$

where $\widehat{z} = z \div \beta$ and $\widehat{z} = z \bmod \beta$. The following arithmetization of the Euler scheme (2) "computed at the scale $\omega = \beta^2$ and interpreted at the intermediary scale β" [1] is then given (without proof):

$$\begin{cases} \widetilde{X}_0 = A \div \beta \\ \widetilde{X}_{n+1} = \widetilde{X}_n + 1 \\[2mm] \widehat{Y}_0 = B \bmod \beta \\ \widehat{Y}_{n+1} = (\widetilde{Y}_n + \widetilde{F}_n) \bmod \beta \\[2mm] \widetilde{Y}_0 = B \div \beta \\ \widetilde{Y}_{n+1} = \widetilde{Y}_n + (\widetilde{Y}_n + \widetilde{F}_n) \div \beta \end{cases} \qquad (9)$$

where $A \equiv \lfloor \beta^2 a \rfloor$, $B = \lfloor \beta^2 b \rfloor$ and

$$\widetilde{F}_n \equiv F_\omega(\widetilde{X}_n \beta + A \bmod \beta, \widetilde{Y}_n \beta + \widehat{Y}_n) \div \beta$$

$$\equiv \left\lfloor \omega F \left(\frac{\widetilde{X}_n \beta + A \bmod \beta}{\omega}, \frac{\widetilde{Y}_n \beta + \widehat{Y}_n}{\omega} \right) \right\rfloor \div \beta$$

$$\equiv \psi_\omega(F(\varphi_\omega(\widetilde{X}_n \beta + A \bmod \beta), \varphi_\omega(\widetilde{Y}_n \beta + \widehat{Y}_n))) \div \beta$$

In (9), the variables of interest are $\widetilde{X}_i$ and $\widetilde{Y}_i$ since the sequence of pairs $(\widetilde{X}_i, \widetilde{Y}_i)$ is the graph of a discrete function $Y(X)$ that is meant to approximate our original continuous function $y = f(x)$.

The next steps in our mechanization therefore involve the formalization of this new scheme and then *proving* that it is an arithmetization of the initial scheme (2). The representation of (9) in Isabelle is slightly more complicated than that of (5) though as it involves mutually recursive functions. Nevertheless, with a little bit of effort, we can capture the step function as:

$$\begin{aligned} &\mathsf{HR\_EulerStepScale} &&:: \mathsf{hypreal} \Rightarrow \mathsf{hypint} \Rightarrow \mathsf{hypnat} \Rightarrow \mathsf{hypint} \\ &\mathsf{HR\_EulerStepScale}\, a\, \beta\, 0 &&= \lfloor \beta^2 a \rfloor \div \beta \\ &|\mathsf{HR\_EulerStepScale}\, a\, \beta\, (\mathsf{Suc}\, n) &&= \mathsf{HR\_EulerStepScale}\, a\, \beta\, n + 1 \end{aligned} \qquad (10)$$

and the quotient and remainder approximations as:

function
HR_EulerSchemeMod   :: (hypreal $\Rightarrow$ hypreal $\Rightarrow$ hypreal) $\Rightarrow$ hypint $\Rightarrow$ hypreal $\Rightarrow$ hypreal $\Rightarrow$ nat $\Rightarrow$ hypint
and
HR_EulerSchemeDiv   :: (hypreal $\Rightarrow$ hypreal $\Rightarrow$ hypreal) $\Rightarrow$ hypint $\Rightarrow$ hypreal $\Rightarrow$ hypreal $\Rightarrow$ nat $\Rightarrow$ hypint
  HR_EulerSchemeMod $F\,\beta\,a\,b\,0 = \lfloor\beta^2 b\rfloor \bmod \beta$
| HR_EulerSchemeDiv $F\,\beta\,a\,b\,0 = \lfloor\beta^2 b\rfloor \div \beta$
| HR_EulerSchemeMod $F\,\beta\,a\,b$ (Suc n) $=$
   (HR_EulerSchemeMod $F\,\beta\,a\,b\,n$
     $+\ \psi_\omega\,(F(\varphi_\omega\,(\text{HR\_EulerStepScale}\,a\,\beta\,n \times \beta + \lfloor\beta^2 a\rfloor \bmod \beta))$
        $(\varphi_\omega\,(\text{HR\_EulerSchemeDiv}\,F\,\beta\,a\,b\,n \times \beta + \text{HR\_EulerSchemeMod}\,F\,\beta\,a\,b\,n))) \div \beta)\bmod\beta$
| HR_EulerSchemeDiv $F\,\beta\,a\,b$ (Suc n) $=$
   HR_EulerSchemeDiv $F\,\beta\,a\,b\,n$
     $+\ (\text{HR\_EulerSchemeMod}\,F\,\beta\,a\,b\,n$
       $+\ \psi_\omega\,(F(\varphi_\omega\,(\text{HR\_EulerStepScale}\,a\,\beta\,n \times \beta + \lfloor\beta^2 a\rfloor \bmod \beta))$
          $(\varphi_\omega\,(\text{HR\_EulerSchemeDiv}\,F\,\beta\,a\,b\,n \times \beta + \text{HR\_EulerSchemeMod}\,F\,\beta\,a\,b\,n))) \div \beta)\div\beta$

$$(11)$$

We can prove the correctness of (9) by formally relating (11) to (7), which we know to be the arithmetized version of (2). The following theorems, all mechanized in Isabelle by inductive proofs, help demonstrate this:

- $\omega = \beta^2 \implies$ HR_EulerStepScale $a\,\beta\,n = $ HR_EulerStep $\beta\,a\,n \div \beta$
- $\omega = \beta^2 \implies$ HR_EulerSchemeDiv $F\,\beta\,a\,b\,n = $ HR_EulerScheme $F\,\beta\,\beta\,a\,b\,n \div \beta$
- $\omega = \beta^2 \implies$ HR_EulerSchemeMod $F\,\beta\,a\,b\,n = $ HR_EulerScheme $F\,\beta\,\beta\,a\,b\,n \bmod \beta$

From the last two theorems, we thus have that:

$$\omega = \beta^2 \implies \text{HR\_EulerScheme}\,F\,\beta\,\beta\,a\,b\,n = $$
$$\text{HR\_EulerSchemeDiv}\,F\,\beta\,a\,b\,n \times \beta + \text{HR\_EulerSchemeMod}\,F\,\beta\,a\,b\,n$$

which verifies the correctness of (11). With this final Euler scheme set up, we can now look at Reveillès' notion of a discrete analytic line [14], which acted as the starting point of a new approach in discrete geometry know as discrete analytical geometry [1].

5   A Verified Arithmetization of the Straight Line

We consider a straight line $\mathcal{L}$ in $\mathbb{R}^*$ with equation $y = cx + d$ where $c \in$ Limited and $d \in$ Limited. This is characterized by the differential equation:

$$y' = F(x, y) = c$$

and the initial condition:

$$y(0) = d$$

By plugging these values in our Isabelle definition(s) (11) and using rewriting, we trivially derive:

$$\omega = \beta^2 \implies \text{HR\_EulerSchemeDiv}\,(\lambda xy.\,c)\,\beta\,0\,d\,\text{Suc}\,n = $$
$$\text{HR\_EulerSchemeDiv}\,(\lambda xy.\,c)\,\beta\,0\,d\,n + $$
$$(\text{HR\_EulerSchemeMod}\,(\lambda xy.\,c)\,\beta\,0\,d\,n + \lfloor\beta^2 c\rfloor \div \beta) \div \beta$$

and

$$\omega = \beta^2 \implies \text{HR\_EulerSchemeMod}\,(\lambda xy.\,c)\,\beta\,0\,d\,\text{Suc}\,n = $$
$$(\text{HR\_EulerSchemeMod}\,(\lambda xy.\,c)\,\beta\,0\,d\,n + \lfloor\beta^2 c\rfloor \div \beta) \div \beta$$

which, along with HR_EulerStepScale $0\,\beta\,0 = \widetilde{X}_0 = 0$, corresponds to the following scheme, computed at the scale β^2 and interpreted at the scale β:

$$
\begin{cases}
\widetilde{X}_0 = 0 \\
\widetilde{X}_{n+1} = \widetilde{X}_n + 1 \\[1mm]
\widehat{Y}_0 = \lfloor \beta^2 d \rfloor \bmod \beta \\
\widehat{Y}_{n+1} = (\widehat{Y}_n + \lfloor \beta^2 c \rfloor \div \beta) \bmod \beta \\[1mm]
\widetilde{Y}_0 = \lfloor \beta^2 d \rfloor \div \beta \\
\widetilde{Y}_{n+1} = \widetilde{Y}_n + (\widehat{Y}_n + \lfloor \beta^2 c \rfloor \div \beta) \div \beta
\end{cases}
\tag{12}
$$

These derivations therefore verify the arithmetized scheme given by Chollet *et al.* for the straight line [1].

Now, assuming that the sequence of pairs $(\widetilde{X}_i, \widetilde{Y}_i)$ corresponds to the discrete function $Y(X)$, then the latter should approximate our original line $y = cx + d$. We demonstrate that this is indeed the case by first mechanizing the following theorem in Isabelle:

$$\omega = \beta^2 \implies$$
$$\text{HR\_EulerSchemeDiv}\,(\lambda xy.\,c)\,\beta\,0\,d\,n = \left\lfloor \frac{\lfloor \beta^2 c \rfloor \div \beta}{\beta} (\text{HR\_EulerStepScale}\,0\,\beta\,n) + \frac{\lfloor \beta^2 d \rfloor}{\beta} \right\rfloor$$

which means that the discrete function $Y(X)$ is given by:

$$
\omega = \beta^2 \implies Y(X) = \left\lfloor \frac{\lfloor \beta^2 c \rfloor \div \beta}{\beta} X + \frac{\lfloor \beta^2 d \rfloor}{\beta} \right\rfloor
\tag{13}
$$

The mechanical proof is similar to the pen-and-paper one [1] and only involves rewriting. As can be seen, the points given by (13) are those that are incident or just below the line $\mathcal{L}_\beta$ given by $y = (\lfloor \beta^2 c \rfloor \div \beta)/\beta x + \lfloor \beta^2 b \rfloor /\beta$, about which we also prove the following theorems:

$$\beta \in \text{HNatInfinite} \implies \frac{\lfloor \beta c \rfloor \div \beta}{\beta} \approx c \text{ and } \beta \in \text{HNatInfinite} \implies \frac{\lfloor \beta^2 d \rfloor}{\beta} \approx \beta d$$

This verifies the claim that for $d = 0$ the line $\mathcal{L}_\beta$ is infinitely close to the original line $\mathcal{L}$ [1] i.e. that the algorithm (12) produces a discrete line that is an infinitely good approximation of the original line. And finally, since

$$\left\{ \frac{\lfloor \beta^2 c \rfloor \div \beta}{\beta} (\text{HR\_EulerStepScale}\,0\,\beta\,n) + \frac{\lfloor \beta^2 d \rfloor}{\beta} \right\} =$$
$$\left(\frac{\lfloor \beta^2 c \rfloor \div \beta}{\beta} \right) \text{HR\_EulerStepScale}\,0\,\beta\,n - \text{HR\_EulerSchemeDiv}\,(\lambda xy.c)\,\beta\,0\,b\,n + \frac{\lfloor \beta^2 b \rfloor}{\beta}$$

we trivially have:

$$\omega = \beta^2 \implies$$
$$0 \le (\lfloor \beta^2 c \rfloor \div \beta)\,\text{HR\_EulerStepScale}\,0\,\beta\,n -$$
$$\beta\text{HR\_EulerSchemeDiv}\,(\lambda xy.c)\,\beta\,0\,b\,n + \lfloor \beta^2 b \rfloor \le \beta$$

This inequality leads to Reveillès original notion of an analytic discrete line [14,1], a concept that set the commencement of discrete analytical geometry and seems a fitting point to end the current detailed description of the formalization.

6   Discussion

Although much of this paper focused on the formalization of the framework for discrete analytical geometry and its properties, it is worthwhile to place these results within the broader context of mechanized geometric reasoning and formal verfication. In mechanizing scheme (9), for instance, we have effectively derived a general algorithm that can be used to discretize geometric objects so that they can be represented by discrete sets of points in a digitized setting such as a computer screen. The nonstandard formulation enables us to verify formally that the algorithm, when applied to particular geometric objects, ultimately yields correct, infinitely-close approximations of said objects i.e. we can obtain near-perfect approximation, if we allow the Euler method to take infinitely small steps.

Moreover, starting from the notion of a discrete line, as given in this paper, various of its properties can be *derived* that are the discrete counterparts of the usual Euclidean geometry axioms (e.g. a version of the axiom that two points determine a unique discrete line and a version of Euclid's Parallel Postulate). In fact, all the usual Euclidean notions and theorems can be recovered for suitably defined discrete objects [14]. This results in a so-called *ideal discrete geometry* – a perfect discretization of Euclidean geometry – which we are (further) formalizing in Isabelle/HOL. In this, for instance, one considers a discrete square screen of width ω and introduces a new, extended notion of points known as *big-points* [14]. The big-point of an integer point A is then a collection of points corresponding to the infinitesimal neighbourhood of A and can easily be shown, through the use of various concepts described in the current paper, to behave like the points encountered in continuous geometry. Big-points, along with other notions (such as the *shadow* of a discrete line), make the foundations that we have presented essential and of immediate relevance to explicit geometric reasoning in the idealized setting and we hope to report on this aspect of the work soon.

7   Conclusion and Further Work

This paper gave an overview of our mechanized treatment of the Harthong-Reeb number system, and of the ensuing geometric approximation algorithms, using the nonstandard analysis theory of Isabelle. In particular, we used the paper by Chollet, Wallet, Fuchs, Largeteau-Skapin and Andres as a blueprint to guide our development. All the results up to their discussion of the constructive nature of the Harthong-Reeb line are fully verified (although we did not present the arithmetization of the exponential function in this paper).

This successful mechanization is not entirely surprising since the original treatment of discrete analytical geometry had NSA at its core and the paper by Chollet *et al.* is meant to revisit this formally. However, it is still pleasing to see how a (pen-and-paper) mathematical framework can be realised rigorously and conservatively in a theorem prover. We hope to use these new theories to investigate the arithmetization of other geometric figures such as the ellipse [2] and formally derive the theoretical results that follow (e.g. connectivity properties). As already remarked in Section 6, we are also mechanizing many of the geometric results presented by Reveillès in one of his early papers [14]. The latter uses an external version of NSA which is identical to the one in Isabelle, making the formalization especially faithful.

As a final remark, we note that a related effort was started in Coq [11] prior to ours and, in fact, provided some of the motivation for starting the current mechanization. However, despite some initial similarities, the two projects have quite different goals: the Coq effort is much more ambitious and closer to the philosophy that underlies the Harthong-Reeb system since it provides hope of an approach that can harness the inherently constructive nature of the model [9,1]. Our mechanization, for its part, is based on a non-constructive approach to NSA and should be viewed more as a mathematical and geometrical exploration rather than one aimed at yielding executable programs via proofs.

Acknowledgements. I wish to thank Laurent Fuchs for introducing me to this research area. His enthusiastic discussion of their approach motivated the current attempt to explore their results and link my own formalization to theirs. I also wish to thank the GALAPAGOS project for providing me with the opportunity to discuss some of the ideas further. Finally, I would like to thank the referees for their useful feedback.

References

1. Chollet, A., Wallet, G., Fuchs, L., Largeteau-Skapin, G., Andres, E.: Insight in discrete geometry and computational content of a discrete model of the continuum. Pattern Recognition 42(10), 2220–2228 (2009)
2. Chollet, A., Wallet, G., Andres, E., Fuchs, L., Largeteau-Skapin, G., Richard, A.: Ω-Arithmetization of Ellipses. In: Barneva, R.P., Brimkov, V.E., Hauptman, H.A., Natal Jorge, R.M., Tavares, J.M.R.S. (eds.) CompIMAGE 2010. LNCS, vol. 6026, pp. 24–35. Springer, Heidelberg (2010)
3. Collected Work of L. Euler, vol. 11 (1913), vol. 12 (1914)
4. Fleuriot, J.D., Paulson, L.C.: A Combination of Nonstandard Analysis and Geometry Theorem Proving, With Application to Newton's Principia. In: Kirchner, C., Kirchner, H. (eds.) CADE 1998. LNCS (LNAI), vol. 1421, pp. 3–16. Springer, Heidelberg (1998)
5. Fleuriot, J.D., Paulson, L.C.: Proving Newton's Propositio Kepleriana Using Geometry and Nonstandard Analysis in Isabelle. In: Wang, D., Yang, L., Gao, X.-S. (eds.) ADG 1998. LNCS (LNAI), vol. 1669, pp. 47–66. Springer, Heidelberg (1999)
6. Fleuriot, J.: On the Mechanization of Real Analysis in Isabelle/HOL. In: Aagaard, M.D., Harrison, J. (eds.) TPHOLs 2000. LNCS, vol. 1869, pp. 145–161. Springer, Heidelberg (2000)

7. Fleuriot, J.: Theorem Proving in Infinitesimal Geometry. Logic Journal of the IGPL 9(3) (2001)
8. Fleuriot, J.: A Combination of Geometry Theorem Proving and Nonstandard Analysis with Application to Newton's Principia. Springer, Heidelberg (2001)
9. Fuchs, L., Largeteau-Skapin, G., Wallet, G., Andres, E., Chollet, A.: A First Look into a Formal and Constructive Approach for Discrete Geometry Using Nonstandard Analysis. In: Coeurjolly, D., Sivignon, I., Tougne, L., Dupont, F. (eds.) DGCI 2008. LNCS, vol. 4992, pp. 21–32. Springer, Heidelberg (2008)
10. Harthong, J.: Une théorie du Continu. La MathéMatique Non Standard. Editions du CNRS, 307–329 (1989)
11. Magaud, N., Chollet, A., Fuchs, L.: Formalizing a Discrete Model of the Continuum in Coq From a Discrete Geometry Perspective. In: Proceedings of the Automated Deduction in Geometry Workshop, Munich (2010)
12. Nelson, E.: Internal set theory: A new approach to nonstandard analysis. Bulletin of the American Mathematical Society 83(6), 1165–1198 (1977)
13. Paulson, L.C.: Isabelle - A Generic Theorem Prover. Springer, Heidelberg (1994)
14. Reveillès, J.-P., Richard, D.: Back and Forth Between Continuous and Discrete For The Working Computer Scientist. Annals of Mathematics and Artifical Intelligence 16, 89–152 (1996)
15. Robinson, A.: Non-standard analysis. North-Holland (1966)
16. Wallet, G.: Integer Calculus on the Harthong-Reeb Line. Revue Arima (9), 517–536 (2008)

A Formalization of Grassmann-Cayley Algebra in Coq and Its Application to Theorem Proving in Projective Geometry[*]

Laurent Fuchs[1] and Laurent Théry[2]

[1] XLIM-SIC UMR CNRS 6172 - Poitiers University, France
Laurent.Fuchs@sic.univ-poitiers.fr
[2] INRIA Sophia Antipolis - Méditerranée, France
Laurent.Thery@inria.fr

Abstract. This paper presents a formalization of Grassmann-Cayley algebra [6] that has been done in the Coq [2] proof assistant. The formalization is based on a data structure that represents elements of the algebra as complete binary trees. This allows to define the algebra products recursively. Using this formalization, published proofs of Pappus' and Desargues' theorem [7,1] are interactively derived. A method that automatically proves projective geometric theorems [11] is also translated successfully into the proposed formalization.

1 Introduction

A well-known application of Grassmann-Cayley algebra is automated theorem proving in projective geometry (see for example [1,4,10]). The usual method is to translate incidence statements of projective geometry into Grassmann-Cayley expressions. These expressions are then translated into bracket polynomials (i.e. the ring of projective invariants [17]). Finally, the bracket polynomial is factorised to get back an equivalent expression in Grassmann-Cayley algebra.

Our motivation in using a proof assistant such as Coq [2] is to capture in a single system all the various aspects of Grassmann-Cayley algebra: we want an abstract generic model on which we can not only reason but also perform both numerical and symbolical evaluations.

Hence, our formalization lets us not only formally check the manipulations of expressions within Grassmann-Cayley algebra but also compute with these very same expressions. De facto, it makes explicit the link between the abstract mathematical object and its applications. In the Coq proof assistant, proofs can be conducted interactively step by step or, using programmed tactics, the expressions can be reduced in a systematic manner. Note that, in our setting, most proofs are parametrized by the dimension of the algebra. So, our development is generic.

Once the formalization of the Grassmann-Cayley algebra is achieved, two kinds of proofs are considered. First some proofs are conducted interactively

---

[*] This work has been supported by the ANR Galapagos.

P. Schreck, J. Narboux, and J. Richter-Gebert (Eds.): ADG 2010, LNAI 6877, pp. 51–67, 2011.

following step-by-step what can be found in the literature, such as the proof of the Pappus' theorem in [7] or the proofs of Desargues' theorem in [1]. The second kind of proofs are conducted automatically following a method published in [11]. All the examples proposed in [11] have been tested successfully.

In future work, we also plan to connect our formalization with other approaches of incidence geometry, such as those based over ranks [12,13,14]. So, our work can be seen as a first step in the study of the formal correctness of automated proof methods in incidence geometry.

This paper is organized as follows. Section 2 introduces Grassmann-Cayley algebra and our choices for the formalization. Section 3 explains how the Grassmann-Cayley is formalized, how the algebra elements are represented and how the products are defined. Section 4 describes how the formalization can be use to prove theorems of incidence geometry, interactively and automatically.

2 Formal Grassmann-Cayley Algebra

Usually, in the literature, the products (join and meet) of the Grassmann-Cayley algebra are introduced by given equations defining their properties. So, they could have been defined in COQ using such an axiomatic approach. However, the main drawback of doing so is that we completely lose the computational aspect of this algebra. In particular, the axiomatic approach gives no hint of how the algebra could actually be implemented on a computer.

For this reason, we favor the definitional approach where the algebra operations are defined as recursive functions over the dimension of the algebra. First, we define a model, i.e. a data-structure that represents elements of the algebra. Then, on this model, we define the usual algebra operations (the join product, the meet product and the duality) and prove that they fulfill the axioms that are used to defined them in the literature. As this representation is quite unusual, we spend some time to detail our data-structure and the related operations.

2.1 The Underlying Vector Space

The Grassmann-Cayley algebra G_n is defined by adding a second product, the meet product, to the Grassmann algebra (or exterior algebra) of a vector space of dimension n, V, over a field K [6,1] where the join product (or the exterior product) is defined.

In order to have a concrete representation of the vectors of V, we need to represent them as n-tuples of K^n. This imposes a basis for V, say the canonical basis $e_n^i = (\delta_{i,0}, \ldots, \delta_{i,i}, \ldots, \delta_{i,n-1})$ where $i = 0, \ldots, n-1$ and $\delta_{i,j}$ is the Kronecker symbol. Then V is seen as the set of n-tuples, K^n.

As we will see this choice also induces the definition of a basis for G_n and this leads to an important change of view in the presentation of the algebra compared to the usual coordinate-free presentation. The elements are represented via their coordinates.

However, in the Coq proof assistant, this does not force us to deal only with numerical computations. As all the axiomatic properties of the algebra operations are proved, we can also reason symbolically using the coordinate-free presentation. Hence, we obtain an abstract generic model on which both numerical and symbolical evaluations can be performed.

2.2   The Join Product

The first step is to define the join product denoted by $\vee$. It is an associative antisymmetric bilinear product and it can be defined axiomatically by:

$$
\begin{array}{ll}
a \vee a = 0 & \lambda a \vee b = \lambda(a \vee b) \\
b \vee a = -a \vee b & (a + b) \vee c = a \vee c + b \vee c
\end{array}
\tag{1}
$$

where a and b are vectors of K^n.

The join product $a \vee b$ of two vectors a and b is non-zero if and only if a and b are linearly independent. If $a \vee b$ is non-zero, it is a grade 2 element of the algebra. More generally, if $\{a_1, \ldots, a_k\}$ are linearly independent vectors of K^n then $a_1 \vee \cdots \vee a_k$ is an element of grade k. Such elements, that are join product of vectors, are called *extensors* or *decomposable k-vectors*. Not all elements of grade k are extensors, they could be linear combination of extensors. In that case they are called *homogeneous vectors* or *k-vectors*. Elements that are linear combination of elements with different grades are the general elements of the algebra. They are called *multi-vectors*.

On the basis elements $\{e_n^0, \ldots, e_n^{n-1}\}$ of K^n, the join product has the following two behaviors:

$$
e_n^i \vee e_n^i = 0 \qquad\qquad e_n^i \vee e_n^j = -e_n^j \vee e_n^i.
$$

This gives the graded structure of G_n. The join product of k basis elements generates the subspace of grade k homogeneous elements. Considering G_3, this means that:

$\{1\}$ generates the elements of grade 0.
$\{e_3^0, e_3^1, e_3^2\}$ generates the elements of grade 1.
$\{e_3^0 \vee e_3^1, e_3^0 \vee e_3^2, e_3^1 \vee e_3^2\}$ generates the elements of grade 2.
$\{e_3^0 \vee e_3^1 \vee e_3^2\}$ generates the elements of grade 3.

Hence, G_n can be seen as a vector space of dimension 2^n. Our model is a representation of this vector space that allows a computational definition of the products of the Grassmann-Cayley algebra.

2.3   The Meet Product

Retrieving the Bracket. Usual presentation of the Grassmann-Cayley algebra [6,1,18] defines a bracket over the vector space V. Given n vectors $a_1, \ldots, a_n$ the bracket $[a_1, \ldots, a_n]$ is a non-degenerate multilinear alternating n-form, taking its values into the field K.

The use of the canonical basis of the vector space K^n defines a bracket implicitly. The set of elements of grade n generated by $e_n^0 \vee \cdots \vee e_n^{n-1}$ is isomorphic to the set of elements of grade 0 via the linear map defined by $i(e_n^0 \vee \cdots \vee e_n^{n-1}) = 1$. This linear map defines a non-degenerate multilinear n-form over the vectors of K^n that is actually a determinant.

Hence, the choice of the canonical basis defines a bracket. This allows us to retrieve the usual definition of the Grassmann-Cayley algebra. This link is used in section 4.2 to introduce automated proof techniques into our formalization.

The Hodge Star. Moreover, as $i(e_n^0 \vee \cdots \vee e_n^{n-1}) = [e_n^0, \ldots, e_n^{n-1}] = 1$, the canonical basis is said to be unimodular [1]. Then, the Hodge star defined as follows:

$$*(e_n^{\rho(0)} \vee_n \cdots \vee_n e_n^{\rho(i)}) = e_n^{\rho(i+1)} \vee_n \cdots \vee_n e_n^{\rho(n-1)}$$

where ρ is an even permutation, satisfies the following properties (see [1]):

(i) $*$ maps extensors of grade k to extensors of grade $n - k$,
(ii) $*(1) = e_n^0 \vee \cdots \vee e_n^{n-1}$ and $*(e_n^0 \vee \cdots \vee e_n^{n-1}) = 1$,
(iii) $*(*(A)) = (-1)^{k(n-k)} A$ if A is of grade k.

The Hodge star realizes the duality between the meet and the join products [1]. Hence, the following definition of the meet product, denoted $\wedge$, can be adopted:

$$*(A \vee B) = *(A) \wedge *(B) \quad \text{and} \quad *(A \wedge B) = *(A) \vee *(B).$$

Thus, in the algebra G_3, the meet product can be defined over the basis elements by the table

$\wedge$	1	e_3^0	e_3^1	e_3^2	$e_3^0 \vee e_3^1$	$e_3^0 \vee e_3^2$	$e_3^1 \vee e_3^2$	$e_3^0 \vee e_3^1 \vee e_3^2$
1	0	0	0	0	0	0	0	1
e_3^0	0	0	0	0	0	0	1	e_3^0
e_3^1	0	0	0	0	0	-1	0	e_3^1
e_3^2	0	0	0	0	1	0	0	e_3^2
$e_3^0 \vee e_3^1$	0	0	0	1	0	e_3^0	e_3^1	$e_3^0 \vee e_3^1$
$e_3^0 \vee e_3^2$	0	0	-1	0	$-e_3^0$	0	e_3^2	$e_3^0 \vee e_3^2$
$e_3^1 \vee e_3^2$	0	1	0	0	$-e_3^1$	$-e_3^2$	0	$e_3^1 \vee e_3^2$
$e_3^0 \vee e_3^1 \vee e_3^2$	1	e_3^0	e_3^1	e_3^2	$e_3^0 \vee e_3^1$	$e_3^0 \vee e_3^2$	$e_3^1 \vee e_3^2$	$e_3^0 \vee e_3^1 \vee e_3^2$

3   Data-Structures

The programming language of CoQ proof assistant [3] is a functional language with dependent types. It is then particularly suitable for the development of abstract algebra. In order to have a generic formalization, our development is parametrized by an abstract field K and its usual operations:

```
Structure FieldParams := {
        K : Set                       ;
        0 : K                         ;
        1 : K                         ;
    _ =? _ : K → K → bool             ;
      - _ : K → K                     ;
    _ + _ : K → K → K                 ;
    _ * _ : K → K → K                 ;
      _ -1 : K → K
}
```

Note that even if every type in Coq is equipped with a propositional equality, i.e. for two elements x and y in K the proposition $x = y$ expresses that they are equal with respect to Leibnitz equality, we have an explicit equality test $x \overset{?}{=} y$ that lets us decide on this equality. This capability is crucial when defining algorithms over elements of K. Along with this parametric definition of K, there is an associated set of axioms that gives the usual basic properties of the operations (associativity, commutativity, distributivity and neutral elements).

From now on, all our definitions are taking this field K and another parameter n for the dimension as parameters. They follow the same pattern: they are defined recursively on the dimension n. For a data-structure D, this means that its version D_{n+1} for the $n + 1$ dimension is going to be expressed in term of D_n. In this work, only the primitive pairing construct of CoQ is used: if a_1 is of type T_1 and a_2 of type T_2, (a_1, a_2) is of type $T_1 \times T_2$.

3.1   Representing the Vector Space K^n

As a first example, here is how the vectors of K^n are defined for $n \neq 0$:

```
Definition K_n := if n = 1 then K else K x K_{n-1}.
```

Compare to traditional programming where vectors would be represented as arrays, here we use recursion and pairing to mimic this data-structure. The type[1] K_1 is equivalent to K, K_2 to $K \times K$ and K_3 to $K \times (K \times K)$ and an element of K_3 is represented by $(x_1, (x_2, x_3))$.

Operations on this data-structure are also defined recursively. For example, addition of two vectors of dimension n is defined recursively as follows:

```
Definition x +_n y := if n = 1 then x + y else
    let (x_1,x_2) := x and (y_1,y_2) := y in
    (x_1 + y_1 , x_2 +_{n-1} y_2).
```

If the parameter n is one, the two elements belong to K so we can add them using the addition on K, otherwise each element can be decomposed into an

---

[1] The exponent n is changed into an index for notational purpose.

element of K and an element of one dimension less and the resulting pair can be composed by adding the elements of K on the left and using a recursive call on the right. To end the vector space structure, scalar multiplication can be defined in a similar way:

```
Definition k ·ₙ y := if n = 1 then k * x else
    let (x₁,x₂) := x in (k * x₁ , k ·ₙ₋₁ x₂).
```

3.2   Representing the Algebra G_n

Representing elements of G_n, the Grassmann-Cayley of dimension n, follows exactly the same schema. This time, instead of a linear data-structure, binary trees are used:

```
Definition Gₙ := if n = 0 then K else Gₙ₋₁ x Gₙ₋₁.
```

The type G_0 is equivalent to K, G_1 to $K \times K$, G_2 to $(K \times K) \times (K \times K)$. Elements of G_n are binary trees of height n. They have 2^n leaves. This corresponds to the fact that G_n is a vector space of dimension 2^n.

The sum and the scalar multiplication for the vector space structure are defined recursively over the dimension as follows:

```
Definition x +ₙ y := if n = 0 then x + y else
    let (x₁,x₂) := x and (y₁,y₂) := y in
    (x₁ +ₙ₋₁ y₁ , x₂ +ₙ₋₁ y₂).

Definition k ·ₙ y := if n = 0 then k * x else
    let (x₁,x₂) := x in (k ·ₙ₋₁ x₁ , k ·ₙ₋₁ x₂).
```

The equality test is defined in the same way:

```
Definition x ≟ₙ y := if n = 0 then x ≟ y
    let (x₁,x₂) := x and (y₁,y₂) := y in
    (x₁ ≟ₙ₋₁ y₁) && (x₂ ≟ₙ₋₁ y₂).
```

In this definition the operator && is a special notation used in this paper for the logical *and* to avoid confusion with the meet product.

Figure 1 explains how the basis components of G_n are mapped to the binary structure. The leaves contain the coefficients. For example, the grade 2 element of G_3,

$$2.(e_3^0 \vee e_3^1) + 3.(e_3^1 \vee e_3^2)$$

is represented as $(((0,2),(0,0)),((3,0),(0,0)))$ and the multi-vector of G_3,

$$2.(e_3^0 \vee e_3^2) + 3.e_3^1 + 4.1$$

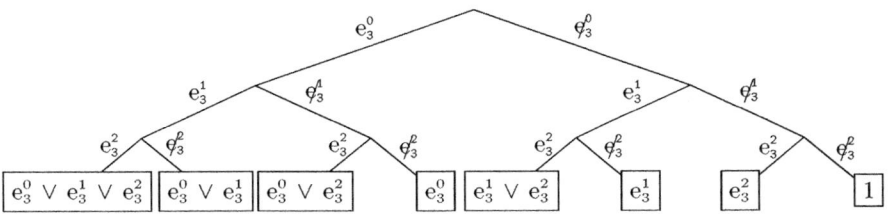

Fig. 1. Mapping of the multi-vector coefficients to the leaves of the binary tree

is represented as $(((0,0),(2,0)),((0,3),(0,4)))$. Here the sign sum indicates that an element of G_n is a linear combination of the basis components.

For a tree of height n, at the level i, a move toward the left child inserts the basis element e_n^i into the join product, while a move toward the right child insures that this element is not present. Then, on a path from the root of the tree to a leaf a move toward the left child increases the grade by one, while a move toward the right child leaves it unchanged. Hence, the left-most leaf of a tree of height n contains the grade n coefficient of an element of G_n while the right-most leaf contains the coefficient of the grade 0 part.

This binary tree structure also allows to increase the dimension of an element x of G_n by injecting it to G_{n+1}. This is done with the function inj_{G_n} by simply pairing the binary tree of height n with all its leaves containing 0, denoted 0_n, and x:

$$\texttt{Definition inj}_{G_n}\ x\ \texttt{:= (}0_n\ \texttt{, x).}$$

This operation does not change the grade of x, but it shifts the basis components. The basis component $e_n^i \vee \cdots \vee e_n^{i+k}$ is mapped to the basis component $e_{n+1}^{i+1} \vee \cdots \vee e_{n+1}^{i+k+1}$.

From the mapping of the multi-vector coefficients and the injection function, the pairing of two elements x and y of G_n can be interpreted in terms of an element of G_{n+1}. The pair (x, y) represents the element

$$e_{n+1}^0 \vee \mathrm{inj}_{G_n}\ x +_{n+1} \mathrm{inj}_{G_n}\ y. \tag{2}$$

This means that pairing two elements x and y inserts the basis component e_{n+1}^0 into the shifted basis component of x. This puts all the coefficients of x into the left part of the tree.

Now, the field K is injected into the binary tree structure representing G_n with the following functions:

$$\texttt{Definition inj}_{n,K}\ k\ \texttt{:= if } n = 0 \texttt{ then } k \texttt{ else (inj}_{n-1,K}\ 0\texttt{,inj}_{n-1,K}\ k\texttt{).}$$

$$\texttt{Definition } 0_n\ \texttt{:= inj}_{n,K}\ 0.$$

$$\texttt{Definition } 1_n\ \texttt{:= inj}_{n,K}\ 1.$$

Hence, the tree representing zero has all its leaves set to 0, while the tree representing one has only its right-most leaf set to 1.

Using the injection of K into G_n, we can define the injection of the elements of K_n into G_n:

```
Definition inj_{K_n} x := if n = 0 then 0 else
    if n = 1 then (x , 0) else
    let (x_1,x_2) := x in (inj_{n-1,K} x_1 , inj_{K_{n-1}} x_2).
```

Note that, as K_0 is not defined, a special case is introduced for $n = 0$ sending any element to zero. Let us take a concrete example to explain how this injection works. An element of K_3 is represented by a triplet $(x_1, (x_2, x_3))$. The element $(((0,0), (0, x_1)), ((0, x_2), (x_3, 0)))$ of G_3 is its image by the injection. For an element of K_n, the coordinates x_i are the coefficients of the basis elements e_n^i.

We can also directly exhibit a base $\{e_n^0, e_n^1, \ldots, e_n^{n-1}\}$ for the vectors of G_n, i.e. the image by the injection inj_{K_n} of the base of K_n induced by the coordinates. Again, this is defined recursively:

```
Definition e_n^i := if n = 0 then 1_n else
    if i = 0 then (1_{n-1},0_{n-1}) else (0_{n-1}, e_{n-1}^{i-1}).
```

If we go back to the relation between K_3 and G_3, the base of K_3 induced by the coordinates is $\{(1,0,0),(0,1,0),(0,0,1)\}$. Then, we have

$$e_3^0 = (((0,0),(0,1)),((0,0),(0,0))) \text{ which corresponds to } (1,0,0),$$
$$e_3^1 = (((0,0),(0,0)),((0,1),(0,0))) \text{ which corresponds to } (0,1,0),$$
$$e_3^2 = (((0,0),(0,0)),((0,0),(1,0))) \text{ which corresponds to } (0,0,1).$$

When they are injected respectively into G_n and G_{n+1}, the basis elements of K_n and K_{n+1} are related by $e_{n+1}^{i+1} = \text{inj}_{G_n} e_n^i$. Hence, in terms of trees, the basis element e_n^i is the right child of the basis element e_{n+1}^{i+1}. In the previous example, we can observe the left zero tree in the representation of e_3^2 and e_3^1 indicating that $e_3^2 = \text{inj}_{G_2} e_2^1$ and $e_3^1 = \text{inj}_{G_2} e_2^0$. This is coherent with interpretation of the pairing of two elements of G_n (see the relation (2)).

At the moment, from the point of view of the properties that can be formally proved, only the usual properties of vector space for K_n and G_n and the properties of morphism of the different injections can be derived.

3.3   Join Product

The next step is to define the join product as a binary tree operation. To explain the definition, we use equation (2) and the mandatory properties expressed by the axioms (1) in section 2.2.

The idea is to define the join product recursively over the dimension. To do so, we decompose the product $x \vee_n y$ in terms of pairing and using the relation (2), we obtain:

$$x \vee_n y = (x_1, x_2) \vee_n (y_1, y_2)$$
$$= (e_n^0 \vee_n \mathrm{inj}_{G_{n-1}} x_1 +_n \mathrm{inj}_{G_{n-1}} x_2) \vee_n (e_n^0 \vee_n \mathrm{inj}_{G_{n-1}} y_1 +_n \mathrm{inj}_{G_{n-1}} y_2)$$

Then, using the axioms (1), we obtain:

$$\begin{aligned}
(x_1, x_2) \vee_n (y_1, y_2) = {} & e_n^0 \vee_n \mathrm{inj}_{G_{n-1}} x_1 \vee_n \mathrm{inj}_{G_{n-1}} y_2 \\
& +_n \mathrm{inj}_{G_{n-1}} x_2 \vee_n e_n^0 \vee_n \mathrm{inj}_{G_{n-1}} y_1 \qquad (3) \\
& +_n \mathrm{inj}_{G_{n-1}} x_2 \vee_n \mathrm{inj}_{G_{n-1}} y_2.
\end{aligned}$$

In the second term in the sum of the right part of this latter expression, the factor e_n^0 needs to be commuted with $\mathrm{inj}_{G_{n-1}} x_2$ in order to be able to factorize the expression with e_n^0 and to get an expression that corresponds to a pairing.

However, the join product is anti commutative and we must pay attention to sign changes into the factors. For example, if x is an homogeneous element of grade k, we have $\mathrm{e}_n^i \vee x = (-1)^k \cdot x \vee \mathrm{e}_n^i$. We want to generalize this property and have a conjugate function, noted $\overline{x}$, such that $\mathrm{e}_n^i \vee x = \overline{x} \vee \mathrm{e}_n^i$ for all x in G_n. Here is the definition of such a function:

```
Definition x̄ⁿ := if n = 0 then x else
    let (x₁,x₂) := x in (-x̄₁ⁿ⁻¹ , x̄₂ⁿ⁻¹).
```

Now expression (3) can be rewritten as:

$$\begin{aligned}
(x_1, x_2) \vee_n (y_1, y_2) = {} & e_n^0 \vee_n (\mathrm{inj}_{G_{n-1}} x_1 \vee_n \mathrm{inj}_{G_{n-1}} y_2 \\
& +_n \overline{\mathrm{inj}_{G_{n-1}} x_2}^n \vee_n \mathrm{inj}_{G_{n-1}} y_1) \qquad (4) \\
& +_n \mathrm{inj}_{G_{n-1}} x_2 \vee_n \mathrm{inj}_{G_{n-1}} y_2.
\end{aligned}$$

Using the definition of the injection inj_G, we get:

$$\begin{aligned}
(x_1, x_2) \vee_n (y_1, y_2) = {} & e_n^0 \vee_n (\mathrm{inj}_{G_{n-1}}(x_1 \vee_{n-1} y_2 +_{n-1} \overline{x_2}^{n-1} \vee_{n-1} y_1)) \\
& +_n \mathrm{inj}_{G_{n-1}}(x_2 \vee_{n-1} y_2).
\end{aligned} \qquad (5)$$

This leads to the following recursive definition of the join product:

```
Definition x Vₙ y := if n = 0 then x * y else
    let (x₁, x₂) := x and (y₁, y₂) := y in
    (x₁ Vₙ₋₁ y₂ +ₙ₋₁ x̄₂ⁿ⁻¹ Vₙ₋₁ y₁, x₂ Vₙ₋₁ y₂)
```

From this definition, we have proved formally that this join product verifies its basic properties (associativity, bilinearity and anti commutativity) defined by the axioms (1).

3.4   Meet Product

In order to define the meet product, we follow exactly the same path than for the join product. We generalize the fact that, for an homogeneous element x of grade k of G_n, we have $e^i_n \wedge x = (-1)^{(n-k)} \cdot x \wedge e^i_n$ and define a dual version of the conjugate function, noted $\overline{x}^d$, such that $e^i_n \wedge x = \overline{x}^d \wedge e^i_n$.

```
Definition x̄^{dn} := if n = 0 then x else
    let (x₁,x₂) := x in (x̄₁^{dn-1} , -x̄₂^{dn-1}).
```

Again, with this auxiliary function, the meet product can be defined recursively as follows:

```
Definition x ∧ₙ y := if n = 0 then x * y else
    let (x₁, x₂) := x and (y₁, y₂) := y in
    (x₂ ∧ₙ₋₁ y₂, x₁ ∧ₙ₋₁ y₂ +ₙ₋₁ x₂ ∧ₙ₋₁ ȳ₁^{dn-1})
```

Note that our recursive approach avoids the use of bracket algebra to define the meet product so that our formalization works internally and independently from the bracket algebra framework. As for the join product, it is quite direct to derive the basic properties of the meet product formally from its definition.

3.5   Duality

The Hodge star operator presented in section 2.3 is also defined recursively over the dimension as follows:

```
Definition *ₙ(x) := if n = 0 then x else
    let (x₁, x₂) = x in (*ₙ₋₁(x̄₂^{n-1}), *ₙ₋₁(x)).
```

Note that in our representation, upto sign flips, the Hodge star just reverses the leaves of the binary tree. For example in G_3, the dual of the element

$$(((x_1, x_2), (x_3, x_4)), ((x_5, x_6), (x_7, x_8)))$$

is the reverse element with two sign flips

$$(((x_8, x_7), (-x_6, x_5)), ((x_4, -x_3), (x_2, x_1))).$$

Because of these sign flips, the Hodge star is not an involution. For homogeneous elements, the following theorem (corresponding to property (iii) in section 2.3) is proved into our formalization using the defined Hodge star:

```
Lemma dual_invo: ∀n k v, if hom^k_n v then *ₙ(*ₙ(v)) = (-1)^{k(n-k)} ·ₙ v.
```

where hom_n^k tests if an element is homogeneous and is defined as follows:

```
Definition homₙᵏ x := if n = 0 then (k = 0 ‖ x ≐ 0) else
    let (x₁,x₂) := x in
    (if k = 0 then x₁ ≐ₙ₋₁ 0ₙ₋₁ else homₙ₋₁ᵏ⁻¹ x₁) && homₙ₋₁ᵏ x₂.
```

As previously, the notation ‖ is a special notation for the logical *or* to avoid confusion with the join product.

Due to our choice of the underlying vector space basis, the Hodge star implements the duality between the join product and the meet product (see section 2.3 and reference [1]). Then the following theorems are proved within our CoQ formalization:

```
Lemma  dual_prod: ∀n v₁ v₂, *ₙ(v₁ ∨ₙ v₂) = *ₙ(v₁) ∧ₙ *ₙ(v₂).
Lemma dual_dprod: ∀n v₁ v₂, *ₙ(v₁ ∧ₙ v₂) = *ₙ(v₁) ∨ₙ *ₙ(v₂).
```

This proves that the join product and the meet product are correctly defined.

At this point, Grassmann-Cayley algebra could already be considered as formalized in the CoQ proof assistant.

4 Theorem Proving in Projective Geometry

In this section, we first show how we can use our formalization of Grassmann-Cayley algebra to model the geometry of incidence. Then, in a second step, we show how proofs in this setting can be fully automatized within CoQ.

4.1 Modeling the Geometry of Incidence

Now that we have Grassmann-Cayley algebra in CoQ, we can use it to represent theorems in projective geometry. All this is standard and can be found by example in [18] or in [15] chapter 3. We just explain how this has been instantiated to our formalization. We work over an arbitrary field K and restrict ourselves to G_3. We take a conservative approach and consider only non-degenerated configurations for constructed points. In this setting, points are vectors, so in our case we are going to use our injection from K_3 to G_3:

```
Definition point K := K₃.
```

To define the fact that a point p_1 is the intersection of the line composed of p_2 and p_3 and the line composed of p_4 and p_5, we simply implement it by saying that using the join to create the line and the meet to perform the intersection:

```
Definition p₁ is the intersection of [p₂,p₃] and [p₄,p₅] :=
    inj₃,ₖ p₁ = (injₖ₃ p₂ ∨₃ injₖ₃ p₃) ∧₃ (injₖ₃ p₄ ∨₃ injₖ₃ p₅).
```

Note that the equality imposes the meet product to be a point, so the lines to be defined and intersecting.

To define the fact that a point p_1 is on the line composed of p_2 and p_3, we simply implement it by saying that the line is well-defined, i.e. the join product of p_2 and p_3 is not zero, and the joint product of the three points is zero:

```
Definition p₁ is free on [p₂,p₃] :=
        (injₖ₃ p₂) ∨₃ (injₖ₃ p₃) ≠ 0₃
   and
        (injₖ₃ p₁) ∨₃ (injₖ₃ p₂) ∨₃ (injₖ₃ p₃) = 0₃.
```

Finally, we consider the collinearity of three points and the concurrency of three lines:

```
Definition {p₁,p₂,p₃} are collinear :=
    (injₖ₃ p₁) ∨₃ (injₖ₃ p₂) ∨₃ (injₖ₃ p₃) = 0₃.

Definition {[p₁,p₂],[p₃,p₄],[p₄,p₅]} are concurrent :=
    ((injₖ₃ p₁) ∨₃ (injₖ₃ p₁)) ∧₃
    ((injₖ₃ p₃) ∨₃ (injₖ₃ p₄)) ∧₃
    ((inj₃,ₖ p₄) ∨₃ (injₖ₃ p₅))     = 0₃.
```

With these definitions, we can start stating some classic theorems of geometry of incidence. First, let us consider Pappus' theorem:

```
Theorem Pappus: ∀ a b c a′ b′ c′ p q r: point K,
if p is the intersection of [a,b′] and [a′,b] and
   q is the intersection of [b,c′] and [b′,c] and
   r is the intersection of [c,a′] and [c′,a] and
   {a,b,c} are collinear and {a′,b′,c′} are collinear
then {p,q,r} are collinear.
```

Introducing the universal quantification and eliminating the points p, q and r, we are left with proving that[2]:

```
if a ∨ b ∨ c = 0 and a′ ∨ b′ ∨ c′ = 0 then
(a ∨ b′ ∧ a′ ∨ b) ∨ (b ∨ c′ ∧ b′ ∨ c) ∨ (b ∨ c′ ∧ b′ ∨ c) = 0
```

Remaining inside the algebra and applying the basic properties it is possible to prove this statement interactively in Coq. For this, we have followed of the proof given in [7]. This requires 10 interactions where the prover is guided in order to apply the symbolic manipulations that leads to the proof.

---

[2] We voluntarily omit the injections and the indices to make the expression more legible.

A more involved proof is Desargues' theorem. It can be stated as:

```
Theorem Desargues: ∀ a b c a' b' c': point K,
if p is the intersection of [a,b] and [a',b'] and
  q is the intersection of [a,c] and [a',c'] and
  r is the intersection of [b,c] and [b',c'] and
then
  {p,q,r} are collinear
iff
  {a,b,c} are collinear or {a,b,c} are collinear or
  {[a,a'],[b,b'],[c,c']} are concurrent.
```

Again, introducing the universal quantification and eliminating the points p, q and r, we are left with proving that:

$$(a \lor b \land a' \lor b') \lor (a \lor c \land a' \lor c') \lor (b \lor c \land b' \lor c') = 0$$
$$iff$$
$$a \lor b \lor c = 0 \text{ or } a' \lor b' \lor c' = 0 \text{ or } a \lor a' \land b \lor b' \land c \lor c' = 0$$

In order to prove this interactively, this time we have followed the paper proof given in [1]. The proof is more intricate and has required 60 interactions with the prover.

4.2   Automating Proofs

Proving the last two theorems is very satisfying because it shows that our algebra can be manipulated symbolically within CoQ but clearly we are at the limit of what is bearable for a user to prove interactively. So, the next step is to automate the proof of such theorems. For this, we are going to introduce bracket algebra and follow the path of [11].

Bracket algebra and its relation with Grassmann-Cayley is a well-known topic [6,1]. Here, we are just going to explain how it has been introduced in our setting. For the moment, this has only been implemented for G_3 but we believe that this could be easy generalised to G_n for an arbitrarily n. In the following, in order to increase legibility we will systematically omit the indices and the injections, so for example $(\text{inj}_{K_3} p_1 \lor_3 \text{inj}_{K_3} p_2)$ will be noted $(p_1 \lor p_2)$ only. A bracket is a function that takes three points and returns an element of our field K. Its definition is the following:

```
Definition [p1,p2,p3] := dC (p1 ∨ p2 ∨ p3).
```

where dC stands for the dual of the constant component, i.e the left-most leaf of the tree-structure given in Figure 1 of page 57. The usual relations between the bracket, the join product and the meet product in G_3 are derived.

Lemma bracket_defE: $\forall p_1 \; p_2 \; p_3$,
  $p_1 \lor p_2 \lor p_3 = [p_1,p_2,p_3] \cdot e^0 \lor e^1 \lor e^2$.

Lemma bracket_defl: $\forall p_1 \; p_2 \; p_3, p_1 \land (p_2 \lor p_3) = [p_1,p_2,p_3] \cdot 1$.

We have also formally proved that it behaves as a determinant:

Lemma bracket0l: $\forall p_1 \; p_2$, $[p_1,p_1,p_2] = 0$

Lemma bracket_swapl: $\forall p_1 \; p_2 \; p_3, [p_1,p_2,p_3] = - [p_2,p_1,p_3]$.

Lemma bracket_swapr: $\forall p_1 \; p_2 \; p_3, [p_1,p_2,p_3] = - [p_1,p_3,p_2]$.

Lemma bracket_free: $\forall \alpha \; \beta \; p_1 \; p_2 \; p_3 \; p_4 \; p_5$,
  if $p_1 = \alpha \cdot p_4 + \beta \cdot p_5$
  then $[p_1,p_2,p_3] = \alpha * [p_4,p_2,p_3] + \beta * [p_5,p_2,p_3]$.

In order to automate as described in [11], we are going to restrict ourselves to a specific skeleton of proofs. The goals we are going to be able to prove automatically have the following shape:

$$\forall p_1 \, p_2 \; \ldots \; p_m, \textit{ if } H_1 \textit{ and } \ldots \; H_n \quad \textit{then} \quad \{p_i, p_j, p_k\} \text{ are collinear}$$

where the H_is are either the construction of a free point on a line

$$p_j \text{ is free on } [p_r, p_s]$$

or the construction of an intersection

$$p_j \text{ is the intersection of } [p_r, p_s] \text{ and } [p_t, p_u].$$

How does the automatic procedure proceed? As the conclusion is a collinearity property, it can be turned into an equality to zero of a bracket expression by the following lemma that is a direct consequence of the lemma bracket_defE:

Lemma collinear_bracket: $\forall p_1 \; p_2$,
  $\{p_1,p_2,p_3\}$ are collinear *iff* $[p_1,p_2,p_3] = 0$

Then, the constructed points are progressively eliminated from the assumptions to obtain a bracket expression. Two lemmas are used corresponding to each construction. In the first case, the assumption is the construction of a free point on a line. The following lemma can be proved thanks to the conservative approach we observed:

Lemma online_def: $\forall p_1 \; p_2 \; p_3$,
  if p_1 is free on $[p_2,p_3]$ *then* $\exists \; \alpha \; \beta, p_1 = \alpha \cdot p_2 + \beta \cdot p_3$

Coupled with the lemma `bracket_free`, this lets us remove the free point from all bracket expressions. In the second case, the assumption is the construction of an intersection then the second rule[3] given in [11] is used to remove the point:

Lemma `bracket_expand`: $\forall p_1 \; p_2 \; p_3 \; p_4 \; p_5 \; p_6 \; p_7$,

  if p_1 is the intersection of $[p_4,p_5]$ and $[p_6,p_7]$ *then*

  $[p_1, \; p_2, \; p_3] = -[p_4, \; p_2, \; p_3] * [p_5,p_6,p_7] + [p_5,p_2,p_3] * [p_4,p_6,p_7]$.

Once all the eliminations of constructed points have been performed, we get an expression that contains sums and products of bracket of initial points and the αs and the βs introduced by the eliminations of the free points. So for the theorem to be true generically, this expression must be equal to zero modulo Plücker relations (see [11]). In order to simplify the obtained expression a contraction rule is used in [11]). In our setting it is stated as:

Lemma `contraction_v0`: $\forall p_1 \; p_2 \; p_3 \; p_4 \; p_5$,

$[p_1,p_2,p_3] * [p_1,p_4,p_5] - [p_1,p_2,p_5] * [p_1,p_4,p_3] = [p_1,p_2,p_4] * [p_1,p_3,p_5]$.

Surprisingly applying this rule unrestrictively as a rewrite rule from left to right as described in [11] is very effective. However, it is not sufficient in our setting to prove all the given examples. To fix this problem, we implement a normalisation method that is very expensive but is known to be complete. This captures the remaining examples. The method is based on an implicit ordering of the initial points $p_1 < p_2 < \cdots < p_i$. Applying some permutation, brackets can always be ordered with respect to this order: $[p_i,p_j,p_k]$ with $p_i < p_j < p_k$.

The order can be lifted to brackets $[p_i,p_j,p_k] \leq [p_{i'},p_{j'},p_{k'}]$ if $p_i \leq p_{i'}$ and $p_j \leq p_{j'}$ and $p_k \leq p_{k'}$. The normalisation proceeds in trying to order the product of brackets from the smallest to the largest. For this, we consider the product of two brackets $[p_i,p_j,p_k] * [p_{i'},p_{j'},p_{k'}]$. Without loss of generality, we can suppose that $p_i \leq p_{i'}$. There are only two situations where this product is not ordered:

1. $p_i < p_{i'}$ and $p_{j'} < p_j$
2. $p_i \leq p_{i'}$ and $p_j < p_{j'}$ and $p_{k'} < p_k$ (or equivalently $p_i < p_{i'}$ and $p_j \leq p_{j'}$ and $p_{k'} < p_k$)

The first rewrite rule takes care of the first case and assures that the resulting expression has every first two elements of brackets in a product properly ordered.

Lemma `split3b_v1`: $\forall p_i \; p_{i'} \; p_j \; p_{j'} \; p_k \; p_{k'}$,

  $[p_i,p_j,p_k] * [p_{i'},p_{j'},p_{k'}] =$
  $[p_i,p_{i'},p_{j'}] * [p_j,p_k,p_{k'}] - [p_i,p_{i'},p_{k'}] * [p_j,p_k,p_{j'}] +$
  $[p_i,p_{j'},p_{k'}] * [p_j,p_k,p_{i'}]$.

---

[3] This rule corresponds to the elimination of the area method [9].

The second rewrite rule takes care of the second case and insures that the resulting products of brackets are all properly ordered.

Lemma split3b_v2: $\forall p_i \; p_{i'} \; p_j \; p_{j'} \; p_k \; p_{k'}$,

$[p_i,p_j,p_k] * [p_{i'},p_{j'},p_{k'}] =$

$\quad [p_i,p_j,p_{i'}] * [p_k,p_{j'},p_{k'}] - [p_i,p_j,p_{j'}] * [p_k,p_{i'},p_{k'}] +$

$\quad [p_i,p_j,p_{k'}] * [p_k,p_{i'},p_{i'}]$.

5   Conclusion

We have described how our formalization of Grassmann-Cayley algebra has been achieved. It is a generic one: it is parametrized both by the underlying field K and by the dimension n. A snapshot of the formalization with a complete zipped archive is available at

$$\texttt{http://www-sop.inria.fr/marelle/GeometricAlgebra}$$

Recursive definitions have played a central role in this formalization. Elements of the algebra are represented as binary trees. With this representation, operations like the meet, the join and the duality can be described as recursive functions in a very direct way. The nice thing about defining these operations in a proof assistant like COQ is that not only can we compute with them like in any programming language but also we can reason about them. This lets us derive all the standard properties of Cayley-Grassmann operations. Our implementation is then verified: we have a certified computational model of Grassmann-Cayley algebra.

One of the most satisfying part of this formalization is, without a doubt, the instantiation that has been done in order to prove Pappus' and Desargues' theorem as proposed in [7] and [1]. We have been capable of justifying formally every step of the paper proofs. Moreover, the method that automatically proves projective geometric theorems proposed in [11] has also be translated successfully into our formalization. An efficient COQ tactic has been developed. This makes us very confident in the potential of our formalization.

In addition to the formalization of the Grassmann-Cayley algebra basics properties, we have also considered other useful operations such as contraction $\langle \phi, v \rangle$ of a linear form ϕ on a vector v and factorisation have also been formalized.

As already mentioned in the introduction, we are very interested in studying the links with other formalized approaches of incidence geometry such as those based over ranks [14,13].

Finally, we also plan to develop our formalization to capture the powerful framework of the geometric algebra [8,5].

References

1. Barnabei, M., Brini, A., Rota, G.C.: On the Exterior Calculus of Invariant Theory. Journal of Algebra 96, 120–160 (1985)
2. Bertot, Y., Castéran, P.: Interactive Theorem Proving and Program Development, Coq'Art: The Calculus of Inductive Constructions. Springer, Heidelberg (2004)

3. Coq development team: The Coq Proof Assistant Reference Manual, Version 8.2. LogiCal Project (2008), http://coq.inria.fr
4. Crapo, H., Richter-Gebert, J.: Automatic proving of geometric theorems. In: White [16], pp. 167–196
5. Dorst, L., Fontijne, D., Mann, S.: Geometric Algebra for Computer Science: An Object Oriented Approach to Geometry. Morgan Kauffmann Publishers (2007)
6. Doubilet, P., Rota, G.C., Stein, J.: On the foundations of combinatorial theory. IX. Combinatorial methods in invariant theory. Studies in Applied Mathematics 53, 185–216 (1974)
7. Hawrylycz, M.: A geometric identity for Pappus' theorem. Proceedings of the National Academy of Sciences U.S.A. 91(8), 2909 (1994)
8. Hestenes, D., Sobczyk, G.: Clifford Algebra to Geometric Calculus: A Unified Language for Mathematics and Physics. In: Fundamental Theories of Physics, vol. 5, Kluwer Academic Publishers (1984)
9. Janicic, P., Narboux, J., Quaresma, P.: The Area Method: a Recapitulation. Journal of Automated Reasoning (2010) (published online)
10. Li, H.: Algebraic Representation, Elimination and Expansion in Automated Geometric Theorem Proving. In: Winkler, F. (ed.) ADG 2002. LNCS (LNAI), vol. 2930, pp. 106–123. Springer, Heidelberg (2004)
11. Li, H., Wu, Y.: Automated short proof generation for projective geometric theorems with Cayley and bracket algebras: I. Incidence geometry. Journal of Symbolic Computation 36(5), 717–762 (2003)
12. Magaud, N., Narboux, J., Schreck, P.: Formalizing Projective Plane Geometry in Coq. In: Sturm, T., Zengler, C. (eds.) ADG 2008. LNCS, vol. 6301, pp. 141–162. Springer, Heidelberg (2011)
13. Magaud, N., Narboux, J., Schreck, P.: Formalizing Desargues' theorem in Coq using ranks. In: Proceedings of the ACM Symposium on Applied Computing SAC 2009, ACM, ACM Press (March 2009),
http://lsiit.u-strasbg.fr/Publications/2009/MNS09
14. Michelucci, D., Schreck, P.: Incidence constraints: A combinatorial approach. International Journal of Computational Geometry & Applications 16(5-6), 443–460 (2006)
15. Sturmfels, B.: Algorithms in Invariant Theory. Springer, New York (1993)
16. White, N.L. (ed.): Invariants Methods in Discrete and Computational Geometry. Kluwer, Dordrecht (1995)
17. White, N.L.: A tutorial on Grassmann-Cayley algebra. In: Invariants Methods in Discrete and Computational Geometry [16], pp. 93–106
18. White, N.L.: Geometric applications of the Grassmann-Cayley algebra. In: Handbook of Discrete and Computational Geometry, pp. 881–892. CRC Press, Inc., Boca Raton (1997)

Automatic Calculation of Plane Loci Using Gröbner Bases and Integration into a Dynamic Geometry System

Michael Gerhäuser and Alfred Wassermann

University of Bayreuth
Department of Mathematics, 95440 Bayreuth, Germany
{michael.gerhaeuser,alfred.wassermann}@uni-bayreuth.de

Abstract. We describe the integration of a well known algorithm for computing and displaying plane loci based on ideal elimination using Gröbner bases in the dynamic geometry software JSXGraph. With our approach it is not only possible to determine loci depending on other loci but it is also possible to extend JSXGraph to deal with loci depending on arbitrary plane algebraic curves. For Gröbner bases calculations we use CoCoa, a computer algebra system with its focus on computations in commutative algebra.

Keywords: dynamic geometry system, gröbner bases, automatic discovery of plane loci.

1   Introduction

The term "Dynamic Geometry System" (DGS) describes software programs which can be used to construct and display geometric configurations. The key characteristic of such software, expressed by the word "dynamic", is that unconstrained elements can be moved freely across the whole drawing board while the other elements are adjusted automatically satisfying existing constraints. One of the features which are present in nearly every modern DGS is the calculation of loci out of a set of free or constrained points, "glued" together by a set of geometric relations. However, all but one of the free points should be fixed to their current location while determining a locus. One of the constrained points is called the *drawer* and it is bound to another object, e.g. a circle or a line or a function plot. However, the implementation of the algorithm described in this paper requires the point to be bound to an algebraic curve. The locus point should depend on this drawer.

A simple way to display the locus of a constrained point is to automatically move the drawer and trace the locations of the locus point. This gives us a set of sample points which can be interpolated to draw a smooth loci curve. This simple algorithm is implemented and used in many DGS, e.g. GEONE$_X$T[1], GeoGebra[2],

---

[1] http://geonext.de
[2] http://geogebra.org

P. Schreck, J. Narboux, and J. Richter-Gebert (Eds.): ADG 2010, LNAI 6877, pp. 68–77, 2011.

and Cinderella[3]. The advantage of this approach is its efficiency. Unfortunately when using this method, in some cases parts of the locus get lost as is shown in Figure 1. Additionally, the locus line equation is not part of the result but there are ways to calculate it as well [13].

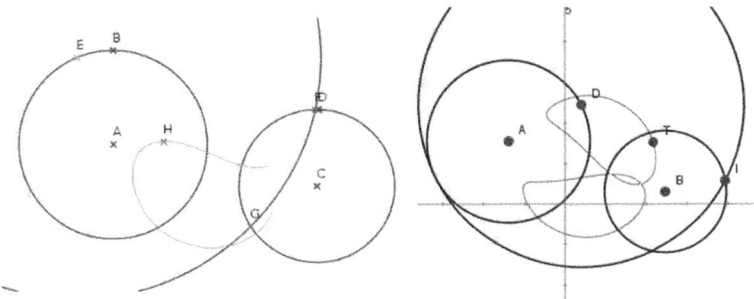

Fig. 1. Some parts of the four-bar-linkage locus get lost in GEONE$_x$T(left hand side) and are preserved in JSXGraph (right hand side)

This behavior is due to the continuity problem [12] and due to possibly occurring complex coordinates in some of the dependent points. In most DGS the calculation of intersection points is done by simply determining the intersection points coordinates. If a circle is involved in an intersection there are two intersection points which can't be algebraically distinguished. Therefore, those two intersection points in fact live only on a semicircle and will jump over the other half of the circle if they get near to one of the transition points.

In the construction shown in Figure 1, I is the intersection point of the circle d and the circle around the drawer D. While we are moving D on circle c, it may occur that I is not real anymore, i.e. its coordinates are complex, but the midpoint T between D and I still is real, but is not shown by most DGS. The only DGS we know that take care of this problem are Cinderella and GeoGebra.

2   Calculating Loci with Gröbner Bases

2.1   The Algorithm

To get around the continuity problem in the successor of GEONE$_x$T which will be based on JSXGraph, we integrated an exact algorithm for the calculation of plane loci. This method uses Gröbner bases to calculate the implicit algebraic equation of the locus and then plots the corresponding curve. Every point bound by geometric constraints fulfills one or two algebraic equations. If we write down all the algebraic equations we get a system of polynomial equations (SPE). By subtracting the right hand sides of the equations we can homogenize the SPE and interpret it as an ideal in an appropriate polynomial ring. Using the Buchberger

---

[3] http://cinderella.de

algorithm we then compute the Gröbner basis of this ideal. From that we obtain an elimination ideal which eventually leads us to the algebraic equation of the variety that contains the locus curve we searched for. For an introduction on Gröbner bases theory see [1,5,7,8,9].

This algorithm itself is not the original work of the authors of this article. It was firstly described in [2] using the ideas in [15], and remotely used in [4]. As the algorithm is described there very well we will give only a short description by example here.

2.2 Example

Given the construction from Figure 1 we first choose coordinates for all points. With the notation in the construction on the right hand side of Figure 1 we get

$$A = (-1.5, 1.5), \ B = (2.5, 0.5), \ D = (d_1, d_2), \ T = (t_1, t_2), \ I = (i_1, i_2) \ . \quad (1)$$

To set up the system of polynomial equations we have to take a look at the geometric constraints of the dependent points, e.g. Point D lives on the circle around A with radius 2, or algebraically spoken

$$(d_1 - a_1)^2 + (d_2 - a_2)^2 = 2^2 \ . \quad (2)$$

The intersection point I lies on the circle d around B with radius $\frac{3}{2}$ and the circle around I with radius 4. Hence, I fulfills the equations

$$(i_1 - b_1)^2 + (i_2 - b_2)^2 = \frac{9}{4} \quad (3)$$

and

$$(i_1 - d_1)^2 + (i_2 - d_2)^2 = 4^2 \ . \quad (4)$$

Doing the same for the locus point T we get

$$d_1 i_2 - t_2 d_1 - i_1 d_2 + i_1 t_2 + t_1 d_2 = 0 \quad (5)$$

as well as

$$(i_1 - t_1)^2 + (i_2 - t_2)^2 = (d_1 - t_1)^2 + (d_2 - t_2)^2 \ . \quad (6)$$

We now can interpret the system of polynomial equations consisting of equations (2) to (6) as an ideal I in the polynomial ring $\mathbb{Q}[d_1, d_2, i_1, i_2, t_1, t_2]$.

Calculating the Gröbner basis results in generators that we will not show here for the sake of brevity. But using this special basis we are able to eliminate the variables d_1, d_2, i_1, and i_2. In other words we can determine easily the fourth elimination ideal which is generated by an algebraic equation in the variables t_1 and t_2:

$$\begin{aligned}
J = (&128t_1^6 + 384t_1^4 t_2^2 + 384t_1^2 t_2^4 + 128t_2^6 - 384t_1^5 - 768t_1^4 t_2 - 768t_1^3 t_2^2 \\
&- 1536t_1^2 t_2^3 - 384t_1 t_2^4 - 768t_2^5 + 128t_1^4 + 2560t_1^3 t_2 + 3328t_1^2 t_2^2 \quad (7) \\
&+ 2560t_1 t_2^3 + 3200t_2^4 - 192t_1^3 - 5232t_1^2 t_2 - 5568t_1 t_2^2 - 8304t_2^3 \\
&+ 2458t_1^2 + 2672t_1 t_2 + 6442t_2^2 - 1170t_1 + 4122t_2 - 3483) \ .
\end{aligned}$$

3   Integration of Locus Computation in JSXGraph

3.1   JSXGraph

JSXGraph[4] is a DGS implemented in JavaScript. It uses either Scalable Vector Graphics (SVG) or Vector Markup Language (VML) or HTML5 Canvas to draw geometric constructions. This enables JSXGraph to run in any major web browser and therefore, it can be used on a wide variety of devices like deskop computers, laptops, tablets and smartphones with many different operating systems [10]. The feature set includes Euclidean geometry, conic sections, function plotting, integration, interactive ode solving, turtle graphics, charts. Importing constructions made with GEONE$_x$T, GeoGebra, or Intergeo is possible, too.

A construction like the one in Figure 1 is done in JSXGraph with the following JavaScript commands:

```
board = JXG.JSXGraph.initBoard('jxgbox', {
    boundingbox:[-4, 6, 10, -4],
    keepaspectratio: true
});

p1 = board.create('point', [0, 0]);
p2 = board.create('point', [6, -1]);
c1 = board.create('circle', [p1, 2]);
c2 = board.create('circle', [p2, 1.5]);
g1 = board.create('glider', [6, 3, c1]);
c3 = board.create('circle', [g1, 4]);
g2 = board.create('intersection', [c2,c3,0]);
m1 = board.create('midpoint', [g1,g2]);

loc = board.create('locus', [m1]);
```

3.2   Calculating Loci with JSXGraph

Because some of the broad range of devices JSXGraph is intended to run on do not have high computing powers and the fact that there are a lot of very efficient algorithms for calculating Gröbner bases already implemented we decided to use one of these implementations remotely via the *XMLHttpRequest* object. The loci calculation in JSXGraph is implemented straightforward:

```
1. Collect the elements the locus depends on
2. Choose a coordinate system
3. Generate polynomial equations from geometric restrictions
4. Upload the equation system to the server
5. Calculate the elimination ideal
6. Calculate the variety generated by the elimination ideal
```

---

[4] http://jsxgraph.uni-bayreuth.de/ and https://sourceforge.net/projects/jsxgraph/

7. Return the variety and the polynomials generating the elimination ideal to JSXGraph
8. Display the variety containing the locus

JSXGraph keeps track of all dependencies between the elements of a construction. We use this knowledge to collect only those elements which are required to calculate the locus. This helps keeping the polynomial equation system generated in step 3 clean and small and thus speeds up the calculations in step 5.

By default JSXGraph uses the actual position of the free points and generates symbolic coordinates for the dependent ones. However, the user can choose one point to be $(0,0)$ and additionally another point either $(x,0)$ or $(1,0)$ to simplify the generated polynomials which in some cases speeds up the calculation:

```
board.options.translateToOrigin = true;
board.options.toOrigin = p1;
```

This will cause point A to be considered as $(0,0)$ all the time even if the user drags A to another place.

If the construction only consists of free points and elements of the following classes, JSXGraph can generate the polynomial equations all by its own. Semi-algebraic elements like a segment will be replaced by their algebraic equivalent:

- Glider on a circle, a line, or another locus
- Intersection points: circle/circle, circle/line, line/line, line/locus, circle/locus, and locus/locus.
- Midpoint
- Parallel line and point
- Perpendicular line and point
- Circumcircle and circumcenter

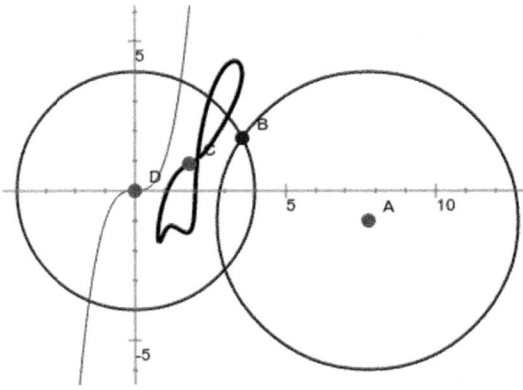

Fig. 2. Gliders on a cubic function can be used to generate a locus

Nevertheless, if further elements are required, JSXGraph can easily be extended by providing an user defined *generatePolynomial()* method. Consider a construction like the one in Figure 2 where we are looking for the locus of a midpoint of a glider living on a cubic function and an intersection point of two circles. To achieve that we have to override the *generatePolynomial()* method of the curve object (called *c* in our example) which represents the cubic function:

```
c.generatePolynomial = function(p) {
   return ['(' + p.symbolic.x + ')^3 - (' + p.symbolic.y + ')'];
};
```

After the system of polynomial equations has been generated successfully it is transferred to a web server where CoCoA [6] is used to calculate the elimination ideal. In a second step the variety generated by the elimination ideal is drawn with *matplotlib* [11]. The resulting set of points is then packed together with the locus equation and sent back to the browser.

Back in JSXGraph the set of points is drawn and the equation is stored for further locus computations. This enables us to intersect the just calculated locus with lines and circles or even other loci or to put a glider on a computed locus and use it for the calculation of other loci. After the calculation the locus is frozen, i.e. it is not recalculated as long as the free points the locus depends on are not moved whereas other points may be moved without triggering an update of the locus.

4   An Idea for Speed Improvements

In Figure 3 we see an extension of the limaçon of Pascal [4]. The extension includes two points which introduce four new variables and these four new polynomials

$$
\begin{aligned}
E = \{ & (u_5 - u_3)^2 + (u_6 - u_4)^2 - 9, \\
& (u_5 - 8)^2 + (u_6 - 8)^2 - 16, \\
& u_4 x - u_3 u_5 + y u_5 - u_3 y + u_3 u_6 - x u_6, \\
& u_3^2 - 2u_3 x + u_4^2 - 2u_4 y - u_5^2 + 2u_5 x - u_6^2 + 2u_6 y \}
\end{aligned}
$$

to the existing set of generators

$$
\begin{aligned}
B = \{ & (u_3 - u_1)^2 + (u_4 - u_2)^2 - 9, \\
& 3u_3 - 3u_1 + u_4 u_1 - 8u_4 + 8u_2 - u_3 u_2, \\
& (u_1 - 8)^2 + (u_2 - 8)^2 - 16 \}.
\end{aligned}
$$

Now we can compute the locus directly eliminating $u_1, \ldots, u_8$ at $(B \cup E)$ or we can eliminate u_1, u_2 from (B) first and use the reduced Gröbner basis L of the resulting elimination ideal and eliminate $u_3, \ldots, u_8$ at $(L \cup E)$.

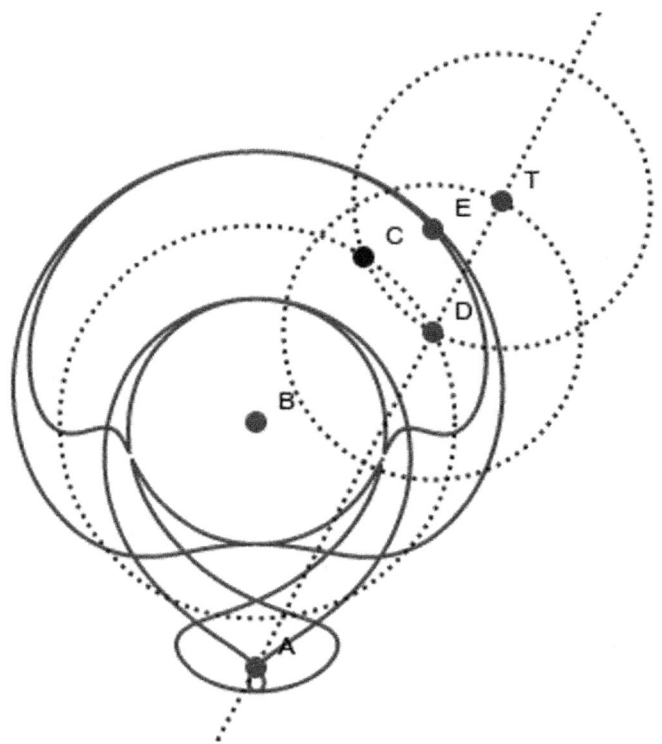

Fig. 3. Extended version of the limaçon of Pascal: Additional to the construction used to compute the limaçon we have a circle around T which intersects in C with the circle around B. The algebraic curve is the locus of the midpoint E of C and T.

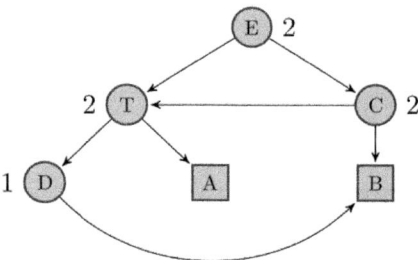

Fig. 4. Dependency graph of the construction seen in Figure 3. The numbers next to some of the nodes show the number of equations that particular point adds to the polynomial equation system.

In general, the results of the direct approach and the split generators approach are not guaranteed to be the same. But in this case we are calculating two loci where the drawer of the second locus is bound to the first locus. Hence, as long as we make sure the intermediate result is a locus, too, we should be fine.

To check if the calculation can be split up we have to take a look at the dependencies of the geometric construction. A dependency graph for the extended limaçon is shown in Figure 4 with the free points in squares and the dependant points in circles. Looking at the dependant points we see that E and C depend on T only. This means we can split the calculation of the locus of E by computing the locus of T first.

The benefit of this approach is the slightly lower time it sometimes takes to calculate the locus which is due to the smaller ideals.

Back to the limaçon example, determining the extended locus consecutively takes about 0.4sec instead of 0.8sec in our setup from above: JSXGraph with CoCoA and the lexicographic order. This measurement also includes the overhead, i.e. the time used to generate and upload the polynomials, to calculate and plot the elimination ideal with matplotlib and in JSXGraph. Additional measures in Singular, Mathematica, and Magma without the overhead generated by JSXGraph confirm this observation. However, in the different computer algebra

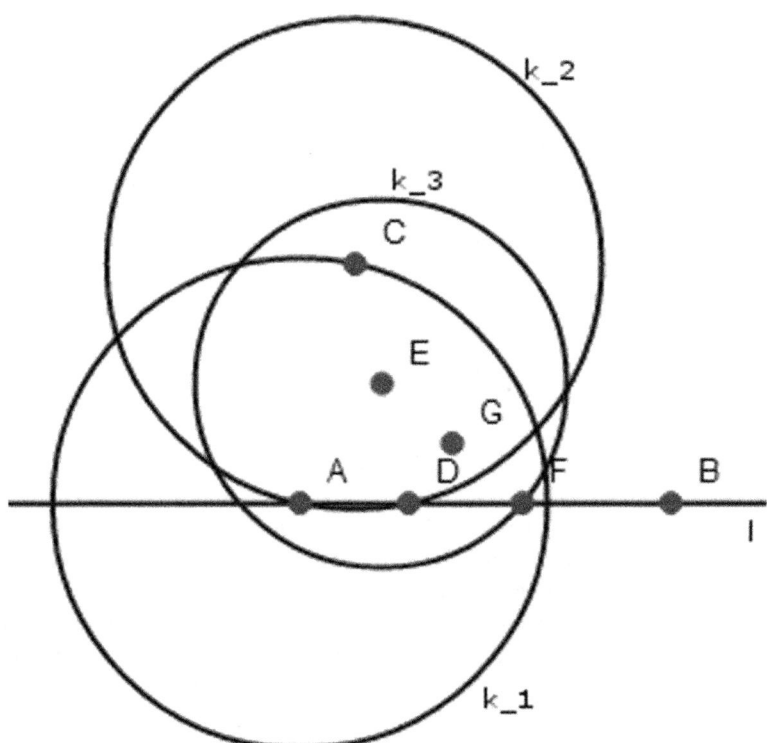

Fig. 5. A counterexample: C is a glider on the circle around A. D is an intersection point of the circle k_2 around C and the line l through A and B. k_3 is a circle around the midpoint E of C and D which intersects with l in F. We are looking for the locus of the midpoint G of E an F.

systems mentioned the gap between the times consumed by the direct and the consecutive computation differs.

	Limaçon	Extension	Total	Directly
JSXGraph + CoCoA	76.5ms	331.8ms	408.3ms	836.3ms
Singular	<=10ms	26.5ms	26.5ms	117.8ms
Mathematica	7.1ms	103.2ms	110.3	418.0ms
Mathematica (EliminationOrdering)	7.9ms	76.7ms	84.6ms	137.8ms
Magma	1.7ms	53.7ms	55.4ms	138.9ms

Unfortunately, it is also possible that the calculation takes much longer when using this approach. An example is shown in Figure 5.

5   Conclusion

The biggest difference between this approach of calculating and displaying plane loci to the relatively simple approach mentioned in the introductory section can be seen in Figure 1. There are no parts of the locus missing, because the construction of the locus is not done by simulating the user dragging around the drawer point, but it is calculated using algebraic methods. Hence, it doesn't matter if any of the dependent points' coordinates are real or not, all the real points of the locus are displayed. The continuity problem is by-passed because in the algebraic equations both intersection points are a solution of the corresponding equations.

In the above example this was an advantage, but sometimes this introduces new branches to our locus which we may not want to appear. One of those cases can be seen in Figure 6 which shows an egg curve. Strictly speaking there are two egg curves in this picture, because the point E was introduced by intersecting the circle around D with the line through A and C. The algorithm considers both intersection points because it simply cannot distinguish them. Or – more generally spoken – semi-algebraic objects are treated as their algebraic equivalent which results in a superset containing the locus but not always being the exact locus. Degenerated conditions in a construction also introduce additional branches which is not always desired.

Another point to mention about this algorithm is the time it consumes. While most of the rather small examples take only a few milliseconds the calculation the egg curve locus seen in Figure 6. takes about 20 seconds on a Intel Pentium Core 2 Duo T5250 which is a rather long time. Unfortunately, loci depending on at least three or four dependant points including the locus point and the drawer usually take way too long to be computed in a reasonable amount of time. However, most of the loci used in educational environments can be determined in a reasonable amount of time.

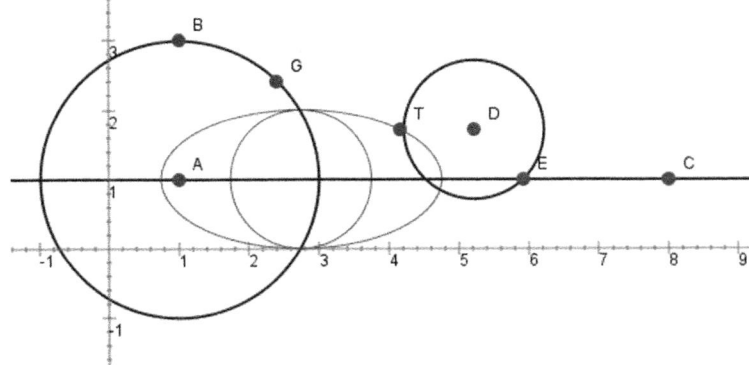

Fig. 6. The egg curve is plotted twice, because the intersection point E originated from an intersection with the circle around D

References

1. Becker, T., Weispfenning, V.: Gröbner bases: a computational approach to commutative algebra. Springer, Heidelberg (1993)
2. Botana, F., Valcarce, J.L.: A software tool for the investigation of plane loci. Mathematics and Computers in Simulation 61(2), 139–152 (2003)
3. Botana, F.: A Web-Based Intelligent System for Geometric Discovery. In: Sloot, P.M.A., Abramson, D., Bogdanov, A.V., Gorbachev, Y.E., Dongarra, J., Zomaya, A.Y. (eds.) ICCS 2003, Part I. LNCS, vol. 2657, pp. 801–810. Springer, Heidelberg (2003)
4. Botana, F., Abánades, M.A., Escribano, J.: Computing Locus Equations for Standard Dynamic Geometry Environments. In: Shi, Y., van Albada, G.D., Dongarra, J., Sloot, P.M.A. (eds.) ICCS 2007, Part II. LNCS, vol. 4488, pp. 227–234. Springer, Heidelberg (2007)
5. Buchberger, B.: Introduction to Gröbner Bases. In: Buchberger, B., Winkler, F. (eds.) Gröbner bases and applications. London Mathematical Society Lecture Note Series. Cambridge University Press (1998)
6. CoCoA: a system for doing Computations in Commutative Algebra, http://cocoa.dima.unige.it
7. Cox, D.A., Little, J., O'Shea, D.: Ideals, Varieties, and Algorithms: An Introduction to Computational Algebraic Geometry and Commutative Algebra. Springer, New York (2008)
8. Fröberg, R.: An Introduction to Gröbner Bases. John Wiley & Sons (1997)
9. von zur Gathen, J., Gerhard, J.: Modern Computer Algebra. Cambridge University Press (1999)
10. Gerhäuser, M., Miller, C., Valentin, B., Wassermann, A., Wilfahrt, P.: JSXGraph: Dynamic Mathematics Running on (nearly) Every Device. To be published in The Electronic Journal of Mathematics and Technology
11. Hunter, J.D.: Matplotlib: A 2D Graphics Environment. In: Computing in Science & Engineering, vol. 9, pp. 90–95. IEEE Computer Society, Los Alamitos (2007)
12. Kortenkamp, U.: Foundations of Dynamic Geometry. Dissertation (1999)
13. Lebmeir, P., Richter-Gebert, J.: Recognition of Computationally Constructed Loci. In: Botana, F., Recio, T. (eds.) ADG 2006. LNCS (LNAI), vol. 4869, pp. 52–67. Springer, Heidelberg (2007)
14. Pilgrim, M.: Dive Into Python. APress (2004)
15. Recio, T., Vélez, M.P.: Automatic Discovery of Theorems in Elementary Geometry. Journal of Automated Reasoning, 63–82 (1999)

Proof Documents
for Automated Origami Theorem Proving

Fadoua Ghourabi[1], Tetsuo Ida[1,*], and Asem Kasem[2]

[1] Department of Computer Science
University of Tsukuba
Tsukuba 305-8573, Japan
{ghourabi,ida}@cs.tsukuba.ac.jp
[2] Faculty of Informatics and Communications Engineering
Yarmouk Private University
Daraa, Syria
a-kasem@ypu.edu.sy

Abstract. A proof document for origami theorem proving is a record of entire process of reasoning about origami construction and theorem proving. It is produced at the completion of origami theorem proving as a kind of proof certificate. It describes in detail how the whole process of an origami construction and the subsequent theorem proving are carried out in our computational origami system. In particular, it describes logical and algebraic transformations of the prescription of origami construction into mathematical models that in turn become amenable to computation and verification. The structure of the proof document is detailed using an illustrative example that reveals the importance of such a document in the analysis of origami construction and theorem proving.

1 Introduction

In this paper, we are interested in computational origami, a discipline of studying mathematical and computational aspects of origami (paper folding). It comprises, among others, the studies of theories of fold, modeling of origami by logical, algebraic and symbolic methods, computer simulation of paper fold, and proving the correctness of geometrical properties of the constructed origami. In particular, we will address the issues of managing automated geometrical origami theorem proving.

It is well known that the set of basic fold operations proposed by Huzita, often referred to as Huzita's axiom set [7,10] or as Huzita's fold principle to be more precise [11], is more powerful than Euclidean tools, i.e. straightedge and compass (abbreviated to SEC hereafter). Huzita's fold principle is more powerful than SEC in the sense that we can construct a larger class of points of coincidence by applying Huzita's fold principle than by SEC [3]. Therefore, the class of the

---

* This research is supported by the JSPS Grants-in-Aid for Scientific Research (B) No. 17300004 and for Exploratory Research No. 19650001.

P. Schreck, J. Narboux, and J. Richter-Gebert (Eds.): ADG 2010, LNAI 6877, pp. 78–97, 2011.

shapes formed by connecting the coincidences is richer than that of the shapes formed by SEC. The trisector of an arbitrary angle is a famous example that is constructible by origami, but not by SEC [15,2]. This shows the importance of deeper and more extensive mathematical study of theories of origami in both foundational and application-oriented domains.

Another interesting aspect of origami is its algorithmic one. The construction is described by a sequence of fold steps, each of which can be specified by Huzita's fold principle. The prescription of the construction of the origami is akin to a program. By obeying the commands of folds, of the form of application of Huzita's fold principle, we can construct a desired shape of an origami. It is not only a human origamist but also the software empowered by high-quality graphics and symbolic and numeric capabilities, that can construct an origami. Subsequently, we should be able to verify the construction, as programs need to be verified.

2   Motivation

Our interest in mathematical aspects of origami coupled with the above observation led us to design and implement a computational origami system called Eos (e-origami system) [9]. While we used Eos for several years as an assistant for studying origami, we found it extremely important to keep record of our interactions with Eos; the record showing which step of constructions creates what shape of the origami, and what geometrical properties are added to the origami; and then how the accumulated geometrical properties are transformed into logical and algebraic expressions. Finally, we should check what algebraic method with specific parameters can be used for proving the origami theorems that we eventually would like to obtain. The construction, reasoning about geometrical and algebraic properties and theorem proving are not separate; they actually are interleaved and the cycle of construction-reasoning-proving is repeated several times with differing parameters. The record of these activities would tend to be scattered, unless we have means to documenting the activities systematically.

For this purpose, we incorporated the functionality of generating the proof document into Eos. The proof document records the entire activities of origami construction and theorem proving with Eos. It also serves as a proof certificate. The development required redesigning and reimplementing several core components of Eos. In this paper we explain the structure and the content of the proof document of Eos.

The organization of the rest of the paper is as follows. In Section 3, we give the overview of the research activities related to origami theorem proving. In Section 4, we explain how reasoning about origami construction and theorem proving is performed in our computational framework using Eos. Then in Section 5, we show the steps of proving whose output constitutes the essential contents of the proof document. In Section 6, we explain in detail the structure of the proof document. Finally, we conclude with remarks on further direction of research.

3   State of the Art of Origami Theorem Proving

Before we start describing the proof document, the discussion about the state of the art is due. Although the history of the research on the automated theorem proving is long, the automated theorem proving focussing on origami is relatively new. One of the earliest attempts, which the second author of the present paper was involved with, was to design a system communicating among origami constructor and reasoner, the geometrical theorem prover and general theorem prover *Theorema* [13]. We computationally construct an origami; then the geometry of the origami is sent to the geometrical prover, which generates the algebraic representation of the origami, and the algebraic representation is sent to *Theorema* [4]. *Theorema* performs a few logical and algebraic manipulations and computes the Gröbner bases. *Theorema* produces a proof text as a sequence of formulas together with statements in a natural language.

As the three components are implemented in *Mathematica* [16], they can easily share the algebraic expressions and their data structures. However, our later experiences show that we need tighter interactions between them, for the sake of analysis of correspondence between symbolic, logical and algebraic structures that are derived from origami geometry.

Although not directly related but influenced, we should mention a geometrical theorem proving system *Geother* [14]. Basically, *Geother* provides an interface to represent dynamic geometrical shapes and a language to prove properties about them. Moreover, the system can detect subsidiary conditions of geometrical theorems (thanks to Wu-Ritt formalism [17]) and transform them into legible statements. The geometrical construction and the proof are two independent parts in *Geother*. As we mentioned in the motivation, we would like to gain the extra advantage from designing a computational origami system where construction and proof can be integrated more tightly.

Our approach is algebraic. We also see the importance of logical approach, exemplified by *GeoProof* system [12]. It allows interactive geometrical constructions and interactive proof assistance. To perform the proof, the user can choose either algebraic methods or *Coq* proof assistance [1]. Employing a proof assistant such as *Coq*, in addition to the algebraic provers, is a possible future extension of our system.

4   Reasoning about Origami

In the generation of a proof document, we are concerned with all facets of reasoning during the entire process of origami construction and theorem proving. Let us briefly discuss the operations of reasoning about origami.

4.1   Algorithm of Origami Construction and Proving

Let i denote the i-th step of the origami construction, $\mathcal{O}_i$ be the origami at step i of the construction, and Ψ_i be the geometrical property associated with

$\mathcal{O}_i$. Suppose that we construct an origami $\mathcal{O}_k$. The origami reasoning process is abstractly described by the following Algorithm **Ori**.

Algorithm Ori

1. [Initialization]
   Let $i := 1$; Define initial origami $\mathcal{O}_i$; $\Psi_i :=$ True;
2. [Construction]

   While $(\mathcal{O}_i \neq \mathcal{O}_k)$ do
   $\quad l_i := f(\mathcal{O}_i, \underline{a_i})$;
   $\quad \Psi_i := \Psi_i \wedge (\overline{l_i = \ulcorner f(\mathcal{O}_i, \underline{a_i}) \urcorner})$;
   $\quad$ Fold $\mathcal{O}_i$ along l_i and obtain $\mathcal{O}_{i+1}$;
   $\quad$ Compute Ψ_{i+1} from $\mathcal{O}_i$ and $\mathcal{O}_{i+1}$;
   $\quad i := i + 1$;

3. [Conclusion formation]
   Define the property Φ to be established.
4. [Additional assumption formation]
   Define an additional assumption Ψ_{i+1} and $i := i + 1$, if needed to establish Φ;
5. [Proving]
   Prove the proposition $\Psi_1 \wedge \cdots \wedge \Psi_i \Rightarrow \Phi$;

The reasoning about origami consists of the above five phases. The first phase is the initialization, where a sheet of origami of specified size and colors is defined. The sheet of origami has two sides, usually with different colors.

The second is the construction phase. It is the iteration of folds or unfolds. The expression $f(\mathcal{O}_i, \underline{a_i})$ describes the operation of finding the line, called *fold line*, along which the origami $\mathcal{O}_i$ is folded or unfolded. Since unfold is a special kind of fold, i.e. fold along the same line as the one for the previous fold in an opposite direction, we will generally use the word *fold* to mean both fold and unfold operations, unless the operation is definitely unfold. The parameter sequence $\underline{a_i}$ is a sequence of points and lines. The result of the evaluation of $f(\mathcal{O}_i, \underline{a_i})$ is an object representing a line. The expression $\ulcorner f(\mathcal{O}_i, \underline{a_i}) \urcorner$ is the unevaluated form of $f(\mathcal{O}_i, \underline{a_i})$. The sequence $\ulcorner f(\mathcal{O}_1, \underline{a_1}) \urcorner$, ..., $\ulcorner f(\mathcal{O}_{k-1}, \underline{a_{k-1}}) \urcorner$ is a program of the construction of the origami $\mathcal{O}_k$. The unevaluated variables in the expression will be treated as universally quantified variables in the proof phase. Each Ψ_{i+1} is a formula of the first-order predicate logic describing the geometrical properties of $\mathcal{O}_{i+1}$ relative to $\mathcal{O}_i$.

The third and fourth phases are usually combined into one [Theorem formation] phase. In certain geometrical constructions, additional conditions may be necessary to establish the conclusion Φ (often called *goal*). For instance, proving certain properties involving a triangle would require the non-collinearity of the vertices of the triangle.

The last phase deals with theorem proving. Since we are interested in computer-assisted (semi-)automated theorem proving, we will use appropriate algebraic methods. Currently, we use Gröbner basis computation (GB) and the cylindrical algebraic decomposition (CAD) methods.

In the following, we show an example of computing each Ψ_i in the process of origami construction.

4.2   A Simple Example of Construction

We show a simple example of construction that follows the way that Eos, and therefore proof document, manages its information for construction and subsequent proving. We start the origami construction with an initial square sheet of origami and fold the origami to bring point A onto point D. Then we unfold the origami. The created origami shapes are shown in Fig. 1. The steps 1 - 3 are a typical prologue of origami constructions, e.g. of origami construction of a regular heptagon [13]. These three steps generate the following Ψ_is. The subscript i is attached automatically to denote the geometrical objects of the origami created at step i, e.g. point A at the step i by A_i.

$\Psi_1 :=$ PPSupQ[A_1, D_1, line1$_1$]
$\Psi_2 :=$ OnLineQ[E_2, line1$_1$] $\wedge$ OnLineQ[E_2, A_1D_1] $\wedge$
        OnLineQ[F_2, line1$_1$] $\wedge$ OnLineQ[F_2, B_1C_1] $\wedge$
        A_2=Reflection[A_1, line1$_1$] $\wedge$ B_2=Reflection[B_1, line1$_1$] $\wedge$
        C_2=C_1 $\wedge$ D_2=D_1 $\wedge$ UnfoldQ[line1$_1$]
$\Psi_3 :=$ A_3=Reflection[A_2, line1$_1$] $\wedge$ B_3=Reflection[B_2, line1$_1$] $\wedge$
        C_3=C_2 $\wedge$ D_3=D_2 $\wedge$ E_3=E_2 $\wedge$ F_3=F_2 $\wedge$ A_3C_3 = line2$_3$

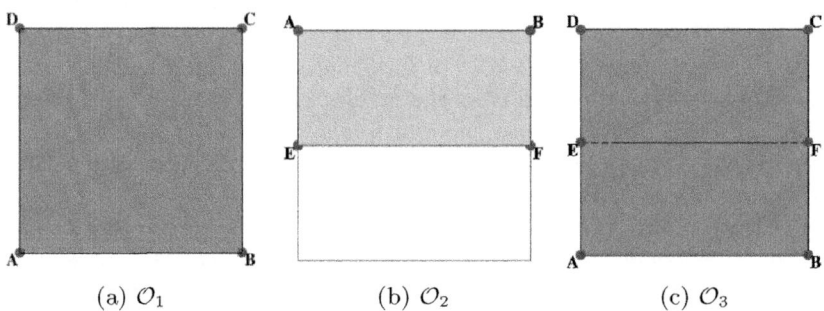

(a) $\mathcal{O}_1$ (b) $\mathcal{O}_2$ (c) $\mathcal{O}_3$

Fig. 1. Fold to bring A to D, and unfold

Formulas Ψ_is are conjunction of predicates (many of them are equalities). They are purely geometrical statements and independent of the coordinate system. The predicate PPSupQ[A_1, D_1, line1$_1$] in Ψ_1 states that line1$_1$ is the fold line to superpose the points A and D. PPSupQ stands for Point-Point Superposition. It corresponds to $f(\mathcal{O}_1, \underline{a_1})$ in Algorithm **Ori**.

Ψ_2 is the conjunction of the predicates whose geometrical meaning is as follows:

- OnLineQ[E_2, line1$_1$] states that point E at step 2 is on the fold line line1$_1$ and the same is true for point F at step 2. Points E and F are constructed at step 2 as the coincidences of the fold line line1$_1$ and the lines obtained by extending the edges AD and BC of origami $\mathcal{O}_1$, respectively. Likewise the other two predicates OnLineQ[E_2, A_1D_1] and OnLineQ[F_2, B_1C_1] are obtained.
- The predicate A_2=Reflection[A_1, line1$_1$] states that point A_2 is the reflection of A_1 over the fold line line1$_1$. The reflections of A_1 and B_1 occur when we fold the origami at step 2.
- Points C and D are not moved by the fold operation. This is expressed by the equalities in the conjunction C_2=C_1 $\wedge$ D_2=D_1.
- The next operation is unfold along the line line1$_1$, which the predicate UnfoldQ[line1$_1$] describes.

The first six predicates in Ψ_3 show the effects of the unfold. Points A and B are reflected back across the fold line line1$_1$ and points C, D, E and F at step 3 are not moved. The last predicate A_3C_3 = line2$_3$, which shows the introduction of the fold line line2, describes the fold operation at step 4. We fold along the line passing through the points A and C, i.e. fold line line2 at step 4, moving the left side of ray $\overrightarrow{AC}$ as shown in Fig. 2.

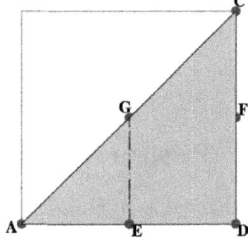

Fig. 2. Fold along the line AC

4.3   Fold Principle

We now explain briefly how Algorithm **Ori** is realized in Eos. We use the *Mathematica* frontend for the interaction with Eos. Users specify the command that folds the origami along the fold line computed by $f(\mathcal{O}_i, \underline{a_i})$ in the syntax of *Mathematica* function calls. $\mathcal{O}_i$s are stored in the global memory of Eos and hence not included in the parameters of the command. The top-level read-eval loop of the *Mathematica* frontend interprets the input command and outputs the next origami.

We adopted Huzita's fold principle to specify the fold operations. Table 1 gives the list of the fold commands. Function HO[1] is the realization of Huzita's seven basic fold operations. Some parameters of HO are syntactically sugared to improve the readability. Those parameters define the fold line. Note that not all the operations are necessary; the first to the fifth, and the seventh fold operations are special instance of the sixth operation. Nevertheless, we provide all of them for the convenience of users and for the ease of proving. For instance, HO[m, n] is equivalent to HO[P, n, Q, n], when m is a line passing through points P and Q.

---

[1] HO stands for Huzita Ori.

Table 1. HO commands

Fold command	Meaning
HO$[H,$ Through $\to \{P, Q\}]$	Fold along the line passing through points P and Q, moving point H.
HO$[P,\ Q]$	Fold to superpose points P and Q.
HO$[m,\ n]$	Fold to superpose lines m and n.
HO$[m,$ Through $\to P]$	Fold along the line passing through point P to superpose line m with itself .
HO$[P, m,$ Through $\to Q]$	Fold along the line passing through point Q to superpose point P and line m.
HO$[P, m, Q, n]$	Fold to superpose point P and line m, and point Q and line n.
HO$[P, m, n]$	Fold to superpose point P and line m, and line n with itself.

4.4   Program of Construction

While constructing the origami, we do not need to reason about the origami in a fundamental level as shown in Algorithm **Ori**. An origami programming language called ORIKOTO is provided for this purpose [5]. Besides the HO commands, ORIKOTO provides commands for managing origamis. Some of them will be explained where they are used.

In ORIKOTO we write a program that constructs the sequence of origamis in Figs. 1 and 2 by executing the following program:

```
BeginOrigami[ ];
HO["A", "D"];
Unfold[ ];
HO["D", Through→ {"A", "C"}];
```

BeginOrigami[ ] produces $\mathcal{O}_1$ in Fig. 1(a). This corresponds to the initialization phase of Algorithm **Ori**. Next, HO["A", "D"] is applied to $\mathcal{O}_1$. This corresponds to the first loop of construction phase. We use double quoted point names, as in "A" and "D", to refer to the points in ORIKOTO to distinguish them from ordinary *Mathematica* variables. In the first loop, $\mathcal{O}_2$, shown in Fig. 1(b), is produced and Ψ_2 is computed. Ψ_2 is saved in the EOS memory and later used to produce the proof document. In the second loop, Unfold[ ] is executed, $\mathcal{O}_3$ in Fig. 1(c) is produced and Ψ_3 is computed. The last predicate in Ψ_3 comes from the execution of the next command HO["D", Through→ {"A", "C"}].

4.5 Program of Proving

Referring to the reasoning about origami of the simple example of subsection 4.2, at the last step of the origami reasoning, we try to prove that the length of segment DG is equal to the length of segment BG. We prove this by issuing the commands:

```
Goal[Distance["D", "G"]^2 == Distance["B", "G"]^2];
Prove["Equal Distance", StrongLineCC -> True];
```

At the end of the execution of `Prove` command, the proof document titled "Proof Document for Equal Distance" will be generated. The keyword `StrongLineCC -> True` is needed in this case to specify the coefficients a, b and c of line equation $ax + by + c = 0$ are constrained to be by the condition $b(-1 + b) = 0 \wedge (-1 + a)(-1 + b) = 0 \wedge a^2 + 1 = 0$. This constraint is discussed further in Subsection 5.7.

5 Method of Proving

5.1 Overview

The geometrical theorem that we prove with Eos is of the form:

$$\forall \underline{U} \ \forall \underline{X} \ \Psi_1 \wedge \cdots \wedge \Psi_k \Rightarrow \Phi \tag{5.1}$$

where $\underline{U}$ is a sequence of independent variables and $\underline{X}$ is a sequence of dependent variables. To prove the theorem given in (5.1), we take an arbitrary but fixed sequences $\underline{U}$, then we prove by contradiction the following:

$$\forall \underline{X} \ \Psi_1 \wedge \cdots \wedge \Psi_k \Rightarrow \Phi. \tag{5.2}$$

This is equivalent to showing that the following is false:

$$\exists \underline{X} \ \Psi_1 \wedge \cdots \wedge \Psi_k \wedge (\neg \Phi) \tag{5.3}$$

The formula (5.3) can be disproved by showing that no instance of $\underline{X}$ exists that satisfy:

$$\Psi_1 \wedge \cdots \wedge \Psi_k \wedge (\neg \Phi) \tag{5.4}$$

5.2 Need for Proof Document

In its purely geometrical form, many of the origami geometrical theorems cannot be proved easily by mere symbolic logical reasoning. In our framework, algebraic transformation of the geometrical formulas is the essential part of the automated theorem proving. Once the geometrical formulas are transformed to algebraic ones, powerful symbolic and algebraic methods, such as GB and CAD, can be employed effectively.

With Eos we proceed in the following way:

1. Geometric reasoning
    (a) local geometrical inference on the formulas of the first-order logic
    (b) collecting all the predicates
    (c) eliminating the formulas that are unnecessary for the proof
    (d) forming equivalence classes of lines and points
    (e) identifying collinearity among points
2. Algebraic manipulation
    (a) assigning coordinates to points, introducing dependent (local) variables
    (b) generating polynomials with appropriate ordering among the variables
    (c) performing algebraic optimization
    (d) forming a system of algebraic equalities (and inequalities)
    (e) calling the appropriate algebraic methods

The generation of the algebraic representation requires insights, and the proof document should be used to assist in the investigation of the process of the algebraic transformation. Also, for better understanding of the process of construction and proof, the reading of complex algebraic computations should be facilitated.

Furthermore, optimization is an important issue for higher performance of the automated proofs of origami geometrical problems. Origami folds lead to the creation of new points and lines on the origami paper, which grow exponentially by the progress of paper folds. For algebraic representation of these points and lines, it is important to keep the number of equations and variables as small as possible. Proving using the Gröbner basis method for instance is highly dependent on the number of variables and polynomials. Besides, the order of variables used to compute the Gröbner bases greatly affects the computation time. We also experienced that the proof process itself might be repeated several times, by modifying small parts in the process, and looking at the final effects. The proof document helps in archiving the proof trials, comparing them, and investigating several aspects of the given problem.

Therefore, the proof document consists of several sections to explain how the origami is constructed, geometrically and algebraically treated and proved. The human-readable document assists us to answer questions like:

– What geometrical properties do the polynomials represent?
– How relations are deduced?
– Which further optimizations can be applied?
– What are the reasons for proof failure?

Roughly speaking, the proof document answers those questions by recording the input/output of the phases of the computation outlined above.

With these observations in mind, we will explain those steps in sequence, in more details in the next subsections.

5.3 Local Geometrical Inference

The inferences on the predicates at construction steps i and $i + 1$ are necessary for the following reason. At construction step i, we compute the fold line. On the transition of step i to step $i + 1$, we fold origami $\mathcal{O}_i$ and create origami $\mathcal{O}_{i+1}$. Only at step $i + 1$ we know which points have been moved. Based on the generated equalities of points at step i and step $i + 1$, we can deduce the following:

1. $P_{i+1} = Q_i$ if P_i and Q_i are specified to be superposed at step i, and P_i is moved.
2. OnLine[P_{i+1}, l_i] if P_i and line l_i are specified to be superposed at step i, and P_i is moved.

Then the above predicates are added, if applicable, to each Ψ_{i+1}.

5.4 Transformation to Logical Formula

Each $\Psi_i, i = 1, \ldots, k$ is the conjunction of predicates, and Φ is a quantifier-free formula of the first-order predicate logic. After specifying independent variables $\underline{U}$, we take all the other points and lines to be quantified variables and let them be variables $\underline{X}$. The result is the formula (5.1). We transform the formula (5.4) to a logically equivalent conjunctive normal form, which is then used for further logical and algebraic transformation. The transformation from this logical formula into the algebraic formula is straightforward. However, most likely this would result in unwieldily large algebraic expressions. If they are used as the input to the Gröbner basis computation or the cylindrical algebraic decomposition, the computation may swamp all the available resources at hand. The elimination of unnecessary predicates is the next step to take.

5.5 Elimination of Unnecessary Predicates

In order to construct a point in the origami to be used in the specification of further fold operations, we first have to make a fold to construct a fold line. The needed point is created as the intersection of the fold line and the existing creases and edges. The fold entails movements of some constructed points, and the non-movements of the other points. As we saw in the example of subsection 4.2, these movements and non-movements generate formulas, some of which may be unnecessary to prove the goal. To eliminate unnecessary formulas, we define a *weakly needed atomic formula*.

Let $\mathcal{T}(A)$ denote the set of geometrical objects (terms) occurring in an atomic formula A (atom for short).

– An atom is needed if it is a sub-formula of the goal.
– An atom is weakly needed if it is needed.
– An equality $P_i = Q_j$ is weakly needed if $j \geq i \wedge Q_j \in \mathcal{T}(A)$ for some weakly needed atom A.

- An equality Reflection(P_i, l) = Q_j is weakly needed if $j \geq i \wedge Q_j \in \mathcal{T}(A)$ for some weakly needed atom A.
- An atom A other than the equalities above is weakly needed if $\mathcal{T}(A) \cap \mathcal{T}(B) \neq \emptyset$ for some weakly needed atom B.

Those atoms that are not weakly needed are eliminated. Let us denote the resulting formula after the elimination by

$$\chi := \phi_1 \wedge \cdots \wedge \phi_j, \text{ where } \phi_i, i = 1, \ldots, j \text{ are literals.}$$

Formula χ is the one that is subjected to the algebraic transformation.

5.6   Forming Equivalence and Collinear Relations

At the end of geometric transformation, we collect the equalities from χ, and generate equivalence classes of points and lines that occur in the equalities. Furthermore, we detect the collinearity relations on the points.

Equivalence classes of points are important to reduce the number of dependent variables. For example, points $\{X_1, \ldots, X_n\}$ that belong to the same equivalence class will be assigned the same coordinate $(x, y)^2$.

5.7   Algebraic Manipulation

Algebraic transformation of the logical formulas was discussed in detail in our previous publication [6]. The overall procedure is as follows. We use the cartesian coordinate system.

- Assigning the coordinates to the points, we use the equivalence relations among the points to ensure that the superposing points are given the same coordinates.
- For each fold line and line extended from the segment, we assign the equality of the form $ax + by + c = 0$ together with the condition

$$b(-1 + b) = 0 \wedge (-1 + a)(-1 + b) = 0 \tag{5.5}$$

where a, b and c are freshly generated variables. For some theorems, we need the following stronger conditions in order to avoid the solutions of a in $\mathbb{C}$:

$$b(-1 + b) = 0 \wedge (-1 + a)(-1 + b) = 0 \wedge a^2 + 1 \neq 0$$

The variables are ordered for the purpose of Gröbner basis computation. The variables are ordered *genetically*. We call a sequence of variables $\underline{X}$ *genetically ordered*, if for every pair of variables v and w in the sequence $\underline{X}$, we have $v \leq w$ if v and w are assigned to geometrical objects V and W (i.e. points and lines in this

---

[2] Later on in the ProofDoc, we see three dimensional coordinates. For the theorem proving two-dimensional ones are all we need, but Eos will also perform (partial) 3D origami folds.

case), respectively, where V is constructed at the step earlier or the same step of the construction of W. This ordering experimentally gives better performance of Göbner basis computation in most of the cases.

The algebraic expressions generated from the predicates can be rational functions of $\mathbb{Q}[\underline{U}, \underline{X}]$. Depending on the Gröbner basis computation libraries, we would need some more algebraic manipulation of the rational functions to obtain the set of polynomials of $K[\underline{X}]$, where K is $\mathbb{Q}[\underline{U}]$.

6 Proof Document

At the end of the proof, EOS generates a proof document, abbreviated to *Proof-Doc* hereafter, that provides detailed information about the construction and proof processes. A ProofDoc is organized as a *Mathematica*[3] notebook and structured as the nested content cells of sections and subsections, which correspond roughly to the items outlined in the previous subsections. The feature of nested cells of *Mathematica* notebooks helps to read lengthy ProofDocs by opening and closing only the nested cells that we are focusing to study.

6.1   Structure of Proof Document

We will explain the structure of ProofDocs using the one produced for the proof of Morley's theorem [8]. Figure 4 shows the ProofDoc that is popped up during the execution of the ORIKOTO program code given in Fig. 3 after the construction of a Morley's equilateral triangle. The colored cells are used for formatting the program as it appears in the ProofDoc. The essential code for proving consists of non-colored cells, in which additional assumption is declared, independent variables are specified as part of coordinate mapping, and the prove command is issued. In the call of Prove, we specify various parameters for Gröbner basis computation.

When the ProofDoc is generated, its cells are not fully open. We see the title and headers of sections, and other short items of information that the readers of the ProofDoc can immediately see without opeing the inner cells. They are about the author, starting time of the computation, software version of *Mathematica*, and the result of computation. Figure 4 is the ProofDoc for Morley's theorem after opening most of section cells to show the headers of the subsections. The title "Proof Document of Morley's theorem by Abe's construction" and the name of the author come from the parameters of Prove command.

6.2   Program, Prover Computation and Result Sections

The information that would immediately interest the reader of the ProofDoc are the construction problem, the method that was used for the proof and whether proof is successful or not. These are the contents of "Program", "Prover Computation" and "Result" sections of the ProofDoc.

---

[3] *Mathematica* 8.0 is used at the time of writing the paper.

■ **Proof**

In[43]:=
```
ProofDocFormat["Proof", "Subsection", 1];
```

In[44]:=
```
ProofDocFormat["Additional assumption", "Subsubsection", 1];
```

In[45]:= **Assume** $\Big[$

$\exists_{t1,\,t1\in Alg}\;\exists_{t2,\,t2\in Alg}\;\exists_{t3,\,t3\in Alg}\;\Big($t1 == ToTangent["L", "A", "B"] $\bigwedge$

t2 == ToTangent["A", "B", "L"] $\bigwedge$ t3 == ToTangent["B", "E", "R"] $\bigwedge$

$(t1 + t2 + t3 - t1\,t2\,t3)^2 == 3\,(1 - t1\,t2 - t1\,t3 - t2\,t3)^2\Big)\Big]$;

In[46]:=
```
ProofDocFormat["Goal", "Subsubsection", 1];
```

In[47]:= **Goal** $\Big[$Distance["L", "R"]^2 == Distance["R", "S"]^2 $\bigwedge$

Distance["L", "R"]^2 == Distance["L", "S"]$^2\Big]$;

In[48]:=
```
ProofDocFormat["Prove", "Subsubsection", 1];
```

In[49]:=
```
Prove["Morley's theorem by Abe's construction",
    Mapping → (StandardMapping[{{"L", Point[u1, u2]}}] &),
    GroebnerBasis → {MonomialOrder → DegreeReverseLexicographic,
      CoefficientDomain → RationalFunctions},
    AuthorName → "Author: Fadoua Ghourabi"];
```

Fig. 3. Proof code for Morley's Theorem by Abe's method

Proof Document of Morley's theorem by Abe's construction

Author: Fadoua Ghourabi

- Computation started at June 20, 2011 12:36:50 PM JST.
- The Mathematica version is 8.0 for Mac OS X x86 (64-bit) (December 2, 2010).
- The constructed origami is as follows:
▽ **Final origami**

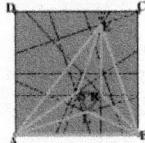

▷ **Program**

▽ **Geometrical reasoning**
  ▷ Geometrical inferences
  ▷ Geometrical operations and relations at each step
  ▷ Elimination of unnecessary predicates
  ▷ Geometrical relations
  ▷ Equivalence
  ▷ Collinearity

▽ **Algebraic transformation**
  ▷ Variable assignment
  ▷ Reflection/Point equality
  ▷ Algebraic relations
  ▷ Notes
  ▷ Occurrence check of inequalities

▽ **Computation of Groebner basis**
  - Preprocessing (elimination of duplicated polynomials) of the input bases is performed.
  - Computation started at June 20, 2011 12:37:41 PM JST.
  - The monomial order is DegreeReverseLexicographic.
  - The variables are ordered genetically.
  ▷ Input polynomials
  ▷ Dependent variables
  ▷ Indepedent variables

▽ **Result**

  ▽ Proof by Groebner basis method is successful.
    - CPU time used for Groebner basis computation is 1762.3 seconds.
    - The computation ended at June 20, 2011 1:11:54 PM JST.

Fig. 4. Structure of ProofDoc

In Fig. 4, "Program" section contains the cells of commands that have been used in the construction and the proof, e.g. HO commands, Prove command and other commands for manipulating origamis. The cells are organized as the usual *Mathematica*. We can reproduce the result once again later by executing all the cells in "Program" section, if necessary. This will reproduce yet another ProofDoc for Morley's Theorem.

Based on the generated algebraic relations, and the user specified options, Eos chooses the suitable proof engine. Currently, it chooses either GB or CAD. As outlined in Fig. 4, the "Computation of Theorem Prover" section records the inputs to the proof engine, in the case of GB, such as the set of polynomials and the genetically ordered variables.

The final section of ProofDoc records the output of the theorem prover; proved or failure to prove, and the time taken for computation. The cell of the Result section of the ProofDoc is open by default to show the result of the proof. The details are kept enclosed in "Geometrical reasoning" and "Algebraic transformation" sections.

6.3   Geometrical Reasoning Section

The results of the computations explained in Sections 5.3 - 5.6 are documented in "Geometrical reasoning" section of ProofDoc. The formulas that are deduced from local geometrical inference are listed in "Geometrical inferences" subsection. The geometrical properties accumulated throughout the construction are recorded in "Geometrical operations and relations at each step" subsection. They correspond to the Ψ_is that define the construction steps. The predicates in Ψ_i are organized in cells. Furthermore, these predicates are expressed in a natural language for readability. In Fig. 5, we show Ψ_7 and Ψ_8, i.e. the geometrical properties that hold at steps 7 and 8 of construction of Morley's equilateral triangle. Table 2 shows some of the statements in English and their original predicate forms.

As we mentioned in Section 5.5, not all the generated predicates are needed for the proof. In the ProofDoc, "Elimination of unnecessary predicates" subsection lists the predicates that are eliminated and "Geometrical relations" subsection lists the predicates that are necessary for the proof, i.e. χ given in Subsection 5.5. Finally, the two remaining subsections list the formed equivalence classes and the inferred relations of collinearity.

6.4   Algebraic Transformation Section

The section "Algebraic transformation" has several subsections. A part of this section is shown in Fig. 6. The subsection "Variable assignment" shows the variables assigned to each geometrical object (i.e. point or line) that is involved in the proof process. The variables on the right hand, i.e. variables of the coordinates of the points and the coefficients of the lines are used to form polynomials to be input to the proof engine (GB or CAD). For instance, we see that line

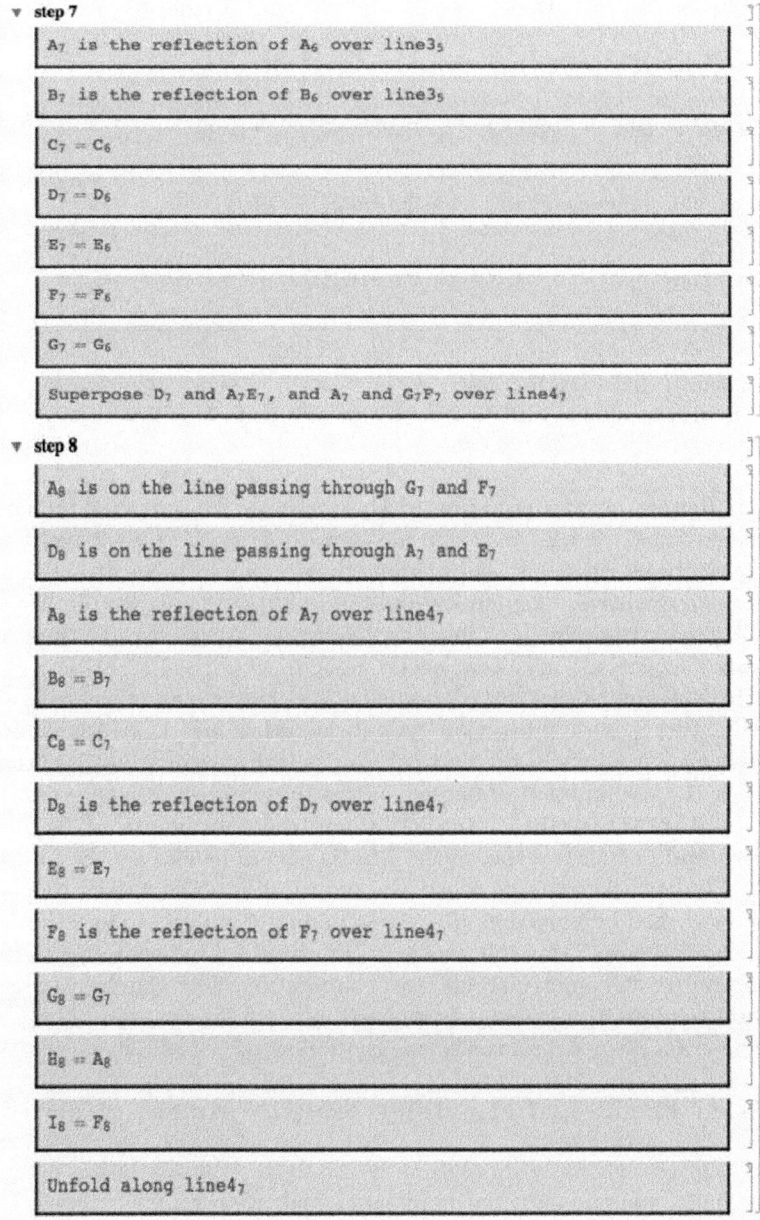

Fig. 5. Part of subsection "Geometrical operations and relations at each step"

Table 2. Translation of predicates in natural language

Statement in ProofDoc	Predicate
P is the reflection of Q over l	P = Reflection[Q, l]
Superpose P and m, and Q and n over l	PLPLSupQ[P, m, Q, n, l]
P is on the line passing through Q and R	OnLineQ[P, QR]
Unfold along l	UnfoldQ[l]

Note:

- Parameter l denotes a fold line.
- The predicate symbol PLPLSupQ stands for two Point-Line Superpositions.

line1$_1$ is defined by the equation $a1x + b1y + c1 = 0$ and the coordinate assignments for A$_1$ and E$_1$ are ($x39$, $y39$) and ($x41$, y41), respectively.

The "Algebraic relations" subsection, shown in Fig. 6, records the algebraic expressions (equalities, inequalities and disequalities) generated from the predicates. We create hyperlinks to link the geometrical properties to their algebraic forms, and vice versa. This allows easier navigation in the ProofDoc and tells the geometrical meanings of the polynomials. We will explain the algebraic forms inserted in the cells of subsection "Algebraic relations" that appear in Fig. 6. The algebraic equations in the first cell comes from predicate A$_1$E$_1$=line1$_1$. The second cell is a conjunction of two equalities that comes from the equality predicate D$_2$ = Reflection[D$_1$, line1$_1$], where the coordinates of D$_1$ and D$_2$ are ($x18$, $y18$) and ($x1$, $y1$), respectively. This is also indicated by the tooltip below the cell. The tooltip pops up when the mouse pointer is placed on the cell to provide a detailed explanation of the polynomial equalities. Moreover, when we click on the content of this cell, our focus of attention hyper-jumps to the corresponding cell in "Geometrical relations" subsection. The algebraic expression of D$_2$ = Reflection[D$_1$, line1$_1$] is derived in the following way. The point that is the reflection of D$_1$ over line1$_1$ has the following coordinates:

$$\left(\frac{-a1^2x18 + b1^2x18 - 2a1(c1 + b1y18)}{a1^2 + b1^2}, \frac{-2b1(c1 + a1x18) + a1^2y18 - b1^2y18}{a1^2 + b1^2} \right)$$

Since point D$_2$ and the reflected point are equal, we obtain:

$$\frac{-a1^2x18 + b1^2x18 - 2a1(c1 + b1y18)}{a1^2 + b1^2} - x1 = 0 \tag{6.1}$$

$$\frac{-2b1(c1 + a1x18) + a1^2y18 - b1^2y18}{a1^2 + b1^2} - y1 = 0 \tag{6.2}$$

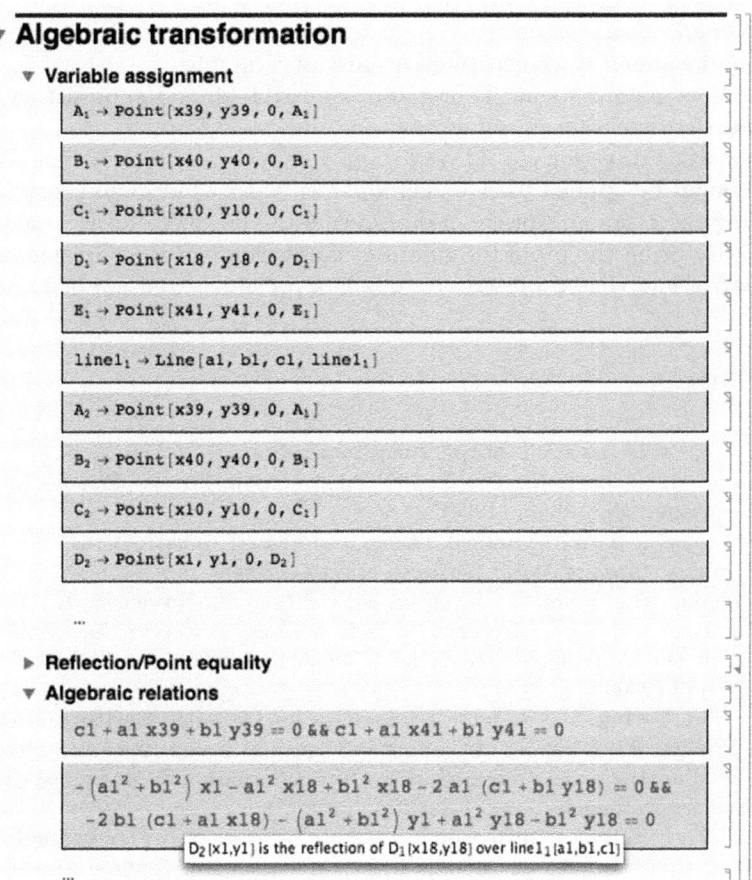

Fig. 6. Parts of the section "Algebraic transformation"

Using the line coefficient conditions (5.5), we multiply both sides of equations (6.1) and (6.2) by $a1^2 + b1^2$, and obtain the algebraic expression in Fig. 6.

In Fig. 4, "Occurrence check of inequalities" subsection shows whether inequalities are generated by the algebraic transformation. The presence of inequalities would requires CAD computation. In Morley's theorem, only equalities are involved, and therefore Gröbner bases method is used.

7 Conclusion

The proof document is a computer-generated document that assists origamists working in computational origami to reason about origami theorems. It documents the whole process of computational origami construction and proving using EOS. It is a *Mathematica* notebook and makes full use of the functionalities of *Mathematica* such as nested cell structures. The functionality of generating the proof document is implemented as part of EOS.

In EOS, we reason about 2D origami. A possible direction of further research would be the extension of our logical and algebraic formalization to cover 3D origami. Other direction would be to apply other proving methods. Wu-Ritt's method could be applied to the algebraic formalism of origami construction. It may bring the extra advantage of discovering the degenerate cases that may be overlooked during the proof formulation. Furthermore, incorporating deductive capability of logic-based proof assistants is a possible extension of EOS system.

References

1. The Coq Proof Assistant, http://coq.inria.fr/
2. Pearson, K.R., Jones, A., Morris, S.A.: Abstract Algebra and Famous Impossibilities. Springer, Heidelberg (1991)
3. Alperin, R.C.: A Mathematical Theory of Origami Constructions and Numbers. New York Journal of Mathematics 6, 119–133 (2000)
4. Buchberger, B., Dupre, C., Jebelean, T., Kriftner, F., Nakagawa, K., Văsaru, D., Windsteiger, W.: The Theorema Project: A Progress Report. In: Symbolic Computation and Automated Reasoning (Calculemus 2000), St. Andrews, Scotland, pp. 98–113 (2000)
5. Ghourabi, F., Ida, T.: *Orikoto*: A Language for Origami Construction and Theorem Proving. In: Frontiers of Computer Science in China (2010); (Submitted, the extended abstract was presented in the fifth International Conference on Origami in Science, Mathematics and Education (5OSME)
6. Ghourabi, F., Ida, T., Takahashi, H., Marin, M., Kasem, A.: Logical and Algebraic View of Huzita's Origami Axioms with Applications to Computational Origami. In: Proceedings of the 22nd ACM Symposium on Applied Computing (SAC 2007), Seoul, Korea, pp. 767–772 (2007)
7. Huzita, H.: Axiomatic Development of Origami Geometry. In: Proceedings of the First International Meeting of Origami Science and Technology, pp. 143–158 (1989)
8. Ida, T., Kasem, A., Ghourabi, F., Takahashi, H.: Morley's theorem revisited: Origami construction and automated proof. Journal of Symbolic Computation 46(5), 571–583 (2011)
9. Ida, T., Takahashi, H., Marin, M., Ghourabi, F., Kasem, A.: Computational Construction of a Maximum Equilateral Triangle Inscribed in an Origami. In: Iglesias, A., Takayama, N. (eds.) ICMS 2006. LNCS, vol. 4151, pp. 361–372. Springer, Heidelberg (2006)
10. Justin, J.: Résolution par le pliage de l'équation du troisième degré et applications géométriques. In: Proceedings of the First International Meeting of Origami Science and Technology, pp. 251–261 (1989)
11. Kasem, A., Ghourabi, F., Ida, T.: Origami Axioms and Circle Extension. In: Proceedings of the 26th Symposium on Applied Computing, pp. 1106–1111. ACM press (2011)
12. Narboux, J.: A Graphical User Interface for Formal Proofs in Geometry. Journal of Automated Reasoning 39(2), 161–180 (2007)

13. Robu, J., Ida, T., Ţepeneu, D., Takahashi, H., Buchberger, B.: Computational Origami Construction of a Regular Heptagon with Automated Proof of Its Correctness. In: Hong, H., Wang, D. (eds.) ADG 2004. LNCS (LNAI), vol. 3763, pp. 19–33. Springer, Heidelberg (2006)
14. Wang, D.: GEOTHER 1.1: Handling and Proving Geometric Theorems Automatically. In: Winkler, F. (ed.) ADG 2002. LNCS (LNAI), vol. 2930, pp. 194–215. Springer, Heidelberg (2004)
15. Wantzel, P.L.: Recherches sur les moyens de connaitre si un problème de géométrie peut se résoudre avec la règle et le compas. Journal de Mathématiques Pures et Appliquées, 366–372 (1984)
16. Wolfram, S.: The Mathematica Book, 5th edn. Wolfram Media (2003)
17. Wu, W.T.: Basic Principles of Mechanical Theorem Proving in Elementary Geometry. Journal of Automated Reasoning 2, 221–252 (1986)

The Midpoint Locus of a Triangle in a Corner

Daniel Lichtblau

Wolfram Research, Inc.
100 Trade Center Dr.
Champaign, IL 61820 USA
danl@wolfram.com

Abstract. We are given an equilateral triangle with vertices constrained to lie in each of the three positive octant coordinate planes (colloquially, "a triangle in a corner"). We wish to describe the locus of points covered by the midpoint of the triangle, as the vertices range over configurations allowed by the above constraint. This locus comprises a solid region. We use numerical and graphical methods, and also computational algebra, to find the boundary surface and visualize this locus.

Keywords: Constraint geometry, implicit surfaces, nonlinear systems.

1   Introduction

A Putnam Exam problem from 1948 [2] asks for the locus of a circle center when the circle is constrained so that it is simultaneously tangent to the non-negative parts of the three coordinate planes (the "penny in a corner" problem). A recent variant on this is the "tile in a corner" problem, as discussed by [3]. In this note I discuss a variant, first proposed in [3], wherein our object is an equilateral triangle. We constrain each vertex to lie in one of the positive octant coordinate planes with no two lying in the interior of the same such plane. We wish to understand the locus spanned by the triangle midpoint. We will show both numeric and algebraic methods for discovering this locus. Our results include direct computation of the polynomials that define its boundary. They comprise the first complete solution to the problem under consideration.

The problem itself is perhaps not of great significance beyond mathematical curiosity. But one can readily imagine constrained geometric problems of a similar nature, but having practical interest. Areas that come to mind include object avoidance in path planning, motion of objects within larger objects (e.g. atoms within molecules), construction/placement of mobile sculptures, etc.. See also [8] for some related applications. Our expectation is that the tools we develop for this might similarly be useful in such more general settings.

I thank Jack Wetzel for introducing me to this problem and inviting me to speak about it at a UIUC geometry seminar. I thank both Jack and Wacharin Wichiramala for email discussions of some of the finer points. In particular, Wichiramala (private communication) has indicated some methods that are of independent interest for finding parts of the locus boundary. I thank Michael

P. Schreck, J. Narboux, and J. Richter-Gebert (Eds.): ADG 2010, LNAI 6877, pp. 98–117, 2011.
© Springer-Verlag Berlin Heidelberg 2011

Trott, Brett Champion, and Yu-Sung Chang for assistance with some of the graphics. I thank the organizers of ADG 2010 for their hospitality and for the overall quality of the workshop. I thank the anonymous referees of both drafts for raising several interesting points and making numerous constructive suggestions. These include, but are not limited to, requests for discussion of important inverse problems.

Mathematica [10] code for our computations may be found in the appendix.

2   Setting up the Problem

The primary contribution of this paper is to phrase the problem in a way that is computationally tractable. To this end we first set up an algebraic problem.

We will denote our midpoint as $M = \{x_m, y_m, z_m\}$. The vertices will be $\{x_j, y_j, z_j\}$ for $1 \leq j \leq 3$. We restrict vertices to coordinate planes by setting appropriate vertex coordinates to zero. Note that we will not enforce non-negativity until later. We use (quadratic) equations for edge lengths that enforce that the triangle be equilateral, and we also have linear equations relating midpoint coordinates to the vertex coordinates. Here are the defining polynomials.

$$-3x_m + x_2 + x_3$$
$$-3y_m + y_1 + y_3$$
$$-3z_m + z_1 + z_2$$
$$x_2^2 + y_1^2 + z_1^2 + z_2^2 - 2z_1 z_2 - 1$$
$$x_3^2 + y_1^2 + y_3^2 - 2y_1 y_3 + z_1^2 - 1$$
$$x_2^2 - 2x_3 x_2 + x_3^2 + y_3^2 + z_2^2 - 1$$

We have six polynomials. We are interested in the midpoint variables, as functions of the vertex coordinates. For this purpose we will regard those latter as parameters. In the remainder of this paper, "variables" refers to the midpoint coordinates and "parameters" to the vertex coordinates unless otherwise stated. We will eliminate three of these parameters so that our midpoint is defined, implicitly, in terms of the remaining three. We obtain the set of polynomials below.

$$-6x_3 x_m + 6y_1 y_m - 18z_m^2 + 18z_2 z_m + 2x_3^2 - 4y_1^2 - 4z_2^2 + 1$$
$$6x_3 x_m - 18y_m^2 + 18y_1 y_m - 6z_2 z_m - 4x_3^2 - 4y_1^2 + 2z_2^2 + 1$$
$$18x_3 x_m - 18x_m^2 - 6y_1 y_m + 6z_2 z_m - 4x_3^2 + 2y_1^2 - 4z_2^2 + 1$$

That we have three equations in the three midpoint coordinates tells us to expect that the locus they span will be a solid. We will confirm this numerically below. Our eventual goal will be to determine the enveloping surface of this solid. We observe here that this solid locus is in contrast to the constrained circle and tile problems, wherein the locus of interest lies on a surface (a sphere, in the case of the classic "penny in a corner") [2,3]. We also point out that the boundary surface pieces will be defined in implicit form as polynomials (which we will compute). This follows from the fact that boundary components are part of the discriminant variety [4] of the polynomial system.

3   Approximating the Locus with a Graphic-Numeric Method

Our secondary contribution is to develop symbolic-numeric and graphical methods to approximate the region of interest. This is useful in its own right, and moreover will help us to see how the algebraic surfaces we later compute actually fit together.

First we will find a large number of solutions to the system of equations. We do this by selecting random values in a suitable range for three of the vertex parameters, and solving from that for all other variables of interest. We will use the original system rather than the above elimination set, even though that latter is in some sense simpler. The reason is that we will need to enforce the non-negativity of our vertex coordinates, and we cannot do so without explicitly solving for them. (We might have all midpoints non-negative, but still have one or more vertices with negative coordinate values. This would violate our configuration constraints. We must be able to remove such solutions, hence we must have values for all vertex as well as midpoint coordinates.)

We can get six times as many valid solutions, and obtain a more accurate picture, simply by enforcing the natural symmetry of the problem. We can permute any vertex pair, or all three vertices, of any given configuration and still have a valid configuration. This means the midpoint coordinates have symmetry under the action of S_3. We utilize this to expand our solution "point cloud" by a factor of six.

It is quite difficult to see detail. In particular one cannot readily visualize contours or get an understanding of whether and where the solid folds into

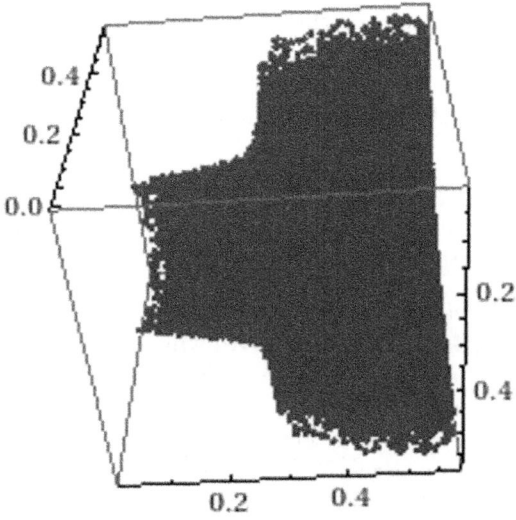

Fig. 1. A point cloud of the midpoint locus

itself. So we will next obtain a better picture using an approximate region plot. We define a point as being inside or outside based on proximity to the points in our region that we computed above.

This is, admittedly, a bit crude, and will have the effect of both expanding the surface and softening any sharp contours. It also carries the risk of omitting "thin" sections, where we perhaps found insufficiently many solutions to accurately sample the region. One could ameliorate this in two ways. One is to recognize that these thin parts tend to be near boundary surface intersections, many of which arise when more than one of our vertex parameter variables are near zero. So we could simply sample more heavily in such regions by constraining some of the random vertex value ranges more tightly. Another way to more reliably enclose the surface would be to use a larger setting for proximity to our computed points. While this will overly inflate the region, in many practical applications that is not a bad thing to do. For example, in collision avoidance, coming "close" is also to be avoided since physical systems always have some looseness in their actual constraints. Hence one would do well to work with a modest enlargement of the danger zone.

All that stated, we will see later that this approximation does in fact provide a quite reasonable picture of the actual region boundary. For purposes of orientation we note that the origin is "pointed to" by the extruding middle in the figure on the left.

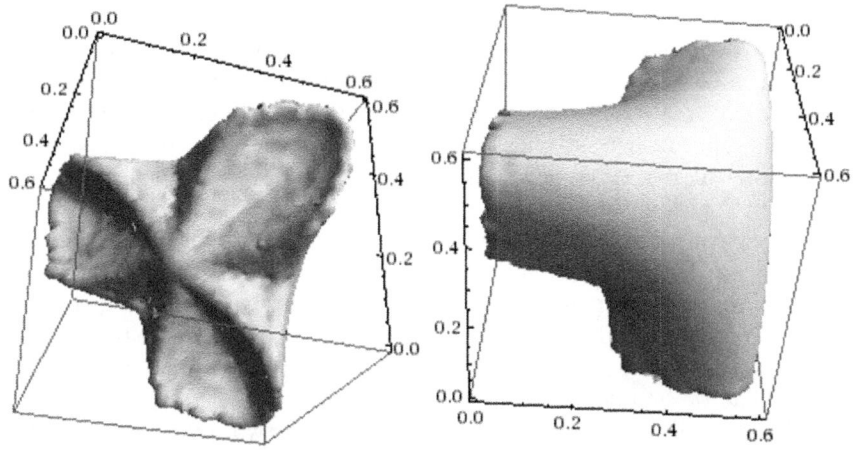

Fig. 2. The approximated region

It looks like something from a sci-fi movie. This is in part from the edge artifact; the solid has boundary seams in the coordinate planes, near which it is quite thin. Our randomization did not suffice to find many solutions in the thin regions. The actual region will not have the jaggedness we see in these approximations.

Perhaps surprisingly, given that we began with but three quadratic and three linear equations, the bounding surface for our region is by no means simple.

4   Solving Algebraically

We now proceed to determine the implicit equations of the algebraic surfaces that bound the triangle midpoint locus. We first ignore positivity constraints on the vertex parameters, and just look for extremal values of a midpoint coordinate when vertices lie in open sets of our parameter space. The reason we pursue this path is as follows. While we might hope to enforce non-negativity of our vertex parameters by replacing them with their squares, we would run into two forms of added computational complexity. One is that we might get spurious surface parts corresponding to complex-valued solutions (we might have this problem anyway, but that turns out not to happen). The second is that, as we will see, the eventual result has several components. Were we to compute them all as one polynomial, the size would be considerably larger, in terms of degree and number of terms, than is the sum of their separate sizes. Not surprisingly, this manifests as increased difficulty in the actual computation; an attempt along these lines showed no success after a day of run time.

We use the reduced system (the one with three vertex coordinates eliminated). We will explicitly regard the midpoint coordinates as functions of the remaining vertex parameters. We then use implicit differentiation of the defining equations in order to form a linear system of the partial derivatives of the midpoint coordinate functions in terms of both the vertex parameters and the midpoint coordinates themselves (see the appendix for code to do this). We solve this system for the partial derivatives of the midpoint coordinates.

We will next create a vector equation involving Lagrange multipliers (see [5] for a similar approach to a constrained geometry problem arising in number theory). This will give three new equations and two new variables (the multipliers), thus cutting dimension by one. To understand what follows, recall that our three vertex parameters–the independent variables in this formulation–are each subject only to the weak inequality constraint of non-negativity. One can find extremal values of the dependent variables by separately considering the dependent variable space boundary i.e. setting one independent variable to zero (which we will do later) and its open interior (where they are all strictly positive). It is this latter that we handle here.

Consider what happens when we fix the values of two midpoint coordinates, say x_m and y_m. The remaining coordinate z_m is now constrained to take values on an interval, and the midpoint moves on a vertical segment. We want to "extremize" these z_m values. This is simply a case of one function in our parameters, z_m, varying subject to (two) equality constraints (since the other two dependent variables, which are also functions of the parameters, are held fixed). It is well known that such a problem can be handled via Lagrange multipliers. Specifically, the gradient of z_m (the vector of partial derivatives of z_m with respect to the vertex parameters) must be a linear combination of scalar multipliers times each of the gradient vectors of x_m and y_m.

We remark that another approach would be to find the vanishing set of the Jacobian of our midpoint variables as functions of the vertex parameters. This is mathematically equivalent to ours, insofar as the Jacobian vanishes exactly when there is a linear dependency among the gradient vectors. We prefer our formulation because it reduces degree, though at the expense of introducing more equations and variables.

We also point out that this relatively simple formulation is related to the setting of discriminant varieties [4]; a difference is that we look at the projection of our system onto the variables of interest rather than the underlying parameter space.

Here is the full set of polynomials. (More correctly, these are rational functions, but we will impose that the common denominator not vanish and this will give the algebraic variety we seek.)

$$2x_3^2 - 6x_m x_3 - 4y_1^2 - 4z_2^2 - 18z_m^2 + 6y_1 y_m + 18z_2 z_m + 1,$$
$$-4x_3^2 + 6x_m x_3 - 4y_1^2 - 18y_m^2 + 2z_2^2 + 18y_1 y_m - 6z_2 z_m + 1,$$
$$-4x_3^2 + 18x_m x_3 - 18x_m^2 + 2y_1^2 - 4z_2^2 - 6y_1 y_m + 6z_2 z_m + 1,$$

$$(y_m x_3^2 - 2z_2 \lambda_2 x_3^2 + 4z_m \lambda_2 x_3^2 + 5x_m y_1 x_3 - 6x_m y_m x_3 - 5y_1 z_2 \lambda_1 x_3 +$$
$$7y_m z_2 \lambda_1 x_3 + 8y_1 z_m \lambda_1 x_3 - 12y_m z_m \lambda_1 x_3 + 5x_m z_2 \lambda_2 x_3 - 12x_m z_m \lambda_2 x_3 -$$
$$6x_m^2 y_1 - x_3^2 y_1 + 9x_m^2 y_m + 9x_m y_1 z_2 \lambda_1 - 15x_m y_m z_2 \lambda_1 - 15x_m y_1 z_m \lambda_1 +$$
$$27x_m y_m z_m \lambda_1 - 3x_m^2 z_2 \lambda_2 + 9x_m^2 z_m \lambda_2)/$$
$$(3(3x_3 y_1 z_2 - 5x_m y_1 z_2 - 5x_3 y_m z_2 + 9x_m y_m z_2 - 5x_3 y_1 z_m +$$
$$9x_m y_1 z_m + 9x_3 y_m z_m - 18x_m y_m z_m)),$$

$$(2x_3 y_1^2 - 4x_m y_1^2 + z_2 \lambda_1 y_1^2 - z_m \lambda_1 y_1^2 - 5x_3 y_m y_1 + 12x_m y_m y_1 - 5y_m z_2 \lambda_1 y_1 +$$
$$6y_m z_m \lambda_1 y_1 - 5x_3 z_2 \lambda_2 y_1 + 8x_m z_2 \lambda_2 y_1 + 7x_3 z_m \lambda_2 y_1 - 12x_m z_m \lambda_2 y_1 +$$
$$3x_3 y_m^2 - 9x_m y_m^2 + 6y_m^2 z_2 \lambda_1 - 9y_m^2 z_m \lambda_1 + 9x_3 y_m z_2 \lambda_2 -$$
$$15x_m y_m z_2 \lambda_2 - 15x_3 y_m z_m \lambda_2 + 27x_m y_m z_m \lambda_2)/$$
$$(3(3x_3 y_1 z_2 - 5x_m y_1 z_2 - 5x_3 y_m z_2 + 9x_m y_m z_2 - 5x_3 y_1 z_m +$$
$$9x_m y_1 z_m + 9x_3 y_m z_m - 18x_m y_m z_m)),$$

$$(-2y_1 \lambda_1 z_2^2 + 4y_m \lambda_1 z_2^2 + x_3 \lambda_2 z_2^2 - x_m \lambda_2 z_2^2 + 5x_3 y_1 z_2 - 7x_m y_1 z_2 -$$
$$8x_3 y_m z_2 + 12x_m y_m z_2 + 5y_1 z_m \lambda_1 z_2 - 12y_m z_m \lambda_1 z_2 - 5x_3 z_m \lambda_2 z_2 +$$
$$6x_m z_m \lambda_2 z_2 - 9x_3 y_1 z_m + 15x_m y_1 z_m + 15x_3 y_m z_m - 27x_m y_m z_m - 3y_1 z_m^2 \lambda_1 +$$
$$9y_m z_m^2 \lambda_1 + 6x_3 z_m^2 \lambda_2 - 9x_m z_m^2 \lambda_2)/$$
$$(3(3x_3 y_1 z_2 - 5x_m y_1 z_2 - 5x_3 y_m z_2 + 9x_m y_m z_2 - 5x_3 y_1 z_m +$$
$$9x_m y_1 z_m + 9x_3 y_m z_m - 18x_m y_m z_m))$$

We now find the implicit surface algebraically, by eliminating from our system the multipliers and the vertex coordinates. We will use a Gröbner basis computation for this task (see appendix for code).

$$16 - 672x_m^2 + 9432x_m^4 - 37800x_m^6 - 155439x_m^8 + 680400x_m^{10} + 3055968x_m^{12} +$$
$$3919104x_m^{14} + 1679616x_m^{16} - 672y_m^2 + 19728x_m^2 y_m^2 - 245592x_m^4 y_m^2 +$$

$$\vdots$$

$$16796160x_m^4 z_m^{12} - 15396480y_m^2 z_m^{12} - 62145792x_m^2 y_m^2 z_m^{12} + 16796160y_m^4 z_m^{12} +$$
$$3919104z_m^{14} + 8398080x_m^2 z_m^{14} + 8398080y_m^2 z_m^{14} + 1679616z_m^{16}$$

We omitted 40 lines for brevity. The full polynomial occupies an entire page. It has total degree 16 and 165 terms. The largest coefficient in magnitude has ten digits. Inspection reveals all exponents are of even degree. If we consider the set of all such polynomials, we realize this one contains every possible monomial. Hence it is dense when restricted to this class.

The first thing one should ask is whether it factors. A simple computation (see the code appendix) will show it does not factor over the rationals. Indeed, it is also absolutely irreducible, that is, irreducible over the complexes (or, equivalently for a polynomial with rational coefficients, irreducible over the algebraic closure of the rationals). This can be shown using, for example, a method described in [7].

As an aside we mention that one might be interested in the singular set, that is, where the gradient vanishes, as this contains the self intersections. A computation of the Gröbner basis takes about four minutes in *Mathematica* to produce a projection curve onto a coordinate plane (that is, we eliminate one variable). It has degree 53 in each of the remaining variables, and total degree 58. The largest coefficient is around 30 digits.

5   Visualizing the Algebraic Surface

We have an irreducible algebraic surface. Parts of it in the positive octant of real space will bound our triangle midpoint locus. We now plot this surface. The broad part in the view on the right is oriented away from the origin.

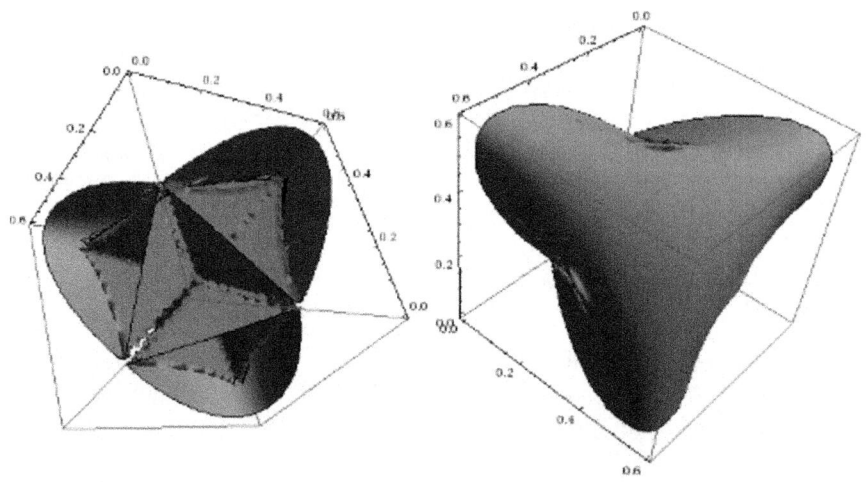

Fig. 3. Two views of one boundary surface

This is a complicated surface; I refer to it as the "Wetzel pretzel". By rotating the algebraic surface, and/or showing it superposed on the region plot, it is not difficult to see how they approximately coincide.

It is clear from geometric considerations that certain parts of the algebraic surface should be correct even for the inequality-constrained problem. That is, the complex algebraic geometry setting should suffice to give the real algebraic geometry bounding surface. Specifically, the part of the surface outermost in the first octant (that is, where all three coordinates are relatively large), is unaffected by the inequalities, because it arises from vertex configurations wherein all three vertices are in open sets in their respective coordinate planes of the first octant.

We remark that as our algebraic approach does not take into account inequality constraints, it does not fully reflect the real algebraic geometry of the problem at hand. This was necessitated by computational considerations. We could, as mentioned earlier, use squares to represent our (non-negative) triangle vertices. This gives rise to a variable elimination problem we were not able to handle. Cylindrical decomposition methods likewise foundered. This is not surprising, as the resulting polynomial would of necessity contain every component of the boundary as factors. We will see below that there are six more, and when multiplied by the one already seen above it becomes considerably larger.

6   The Rest of the Boundary

We used Lagrange multipliers to extremize a coordinate of our midpoints, where midpoint coordinates are all defined implicitly as algebraic functions of some vertex parameters. We can regard those latter as living in a "parameter space". In enforcing inequality constraints–specifically, that vertices lie in the positive octant–we have created a boundary for our parameter space. So we need to also allow for the possibility that some of the extreme coordinates come from the boundary parts of the parameter space. We now handle this case.

This part of the boundary arises exactly when a second vertex coordinate is zero, that is, when it lies on a positive coordinate axis rather than the interior of a positive coordinate plane. For example, the set of polynomials we would use to find the part of the boundary where x_2 vanishes can be obtained from the initial polynomial system simply by setting that variable to zero. Note that such sets themselves have a lower dimensional boundary, where the remaining coordinate goes from positive to negative at the origin. It is easy to see that these boundary edges are arcs traced by the midpoint when one vertex is pinned to the origin.

We remark that algebraically we are simply restricting to that part of the parameter space where weak inequalities become equalities. This could also be viewed in the context of a discriminant variety [4].

We eliminate the parameters (that is, the non-midpoint coordinates) to get our polynomial defining a new piece of the boundary. Straightforward code for this is in the appendix. This part of the boundary has six equivalent components. We show the equation defining one of them. The others are obtained by permuting the variables.

$$20736x_m^8 + 41472y_m^2x_m^6 + 10368z_m^2x_m^6 - 6912x_m^6 + 20736y_m^4x_m^4 +$$
$$1296z_m^4x_m^4 - 10368y_m^2x_m^4 + 10368y_m^2z_m^2x_m^4 - 5184z_m^2x_m^4 +$$
$$864x_m^4 - 3456y_m^4x_m^2 - 864z_m^4x_m^2 + 864y_m^2x_m^2 + 864y_m^2z_m^2x_m^2 +$$
$$648z_m^2x_m^2 - 48x_m^2 + 144y_m^4 + 144z_m^4 - 24y_m^2 - 144y_m^2z_m^2 - 24z_m^2 + 1$$

As noted above, this is still not faithful to the inequality constraints of having vertices in the first octant. Indeed, it gives something that has symmetry across all octants. Nonetheless, we know that parts of it will correspond to our desired surface, because parts come from an open set in the restricted parameter space wherein one vertex is on a positive coordinate axis, and the other two are in the interior of their respective positive coordinate planes. This is why we did not have to enforce non-negativity. The price we pay is that we have, in our algebraic surface, parts that we do not want. The advantage, however, was that the computations were quite simple to perform.

We now graph this. The indented part points toward the origin. One might see that this surface intersects the coordinate planes in a way that appears similar to that of the boundary surface we found earlier. This is no accident. These intersection curves correspond to midpoint loci wherein two triangle vertices lie on coordinate axes and the third is strictly inside its designated coordinate plane. As such sets form curves, and provide the only means by which the midpoint can lie on a coordinate plane, the implication is that the surfaces from figures 3 and 4 must meet at these curves.

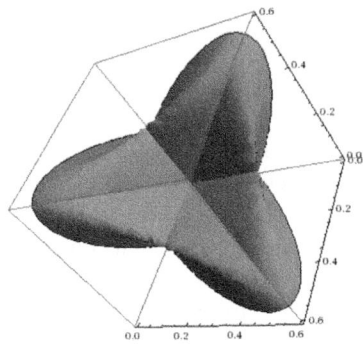

Fig. 4. The rest of the midpoint locus boundary

We now show slices of both the algebraic surfaces and the geometric region. While the latter is approximated, this will still give a good idea of how the two fit together. The slices are planar cuts with x_m taking constant values from 0.015 to about 0.6 in increments of 0.07. The top slices show the region nearest the $y - z$ coordinate plane, with increasing values of x_m as we proceed left-to-right and downward. The dashed blue curves demarcate the extremal surface we computed, and the solid red curves the six boundary components found by setting vertex coordinates individually to zero. That is to say, these curve sets respectively correspond to the first and second boundary surfaces we found and graphed earlier.

The slice views of figure 5 indicate two things of importance. One is that the approximation function we derived from the point cloud gives a reasonably faithful rendering of the region. This we deduce from seeing the green regions

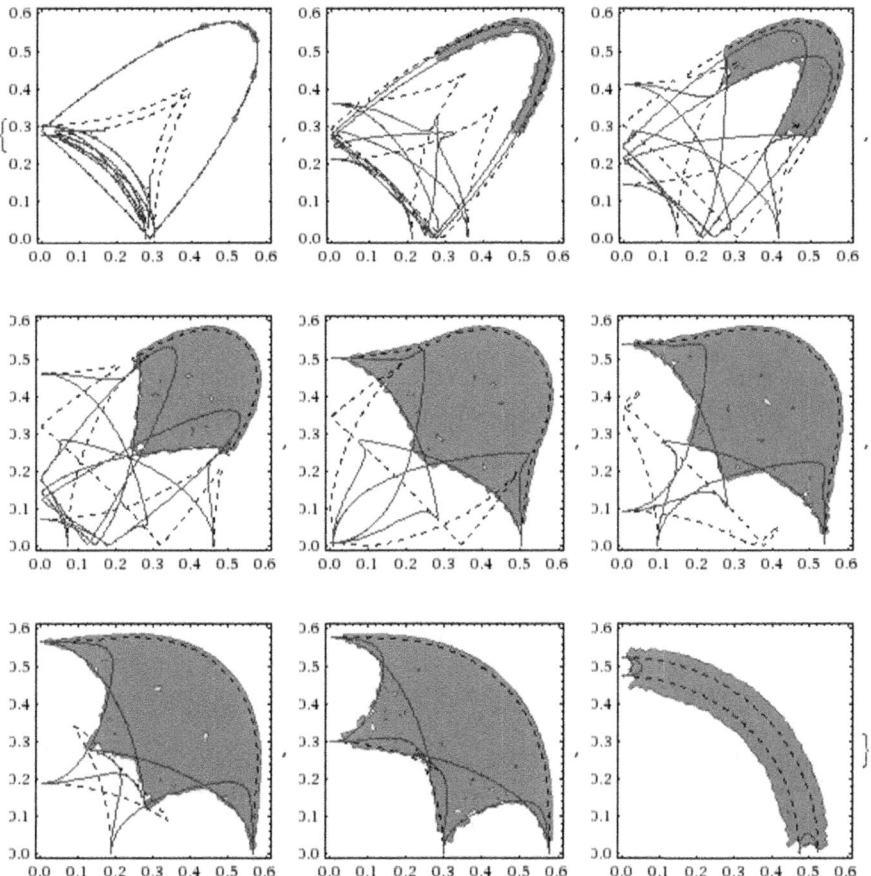

Fig. 5. Several midpoint locus slices along the x axis

mostly fill in, and only barely overflow, the boundary curves. The other important item is that the way in which the boundary surfaces fit together to define
the region is quite complicated. In the bottom middle slice, for example, we see
green beneath the blue boundary at the top (expected), and bounded by red
on each side (unsurprising). But then it dips under red curves and hits the blue
again along the bottom left. This means the extremal surface (the one found via
Lagrange multipliers) contains both maximal and minimal values. It also means
the region is above some, but not everywhere all, of the six surface pieces that
came from setting vertex coordinates separately to zero (that is, forcing a vertex
onto an edge). An implication is that it will be difficult to determine an exact
region from inequality predicates based on the polynomials defining the surface
boundaries.

Finally we show the two boundaries together, using opacity to allow us to see
through one to the other. This is at best an imperfect rendition, but gives an
idea of how they fit together to enclose the midpoint locus.

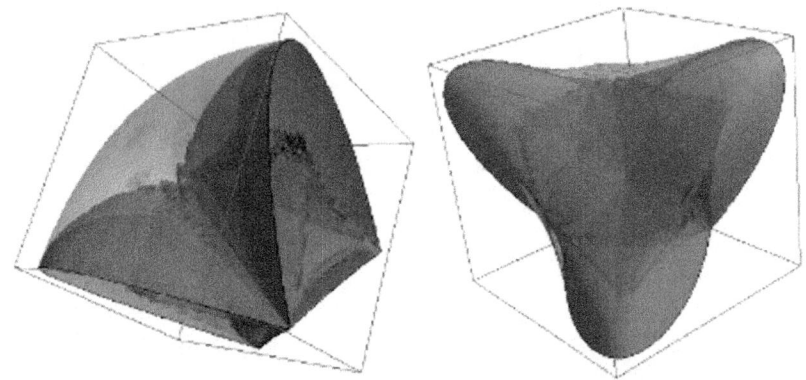

Fig. 6. Two views of the outer and inner boundary surfaces superimposed

7   Inverse Problems

One inverse problem relevant to this study is to understand the vertex locus that gives rise to the midpoint locus. It suffices for this purpose to find those vertex values that give the extreme midpoint values, that is, the boundary set we have found above. This comprises an important part of the discriminant variety. It might be of interest in, say, the setting of motion planning where we might need to be concerned about such parameter values.

It turns out that this can be done using the same methods as for finding the midpoint extremal values. Recall that to find the lower boundary of the midpoints, we simply fixed in turn each of the (generally positive) vertex parameters to zero. We then eliminated the remaining parameters to obtain a polynomial in the midpoint variables alone. To find corresponding "surfaces" in the space of vertex parameters, one would instead eliminate the three midpoint variables from the same systems of equations, retaining now some vertex parameters.

A small subtlety is that we still eliminate those vertex coordinate parameters we had effectively removed in our preliminary computations (when we solved for three in terms of the other three). This has the beneficial effect of keeping our parameter region to three dimensions, making it thus amenable to graphical analysis. It is also useful because this reduced space aptly captures the fact that, due to algebraic constraints, there are only three degrees of freedom. When all is done we have six surfaces. Three of them are trivial, occurring when each in turn of the retained vertex variables is zero. The other three surfaces are modest sized polynomials in the retained vertex variables, and correspond to setting each in turn of the eliminated vertex variables to zero.

One can similarly obtain a description of the vertex parameter space that gives midpoint values on that part of the boundary surface found via Lagrange multipliers. Again, one simply changes roles of vertex parameters and midpoint variables, eliminating the latter instead of the former (of course we still eliminate the Lagrange multipliers). This computation is about as strenuous as its midpoint surface counterpart. It gives a result comparable in size, and with an eerily similar surface plot.

A second inverse problem of interest is to compute vertex coordinates when given a midpoint location. A more general, and more important, form of this problem is to compute a one dimensional locus of vertex coordinates that takes a midpoint from one given position to another. This is the sort of computation that arises in collision avoidance and path planning. It is referred to in the literature as the "reachability problem".

We start with a simple example of finding vertices that give a particular midpoint location. We solve the initial polynomial system by plugging in specific values for the midpoint and solving for the vertex coordinates. If we started with a midpoint in the locus then at least one solution must be real-valued. We remark that generically there are eight solutions in $\mathbb{C}^3$ (since there are 8 for random choices of midpoint), so in fact there must be at least two that are real-valued (though one such need not have all vertex coordinates non-negative). It seems moreover that for all midpoint choices there are exactly two such real solutions to the vertices, one of which might have negative values. It would be interesting to have a geometric understanding of this phenomenon.

For the midpoint $(0.3, 0.4, 0.5)$ the vertex coordinates $(x_2, x_3, y_1, y_3, z_1, z_2)$ (in this order) can occupy either of two sets of values. They are, respectively (to three digits), $(0.324, 0.576, 0.886, 0.314, 0.584, 0.916)$ and $(0.716, 0.184, 0.536, 0.665, 0.974, 0.526)$. The first thing to realize is that both are valid vertex coordinate positions, so there are indeed two distinct ways to place the triangle and have its midpoint at $(0.3, 0.4, 0.5)$. The midpoint location $(0.2, 0.3, 0.4)$, in contrast, gives rise only to the positive vertex $(0.872, 0.328, 0.063, 0.837, 0.542, 0.058)$. The second real-valued solution has one coordinate slightly negative, hence that vertex lies in the wrong octant.

Next we describe a simple way to find a path from one midpoint to the other. We use the two midpoints above to illustrate. We simply subdivide the segment between the two into several parts. We solve the inverse problem to find vertex values, then connect them to form a path between vertices. We remark, first, that this will not work at all if the segment connecting the midpoints does not lie in the valid locus (recall that said locus is not convex). Assuming the segment is in the range of validity, there remains a more subtle issue. A vertex set might not connect smoothly from source to target. We show this below. Using one set of vertex values for midpoint $(0.3,0.4,0.5)$ we get a smooth path taking the triangle to the unique one with midpoint $(0.4,0.3,0.2)$ (figure 7). The other triangle with midpoint $(0.3,0.4,0.5)$ does not connect smoothly in this manner: one sees jumps in the vertex paths on the right side.

The code used to compute and graph these paths follows the ideas described above. It is omitted from the code appendix for brevity.

A natural question is how to find paths in vertex space that connect an arbitrary pair of triangles. Such a computation would arise, for example, in motion planning when the target is not necessarily near to the starting point. We

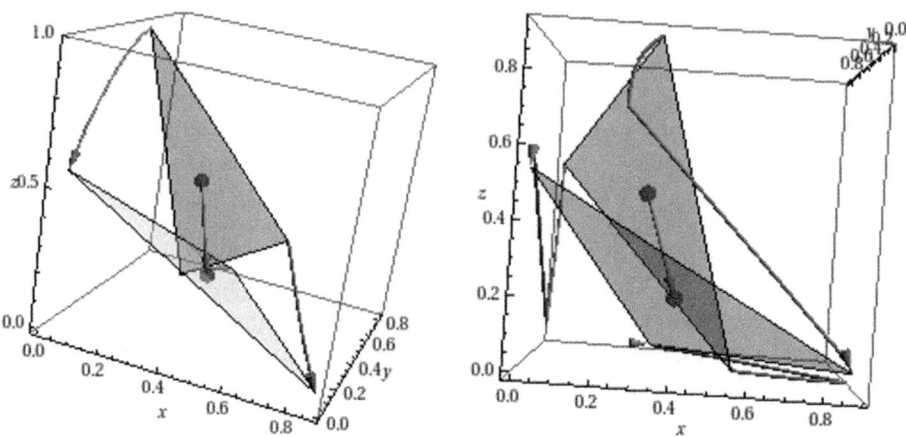

Fig. 7. Continuous and discontinuous paths from one midpoint to another

outline a method that uses differential equations. See [1] for a similar approach to motion planning.

We are given the triangle initial and destination vertices. We will move the former to the latter while enforcing that the vertices stay in the non-negative parts of their respective coordinate planes, and that the triangle remains equilateral at all times. One aspect of keeping the vertices in legal ranges is automatic: by our formulation, a component of each will remain at zero. The non-negativity will be influenced by a repulsion term that grows as a vertex approaches a coordinate axis, but also shrinks as it approaches its destination point. This is important since that destination might actually be on or near an axis, hence it must be possible for the vertex to arrive there.

We will enforce the length constraints by insisting that, for any pair of vertices, the difference in their velocities along the direction of the connecting edge must vanish. This gives three differential conditions. As we have six generically non-zero vertex parameters we must provide exactly three more equations.

We will again work with first derivatives (velocities, in physical terms) as this makes the task easier for the numerical ODE solver than, say, working with second derivatives (physically manifesting as forces). Since we cannot specify the velocities of all six parameters we will only use, for each vertex, its velocity in the direction between that vertex (at a given time) and its destination point. We provide velocity components between vertices and their destinations proportional to the distance between them. These serve as "attractors". In addition we consider velocities that in effect repel vertices from the coordinate axes so they cannot become negative. We take the components of these in the direction we use (between vertex and target position) and add that to the attractive velocity already described. This gives our other three differential conditions. As the system uses six first derivatives, we only require six initial conditions and they of course are provided by the initial position of the six vertex parameters.

We illustrate this method by computing a smooth transition between the second pair of configurations used above. We will show the path a bit differently this time, using a succession of points to give the path. This is useful because it is not the individual paths that matter so much as how they are synchronized. That is to say, if we are to plan a constrained motion then we not only must know the path on which each vertex will travel but also need to keep track of where it is at given times.

We show the starting midpoint as a red point and the termination midpoint as green. One sees that the paths are by no means straight, and vertex components may at times move in the "wrong" direction. It is readily checked that the algebraic constraints (that the triangle keep its shape at all times) are met to within the tolerance of machine precision ODE solver (around 10^{-8} in this case).

We remark that variations on this could be considered. One might, for example, work in a differential algebraic system setting, enforcing the equilateral triangle constraints via algebraic equations rather than vanishing velocity components. Also one might in some problems wish to work with forces rather than velocities, and this could entail either finding consistent initial velocities, or working with a boundary value problem since the target values are known in advance. Finally we observe that the repelling terms might, in extreme cases, not suffice to enforce the con-negativity constraints. A robust solver would thus need to use "event detection" to catch vertices going into a disallowed region. It might

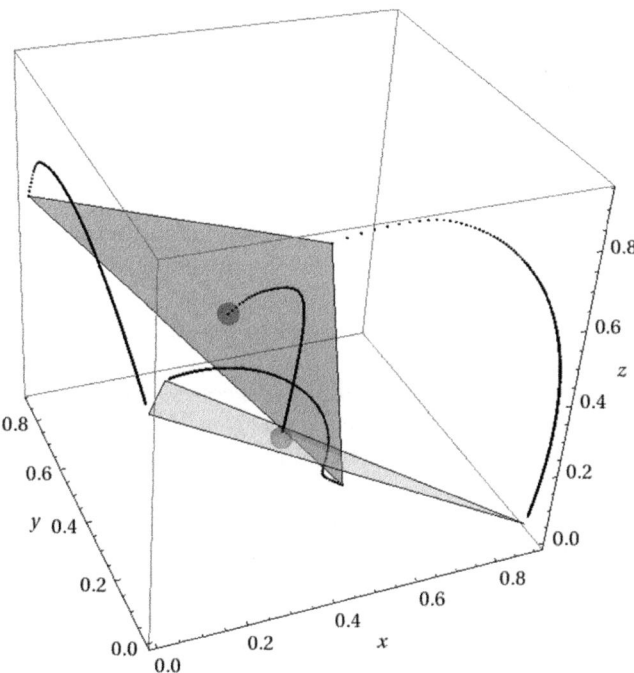

Fig. 8. Continuous and discontinuous paths from one midpoint to another

then either restart with stronger repulsion terms, or continue with a constraint that the appropriate vertex coordinate remain at zero until the corresponding velocity component becomes positive (which would also require event detection).

8   Summary and Open Questions

We presented several methods of visualizing the locus of points of an equilateral triangle with vertices constrained to lie in positive octant, and one on each of the three coordinate planes. Using symbolic methods we derived algebraic expressions corresponding to the boundaries of that region. Taking a different approach, with hybrid (symbolic-numeric) methods for equation solving, we found numerous solutions, retaining the ones that satisfied the appropriate constraints. We then used computational geometric and numeric methods to construct an approximation of the geometric region from this large point cloud of midpoint positions. With all this we were able to show how the algebraic surfaces matched the approximated region. We also addressed some interesting inverse problems related to this study.

To keep this to reasonable size, we did not show some of the other possible ways to visualize the algebraic surfaces together with the approximated geometric region. For example, one can get pictures of the "fit" by cutting away surface parts, and showing the result together with the point cloud or the approximated region derived therefrom. Another direction might be to punch holes in the algebraic surfacein order to see through to regions of interest.

We mention that one might wish to apply more powerful tools for handling such problems. For example, computational geometry methods could perhaps to "take apart" the algebraic surface, removing pieces at self intersections that do not correspond to the desired region boundary. Another nice capability would be to form a bounding surface directly from the point cloud. This might provide a better approximation than we were able to obtain using a proximity approximation. Such graphical methods utilize both symbolic and numeric processing and therefore present interesting opportunities for future development in the area of geometric computation.

We now pose several questions that lead to related areas of inquiry.

How might one show algorithmically that the region we found is topologically equivalent to the three dimensional ball? It is of course obvious from the graphics, but it does not seem to follow immediately from the way the geometry is set up. It would be nice to have a proof based either on computational or geometric methods.

Is the region star-shaped? (If so, this would suffice to show the topology is trivial.) We conjecture that the region is star-shaped with respect to the point obtained when the three vertices are all placed symmetrically in their respective octant planes.

Is there a simple way to obtain a good outer approximation to the convex hull? This might be of use e.g. in motion planning problems where one wishes to avoid the region.

Is there a convenient way, using algebraic inequalities, to show where the various components of the boundary surface end? This would involve computing the intersection curves. One would then need to compute points on those curves which delimit where they are "active" as surface boundaries. Finding these algebraically, and putting together all the information to see only the surface parts that are themselves "active" as midpoint locus boundaries, might provide an interesting symbolic computation counterpart to the graphical or computational geometry possibilities alluded to above. We remark that the task of finding the surface intersection curves is computationally feasible for this particular problem. But it is an open question as to how useful they will be, in implicit form, for the task of better delimiting the boundary surface components. As mentioned in the code appendix, at least one such curve is quite unwieldy.

9   Appendix: *Mathematica* Code

We now provide explicit *Mathematica* code used to set up the computations, create the graphics, etc.

These first lines set up the polynomial system.

```
midpt = {x_m, y_m, z_m};
coords = {x, y, z};
ptcoords = Map[Table[Subscript[#, j], {j, 3}]&, coords];
pts = Transpose[ptcoords];
vals = MapIndexed[(Subscript[#1, #2[[1]]] = 0)&, coords];
polys = Numerator[Together[Flatten[{Mean[pts] - midpt, Map[#.# - 1&,
   Flatten[Table[pts[[j]] - pts[[k]], {j, Length[pts] - 1},
   {k, j + 1, Length[pts]}], 1]]]]]]; }
```

We compute the elimination basis as below.

```
elims = {z_1, x_2, y_3};
params = Complement[DeleteCases[Flatten[pts], 0], elims];
gbrat = GroebnerBasis[polys, midpt, elims,
  CoefficientDomain → RationalFunctions,
  MonomialOrder → EliminationOrder]
```

We find random points in the region using the code below. The idea is to select random values for the parameters, and then solve for the triangle midpoint values. As discussed earlier, we use the original set of polynomials so we can readily check vertex values from eliminated parameters to make sure the result comes from a positive octant configuration.

```
vertices = RandomReal[{0, 1}, {25000, 3}];
substs = Map[Thread[params → #]&, vertices];
vars = Complement[Variables[polys], params];
```

First we find all possible solutions. This takes several minutes to solve 10,000 systems of nonlinear polynomial equations, using version 8 of *Mathematica*.

Timing[solns = Flatten[Map[NSolve[polys/.#, vars]&, substs], 1];]
    {785.884, Null }

Next we select only the ones that satisfy our configuration constraints.

mdptsols = Pick[midpt/.solns, Apply[And,
  Map[(Head[#]===Real&&# ≥ 0)&, vars/.solns, {2}], {1}]];

We symmetrize to expand our valid points by a factor of six, then plot the resulting point cloud.

symmdptsols = Flatten[Map[Permutations, mdptsols], 1];
ptreg = ListPointPlot3D[symmdptsols, BoxRatios → 1]

Here we form our region function, for use in plotting a reasonable approximation to the boundary. For computational speed we use an octree structure This is provided in *Mathematica* by the function Nearest. It will allow a plotting function to determine quickly whether a given point is to be regarded as inside the region under consideration.

nf = Nearest[symmdptsols];
inRegion[pt : {_Real, _Real, _Real}, eps_Real]:=
  TrueQ[Norm[nf[pt, 1][[1]]] − pt] < eps]
reg = RegionPlot3D[inRegion[{x, y, z}, .015], {x, 0, .6}, {y, 0, .6}, {z, 0, .6},
  Mesh → False, PlotPoints → 30]

We now set up the Lagrange multiplier system of polynomials.

varsub = Map[# → #[Sequence@@params]&, midpt];
fullrats = gbrat/.varsub;
reversesub = Map[Reverse, varsub];
ratderivs = Flatten[Map[D[fullrats, #]&, params]];
derivvars = Cases[Variables[ratderivs], Derivative[_][_][_]];
derivs = First[Solve[(ratderivs/.reversesub) == 0, derivvars]];

We form our gradients and use them in the Lagrange multipliers.

grad[v_]:=Map[D[v/.varsub, #]&, params]/.derivs
lambdas = {λ₁, λ₂};
auxpolys = Flatten [Together [grad [z_m] − lambdas. {grad [x_m] , grad [y_m]}]] ;
fullpolys = Join[gbrat, auxpolys];

We find the implicit surface by an elimination of variables, using GroebnerBasis. The settings are based on a method described in [9,6].

Timing[implicit = First[
  GroebnerBasis[fullpolys, midpt, Join[lambdas, params], Sort → True,
    Method → { "GroebnerWalk", "EarlyEliminate" → True}]];]
      {3.14852, Null }

This is quite a large polynomial. The first thing one should ask is whether it factors. We check below that it is irreducible over the rationals. In *Mathematica* one can readily use the function FactorList for this purpose. It gives a result as a list of elements of the form {poly, expon}, where the first factor is explicitly numeric. Thus if the length is two, and the second exponent is one, then the polynomial did not factor. We check this explicitly below.

Length[fax = FactorList[implicit]]
{fax[[1]], fax[[2, 2]]}
  2
  $\{\{1, 1\}, 1\}$

We can graph the algebraic surface as below. As it is complicated, we use some non-default options to speed the process.

g = Compile [{{x_m, _Real}, {y_m, _Real}}, {z_m, _Real}}, Evaluate[implicit]] ;
g1[xm_Real, ym_Real, zm_Real]:=g[xm, ym, zm];
cp1 = ContourPlot3D[
  g1 [x_m, y_m, z_m] == 0, {x_m, 0, .6}, {y_m, 0, .6}, {z_m, 0, .6},
  MaxRecursion → 0, PlotPoints → 40, Mesh → False,
  ContourStyle → RGBColor[0.137255, 0.913725, 1]];

We now produce the polynomials that define the other parts of the boundary. These arise, as explained above, from forcing one vertex coordinate to lie on one or the other boundary axis of its octant face.

allparams = Complement[Variables[polys], midpt];
boundary = Table[First[GroebnerBasis[
  Join[polys, {allparams[[j]]}], midpt, allparams,
  MonomialOrder → EliminationOrder]],
  {j, Length[allparams]}];

We can plot this as below.

cp2 = ContourPlot3D[
  Evaluate[boundary == 0], {x_m, 0, .6}, {y_m, 0, .6}, {z_m, 0, .6},
  MaxRecursion → 0, PlotPoints → 30, Mesh → None,
  ContourStyle → Lighter[Red], Lighting → "Neutral"];

We remark that one can compute projections of the curves where the bottom surfaces intersect the top one. For example:

Timing[intersect1 = GroebnerBasis[
  {boundary[[1]], implicit}, {x_m, y_m}, {z_m},
  MonomialOrder → EliminationOrder];]
  {5.73113, Null}

The result is quite large, having total degree 62 and coefficients as large as 30 digits or so.

  Here we prepare two dimensional slices

bottomslices = Table[boundary, {x_m, .015, .6, .07}];
bottomplots = Map[

```
ContourPlot[# == 0, {y_m, 0, .6}, {z_m, 0, .6},
  MaxRecursion → 1, PlotPoints → 25,
  ContourStyle → {Thickness[.004]}
  ColorFunction → Function[{x, y, f}, Red]]&,
 bottomslices];
```

```
topslices = Table[implicit, {x_m, .015, .6, .07}];
topplots = Map[
 ContourPlot[# == 0, {y_m, 0, .6}, {z_m, 0, .6},
  MaxRecursion → 1, PlotPoints → 25,
  ColorFunction → Function[{x, y, f}, Blue],
  ContourStyle → {Thickness[.004], Dashed}]&,
 topslices];
```

```
midslices = Table[x, {x, .015, .6, .07}];
midplots = Map[
 RegionPlot[inRegion[{#, y, z}, .015], {y, 0, .6}, {z, 0, .6},
  ColorFunction → Function[{x, y, z}, Green],
  MaxRecursion → 0, PlotPoints → 50]&,
 midslices];
```

This next line generates the actual slices we showed earlier.

```
Map[Show, Transpose[{midplots, topplots, bottomplots}]]
```
This last graphic puts the surfaces together, using opacity to allow one to better see how they bound the region of interest.

```
Graphics3D[{{Opacity[0.35], First[cp1]}, {Opacity[0.55], First[cp2]}}]
```

Here we solve the inverse problem of locating vertex values that correspond to a given midpoint location.

```
vsols = NSolve[polys/.Thread[midpt → {.3, .4, .5}]];
vertexsols = Pick[allparams/.vsols,
 Apply[And, Map[(Head[#]===Real&&# ≥ 0)&, allparams]
```

References

1. Chibisov, D., Mayr, E.W., Pankratov, S.: Spatial Planning and Geometric Optimization: Combining Configuration Space and Energy Methods. In: Hong, H., Wang, D. (eds.) ADG 2004. LNCS (LNAI), vol. 3763, pp. 156–168. Springer, Heidelberg (2006)
2. Gleason, A.M., Greenwood, R.E., Kelly, L.M.: The William Lowell Putnam Mathematical Competition Problems and Solutions: 1938-1964. Mathematical Association of America (1980)
3. Jerrard, R.P., Wetzel, J.E.: Tile in a corner. Mathematics Magazine 82, 300–309 (2009)
4. Lazard, D., Rouillier, F.: Solving parametric polynomial systems. Journal of Symbolic Computation 42, 636–667 (2007)

5. Lichtblau, D.: Computing Curves Bounding Trigonometric Planar Maps: Symbolic and Hybrid Methods. In: Hong, H., Wang, D. (eds.) ADG 2004. LNCS (LNAI), vol. 3763, pp. 70–91. Springer, Heidelberg (2006),
   http://library.wolfram.com/infocenter/Conferences/7516/
6. Lichtblau, D.: Implicitization via the Gröbner walk (2007), slides:
   http://library.wolfram.com/infocenter/Conferences/7512/
7. Lichtblau, D.: Polynomial GCD and factorization via approximate Gröbner bases. In: SYNASC 2010: 12th International Symposium on Symbolic and Numeric Algorithms for Scientific Computing, pp. 29–36. IEEE Press (2010)
8. Ruiz, O.E., Ferreira, P.M.: Algebraic geometry and group theory in geometric constraint satisfaction. In: Proceedings of the International Symposium on Symbolic and Algebraic Computation, ISSAC 1994, pp. 224–233. ACM, New York (1994)
9. Tran, Q.-N.: Efficient Groebner walk conversion for implicitization of geometric objects. Computer Aided Geometric Design 21, 837–857 (2004)
10. Wolfram Research. Mathematica 7 (2008)

Some Lemmas to Hopefully Enable Search Methods to Find Short and Human Readable Proofs for Incidence Theorems of Projective Geometry

Dominique Michelucci

LE2I, UMR CNRS 5158, 9 av Alain Savary, BP 47870, 21078 Dijon cedex, France
Dominique.Michelucci@u-bourgogne.fr

Abstract. Search methods provide short and human readable proofs, *i.e.* with few algebra, of most of the theorems of the Euclidean plane. They are less succeful and convincing for incidence theorems of projective geometry, which has received less attention up to now. This is due to the fact that basic notions, like angles and distances, which are relevant for Euclidean geometry, are no more relevant for projective geometry. This article suggests that search methods can also provide short and human readable proofs of incidence theorems of projective geometry with well chosen notions, rules or lemmas. This article proposes such lemmas, and show that they indeed permit to find by hand short proofs of some theorems of projective geometry.

1 Introduction

What is a proof? In a first acceptation, a proof is a guarantee, a certificate that some theorem holds; these proofs are very detailed and rigorous, *e.g.* they account for degenerate cases. These proofs are not intended to be read or understood by a human: they can be tedious computations, resorting to some algorithms in Computer Algebra (Wu-Ritt method, Gröbner bases); the guarantee is due to the correctness of the Computer Algebra program. Such proofs provide certitudes, but do not always bring understanding or enlightenment. In a second acceptation, a proof brings us (*i.e.* human beings) explanation, knowledge, understanding, and even enlightenment: *e.g.* these proofs may suggest generalizations, or apply to more general theorems. They must be easy to read and understand. The shorter the proof, the better. Visual proofs are extreme examples of such proofs. Details such as degeneracies must not occlude the main arguments of these proofs.

This paper considers the possibility for search methods to produce short and human readable proofs (*i.e.* with as few algebra as possible) for theorems of projective geometry, mainly for the projective plane. The hypothesis and the conclusion of these theorems are point-line or point-conic incidences. An avenue to compute short proofs is to apply powerful lemmas, in contrast to proofs of the first kind which relies on long and tedious algebraic computations.

P. Schreck, J. Narboux, and J. Richter-Gebert (Eds.): ADG 2010, LNAI 6877, pp. 118–131, 2011.

Previous works: search methods [7,8,6,2,3,5,17] give human readable proofs in Euclidean geometry with very few algebra. Many consider a typical figure to prune the search combinatorial space and discard irrelevant degenerate cases in the wake of 1959 Gelertner's pioneering work [7,8,6]. However the ruleset of current search methods is not well suited to prove incidence theorems of projective geometry; for example angles, distances, similitudes, isometries are relevant for Euclidean geometry, but not for projective geometry. Raymond Pouzergues [13] (unfortunately in French only) proves by hand 2 dozens of incidence theorems in the projective plane, relying on a variant of Pascal's Mystical Hexagram theorem as a main lemma, which he calls the hexamys theorem. Michelucci and Schreck [11] automatize the search of hexamys. But they do not rely on a typical figure to prune the combinatorial space and to discard irrelevant degenerate cases, and (with hindsight) their ruleset is not powerful as it could be; for instance they do not use brianchons (hexamys duals, defined below). Richter-Gebert et al proposed combinatorial-algebraic proofs [1,14] called binomial proofs.

This article proposes lemmas and rules which could make search methods able to provide short and human readable proofs of incidence theorems of projective geometry.

Only some of the proofs given below have been found after a computer search (with an ad-hoc program). The proofs given below must be considered as an empirical evidence that the rules or lemmas which are proposed indeed permit to find short and human readable proofs of incidence geometry. This article does not focus on the algorithmic part of combinatorial search methods. I will only mention that a feature of theorems in projective geometry, and an issue (or an opportunity?), is their big number of symmetries.

The plane of this article is as follows: §2 presents the main lemmas usable in short proofs. Then §3 proves Desargue's theorem, §4 proves Desargue's theorem in the Cevian case, §5 proves the 3 chords theorem and a generalization, §6 explicits the dual of this theorem, §7 proves the 3 circles theorem and its generlization, §8 proves the 4 circles theorem and a generalization. §9 gives some algorithmics to compute automatically this kind of proofs. §10 concludes.

2   Chasles, Pascal, Brianchon

Powerful lemmas enable short proofs. A main lemma which seems able to prove a significant number of theorems involves cubic curves; it was first proved by Michel Chasles and later generalized to curves of higher degree by Cayley and Bacharach.

Lemma 1. *In the projective complex plane, all cubic curves C which pass through 8 of the 9 (distinct) intersection points of 2 other cubic curves C_1 and C_2 (without common component) also pass through the 9th point.*

This theorem solves an apparent contradiction. On one hand, a cubic curve is defined by 9 different points, under some genericity conditions (for instance, no 4 of the points lie on a common line, and no 7 of the points lie on a common conic). On the second hand, after Bézout theorem, two cubic curves (without common component) C_1 and C_2 in the complex projective plane intersect in exactly 9 different points (in the generic case); but these 9 intersection points do not define an unique cubic curve, because the two cubic curves C_1 and C_2 (and all linear combinations $C(x, y, h) = tC_1(x, y, h) + (1 - t)C_2(x, y, h)$ where $C_i(x, y, h) = 0$ is the homogeneous equation of the curve C_i) are different and pass through the 9 points. The solution to this apparent dilemma is to realize than the 9 intersection points are not independent. Actually they have rank 8, in a sense precised in the following proof of Chasles' theorem:

Proof. Assume points have homogeneous coordinates (x, y, h) in $\mathbb{C}^3 \setminus (0, 0, 0)$. Define $\phi : \mathbb{C}^3 \to \mathbb{C}^{10}$,

$$\phi(x, y, h) = (x^3, y^3, h^3, x^2y, x^2h, y^2h, xy^2, xh^2, yh^2, xyh)$$

Then every cubic curve has equation $Q \cdot \phi(x, y, h) = 0$, where $\cdot$ denotes the Hermitian scalar product, and Q is a non zero vector in $\mathbb{C}^{10}$. Each cubic curve is represented with an hyperplane in $\mathbb{C}^{10}$. Hyperplanes in $\mathbb{C}^{10}$ have rank 9. The intersection of 2 hyperplanes (representing the intersection of 2 cubic curves) has rank 8. Now, after Bézout' theorem, two cubic curves intersect in 9 different points in generic case. Thus the 9 intersection points $\phi(p_1), \phi(p_2), \dots \phi(p_9)$ between the 2 cubics have rank 8: only 8 of the 9 points are independent, and the 9th lies in the vector space spanned by the 8 others.

Remark 1. Rank 10 matroids capture Chasles' theorem. A method to prove incidence theorems searches the matroids compatible with the hypothesis incidences [11].

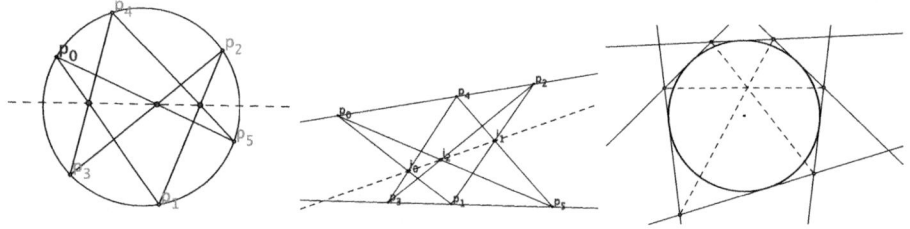

Fig. 1. From left to right: Pascal', Pappus', Brianchon's theorems

Chasles' theorem permit to prove the Pascal mystical hexagram theorem (Fig.1):

Theorem 1 (Pascal's mystical hexagram). *The opposite sides of an hexagon inscribed in a conic curve meet in 3 colinear points.*

Proof (with Chasles theorem). Let $p_0, p_1, \ldots p_5$ the 6 points on a conic. The 3 intersection points of opposite sides are $i_0 = p_0p_1 \cap p_3p_4$, $i_1 = p_1p_2 \cap p_4p_5$ and $i_2 = p_2p_3 \cap p_5p_0$. Call C_1 the cubic curve which is the union of the 3 lines p_0p_1, p_2p_3 and p_4p_5. Call C_2 the cubic curve which is the union of the 3 lines p_1p_2, p_3p_4 and p_5p_0. C_1 and C_2 meet at the 9 intersection points $p_0, \ldots p_5, i_0, i_1, i_2$. The cubic curve C is the union of the conic curve through the p_is and of the line i_0i_1. C passes through 8 of the 9 points (namely the p_is and i_0 and i_1). Thus after Chasles' theorem, C also passes through the 9th point i_2. Admitting i_2 does not lie on the conic (an example, *i.e.* a figure –also called a witness– is a visual proof sufficient and very convenient for a human), i_2 must lie on the line i_0i_1.

Chasles' theorem permits to prove Pappus' theorem (Fig.1):

Theorem 2 (Pappus). *3 points p_0, p_2, p_4 lie on a first line, and 3 points p_1, p_3, p_5 lie on a second line. Then the 3 intersection points $i_0 = p_0p_1 \cap p_3p_4$, $i_1 = p_1p_2 \cap p_4p_5$, $i_2 = p_2p_3 \cap p_5p_0$ are colinear.*

Proof (with Chasles theorem). Define C_1 and C_2 as before: C_1 is the cubic curve which is the union of the 3 lines p_0p_1, p_2p_3 and p_4p_5. C_2 is the cubic curve which is the union of the 3 lines p_1p2, p_3p_4 and p_5p_0. C_1 and C_2 meet at the 9 intersection points $p_0, \ldots p_5, i_0, i_1, i_2$. The cubic curve C is the union of the line $p_0p_2p_4$, the line $p_1p_2p_3$, and the line i_0i_1. C passes through 8 of the 9 points, thus it passes through the 9th point which is i_2. Admitting i_2 does not lie on the lines $p_0p_2p_4$ nor $p_1p_3p_5$ (an example, *i.e.* a figure, is sufficient), i_2 must lie on the line i_0i_1.

Remark 2. This line of thought was introduced by Chasles. It has been somewhat forgotten for the benefit of Bourbaki style. It is today revisited, for instance in Richter-Gebert's book [15].

Pouzergues reformulates Pascal' theorem as follows:

Definition 1. *An hexamys is an hexagon $p_0p_1p_2p_3p_4p_5$ such that opposite sides meet in 3 colinear points (either 3 distinct colinear point, or 2 distinct points) $i_0 = p_0p_1 \cap p_3p_4$, $i_1 = p_1p_2 \cap p_4p_5$ and $i_2 = p_2p_3 \cap p_5p_0$.*

Theorem 3 (Hexamys). *All permutations of an hexamys are hexamys.*

Proof. Trivially, the 6 points of an hexamys lie on a conic, whatever the permutation of the 6 points.

Pouzergues [13], then Michelucci and Schreck [11], use hexamys to prove incidence theorems in the projective plane: a colinearity between 3 points i_0, i_1, i_2 (together with 6 lines: d_0, d'_0 through i_0, d_1, d'_1 through i_1, d_2, d'_2 through i_2) generates an hexamys, every permutation of which imply new colinearities. Hexamys also permit to prove concurrences of 3 lines.

Instead or together with hexamys, it is possible to use Brianchons, from Brianchon's theorem. Brianchons permits to prove concurrence of lines. Brianchon's theorem (Fig.1) states that

Theorem 4 (Brianchon). *If a conic is inscribed in an hexagon with vertices $p_0p_1p_2p_3p_4p_5$ (i.e. the 6 lines p_0p_1, ... p_4p_5, p_5p_0 of the hexagon are tangent to the conic) then the 3 diagonal lines of the hexagon, namely p_0p_3, p_1p_4, p_2p_5, are concurrent.*

Proof. with Chasles. Omitted for conciseness.

It is possible to cancel all references to conics in Brianchon's theorem, as we do for Pascal's.

Definition 2 (brianchon). *A brianchon is an hexagon with lines $d_0d_1d_2d_3d_4d_5$ and vertices $p_i = d_i \cap d_{(i+1)mod\ 6}$ and such that the 3 diagonal lines p_0p_3, p_1p_4, p_2p_5 are concurrent.*

Brianchon's theorem can be restated as:

Theorem 5. *Every permutation of the lines of a brianchon is a brianchon.*

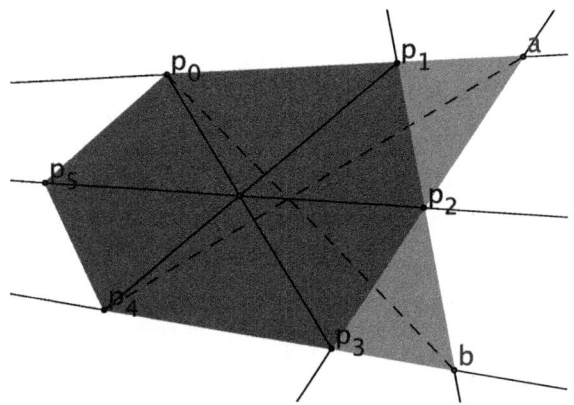

Fig. 2. Brianchon's theorem: if $p_0, p_1, p_2, p_3, p_4, p_5$ is a brianchon, then p_0, $a = p_0p_1 \cap p_2p_3, p_2$, $b = p_1p_2 \cap p_3p_4$, p_4, p_5 is a brianchon as well

It suffices to prove this theorem for a transposition (an exchange), since transpositions generate the group of permutations (Fig.2).

Proof. with Pappus. A brianchon has vertices p_0, p_1, p_2, p_3, p_4, p_5 and lines $d_0 = p_0p_1,\ldots d_5 = p_5p_0$. Let us exchange lines d_1 and d_2, and prove that the hexagon with lines d_0, d_2, d_1, d_3, d_4, d_5, and with vertices p_0, p_1, $a = d_0 \cap d_2 = p_0p_1 \cap p_2p_3$, p_2, $b = d_1 \cap d_3 = p_1p_2 \cap p_3p_4$, p_4, p_5 is a brianchon. So we need to prove that the 3 diagonal lines p_0b, ap_4, p_2p_5 are concurrent. By hypothesis, p_0p_3, p_1p_4, p_2p_5 concur in some point o. Apply Pappus' theorem on the 3 colinear points: p_0, p_1, a and on the 3 colinear points p_4, p_3, b; it implies that the 3 points: $p_0p_3 \cap p_1p_4 = o, p_1b \cap ap_3 = p_2, ap_4 \cap p_0b = x$ are colinear. Thus the point x lies on ap_4, on p_0b and on $op_2 = p_5p_2$. Thus the hexagon with vertices p_0, a, p_2, b, p_4, p_5 is a brianchon.

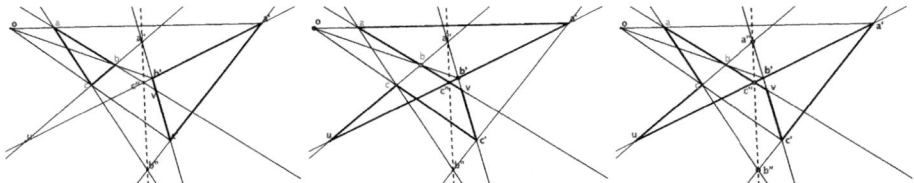

Fig. 3. The 2 triangles in perspective of Desargue's theorem; the first hexamys with points o, b', b colinear by hypothesis; the second hexamys which proves that b'', a'', c'' are colinear

Proof. by hexamys. Omitted for conciseness.

Proof. by Chasles. In the previous proof by Pappus' theorem, replace Pappus' theorem with its proof by Chasles.

Remark 3. A combinatorial search for brianchons (find 3 concurrent lines, and 2 points on each line) in a specified configuration permits to deduce new brianchons, and thus new triples of concurrent lines. It also permits to prove colinearities.

Another short proof of Brianchon's theorem is

Proof. By duality: Brianchon's theorem is the dual of Pascal's theorem.

Indeed duality is another powerful lemma which yields short proofs. Duality exchanges the roles of points and lines, preserving incidences. Gergonne [4] realized first that all the theorems in the projective plane can be dualized. Duality exchanges circles (conics, cubics) with dual circles (conics, cubics). A dual circle (conic, cubic) is a set of lines tangent to a circle (conic, cubic). Duality is used in §6.

3   Desargue's Theorem

Desargue's theorem (Fig. 3) is a combinatorial property of 5 planes in 3D, which still holds after projection on any plane:

Theorem 6 (Desargue theorem). *Let a, b, c and a', b', c' be 2 triangles in perspective, i.e. the 3 lines aa', bb', cc' concur in a point o. Then the 3 intersection points between homologous sides: $c'' = ab \cap a'b'$, $a'' = bc \cap b'c'$, and $b'' = ca \cap c'a'$ are colinear.*

Proof (with dimension lifting). Assume the triangles abc and $a'b'c'$ lie in 2 distinct planes, in 3D, and are still in perspective when viewed from point o. Points a, b, a', b' are coplanar (since lines aa' and bb' cross at o). Thus lines ab and $a'b'$ are coplanar and intersect at some point c'' (possibly at infinity). Now, line ab lies on plane abc, line $a'b'$ lies on plane $a'b'c'$, thus these 2 lines must intersect

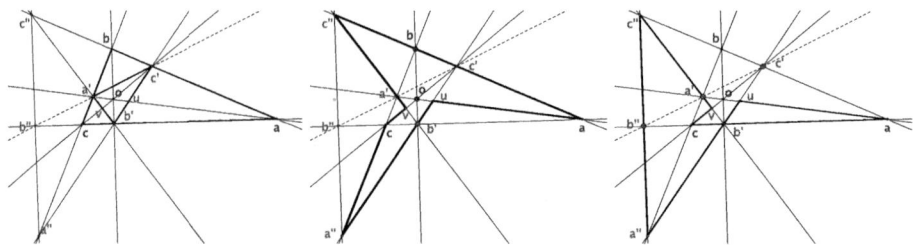

Fig. 4. Left to right: the 2 triangles in perspective in the Cevian case (a' lies on bc, etc); c, a'', u, a, c'', v is an hexamys because opposite sides cross in 3 points b, b', o colinear by hypothesis; thus c, a, u, a'', c'', v is also an hexamys; thus b'', a', c' are colinear

somewhere along the intersection line l of planes abc and $a'b'c'$. The same holds for a'' and b'': they lie on l (here we use symmetry to factorize and shorten the proof). Thus a'', b'', c'' are colinear.

Remark 4. This proof is captured by rank 4 matroids. This kind of proof is used in [9] (in [16]).

Proof (By hexamys (thus by Chasles)). Define $u = a'b' \cap bc$ and $v = ab \cap b'c'$. (a, a', u, c, c', v) is an hexamys because its opposite sides meet in points o, b', b, aligned by hypothesis. Thus (a, c, u, a', c', v) is another hexamys, the opposite sides of which meet in 3 aligned points: b'', a'', c''. $\quad\blacksquare$

4   Desargue's in Cevian Case

In the cevian case of Desargue's theorem (Fig.4), the two triangles are still in perspective, but the vertices of one triangle lie on the edges of the second triangle.

Theorem 7 (Desargue in Cevian case). *Again, 2 triangles abc and $a'b'c'$ are in perspective viewed from point o. Moreover each of the vertices a', b', c' lies on the corresponding side bc, ca, ab. As in the generic case, homologous sides intersect at colinear points $a'' = bc \cap b'c'$, $b'' = ca \cap c'a'$, $c'' = ab \cap a'b'$.*

Proof (with hexamys). The following proof (see Fig.4) needs only one hexamys and is much simpler than the proof in [11]. It was found with a computer search. Points a, b, c, o are given. As usual define $a' = oa \cap bc$, $b' = ob \cap ac$, $c' = oc \cap ab$. Then define $a'' = bc \cap b'c'$, $c'' = ab \cap a'b'$, and here comes the unusual thing: $b'' = a''c'' \cap ac$; thus a'', b'', c'' are colinear but we have now to prove that b'' indeed lies on $a'c'$. Pose $u = oa \cap b'c'$, and $v = oc \cap a'b'$. Then c, a'', u, a, c'', v is an hexamys because its opposite sides intersect in b, b', o colinear by hypothesis; thus after permutation, c, a, u, a'', c'', v is also an hexamys; its opposite sides intersect at points b'', a', c', thus b'' indeed lies on line $a'c'$. $\quad\blacksquare$

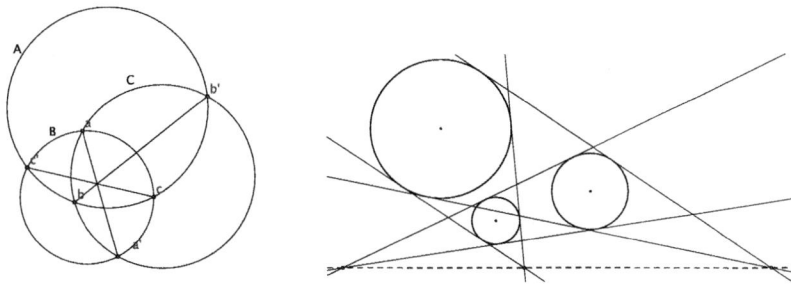

Fig. 5. Left: The 3 pairwise common chords concur. Right: the dual theorem: The 3 homothety centres of pairwise circles are colinear. For readability, only 3 centres and 1 line are displayed. Actually there are 6 centres, forming 4 lines.

5   The 3 Chords Theorem

Theorem 8 (The 3 chords theorem). *Let A, B, C be 3 intersecting circles. Apart cyclic points, A and B meet in points c, c', A and C meet in points b, b', B and C meet in points a, a'. Then the 3 chord lines aa', bb', cc' concur.*

Remark 5. Circles are objects living in the Euclidean plane, not in the projective plane. But we will replace circles by conic in a moment.

Proof. By Chasles. The cubic curve A' is the union of circle A and line aa'. The cubic curve B' is the union of circle B and line bb'. The cubic curve C' is the union of circle C and line cc'. The 2 cubic curves A' and B' intersect in 9 points: $a, b, c, a', b', c', I, J, aa' \cap cc' = o$, where I and J are the two cyclic points (they have homogeneous coordinates $(1, \pm\sqrt{-1}, 0)$ and belong to all circles). The cubic curve C' passes through the first 8 of these points. By Chasles' theorem, $C' = C \cup (cc')$ passes also through the 9th point $aa' \cap cc' = o$. Since o' does not lie on C (a witness, *i.e.* a figure, is a sufficient visual proof for a human), o' lies on line cc'. Thus the 3 chords are concurrent.

The usual proof is as follows: first the power of a point p relatively to a circle C is defined; let l an arbitrary line through p which cuts C in points c and c'. Then the power of p relatively to C is the product $(\bar{c} - \bar{p})(\bar{c'} - \bar{p})$ where $\bar{c}, \bar{c'}, \bar{p}$ are abscissas of points c, c', p along the line. Then it is proved that the power is independent on the line l, and that the line of the common chord of two circles is the locus of points with equal power relatively to the 2 circles. Finally, if o lies on the common chords of circle A and B, and of A and C, then o has equal power relatively to circles A, B and C, thus o lies on the third common chord of B and C. The proof is partly algebraic, but short enough to be human readable. But it is hard to generalize this theorem. The proof by Chasles proves more than this theorem. Actually, the proof by Chasles' theorem proves the more general theorem:

Theorem 9. *Let A, B, C be 3 conics. All 3 conics pass through 2 common distinct points (called I and J in the initial 3 chords theorem). A and B also intersect in c and c', B and C also intersect in a, a', and A and C intersect in b, b'. Then after the previous Chasles' proof, the lines aa', bb' and cc' concur.*

We mention yet another proof of the 3 chords theorem: Chasles' theorem lifts in dimension 10, but dimension 3 is sufficient (and more intuitive):

Proof. Dimension lifting. Lift the Euclidean plane on the parabolic sheet $z = x^2 + y^2$: $L(x,y) = (x, y, z = x^2 + y^2)$. Cocyclic points in the plane become coplanar points after lifting. The common chord of 2 circles A and B is the projection on the plane Oxy of the intersection line between the 2 planes $L(A)$, $L(B)$ of the lifted circles. Now, the 3 planes of $L(A), L(B), L(C)$ in 3D intersect in one common point.

6   The Dual of 3 Chords Theorem

Duality is illustrated with the dual of the 3 chords theorem, in Fig 5. Both theorems and their proofs can be dualized.

Theorem 10 (Dual of the 3 chords theorem). *Let A, B, C be 3 circles. Lines a, a' are common tangents to B and C, Lines b, b' are common tangents to A and C. Lines c, c' are common tangents to A and B. Then the 3 intersection points $a \cap a', b \cap b', c \cap c'$ are colinear.*

Proof (Usual proof). $a \cap a'$, etc is the centre of the scaling (homothety, or homothecy, a non-rotating dilation) which maps circle B to C. This scaling is equal to the composition of the scaling which maps circles B to A (with centre $c \cap c'$), and the scaling which maps circles A to C (with centre $b \cap b'$). These 2 scalings leave globally invariant the line joining their centres $c \cap c'$ and $b \cap b'$. Thus $c \cap c'$ lies on this line, using the lemma: lines globally invariant through a scaling all pass through the centre of the scaling.

For conciseness, the dualization of other theorems (Chasles', theorem 9, etc) and their proofs are left to the reader.

7   The 3 Circles Theorem

Theorem 11. *The 3 circles theorem. Let a, b, c be the 3 vertices of a triangle. Let a' be any point on line bc, let b' be any point on line ac, and c' any point on line ab. Let A be the circle through points a, b', c', let B be the circle through points b, a', c', and C be the circle through points c, a', b'. Then the 3 circles A, B, C have another common point ω.*

Proof. by Chasles. See Fig.6. Let A' be the cubic which is the union of circle A and line $a'bc$, B' the cubic which is the union of circle B and line $ab'c$, and C' the cubic which is the union of circle C and line abc'. The 2 cubic curves A'

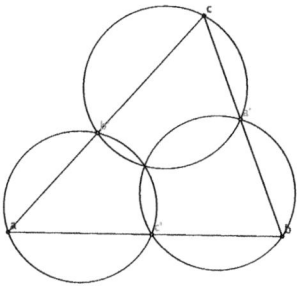 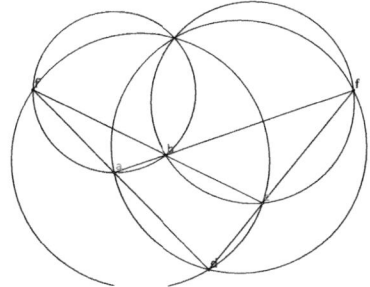

Fig. 6. Left: Three circles theorem: the 3 circles share a common point (other than the 2 cyclic points). Right: the 4 circles theorem, the four circles share a common point.

and B' meet in 9 different points $a, b, c, a', b', c', I, J, \omega$ where I, J are the 2 cyclic points common to all circles, and ω is the intersection point of $A \cap B$ which is not c'. The third cubic C' passes through the first 8 of these 9 points. Thus after Chasles' theorem, it also passes through the 9th point ω. Thus the 3 circles share a common point, ω.

Remark 6. Again, the proof by Chasles' theorem proves more than the 3 circles theorem, because the cyclic points I and J can be replaced with any generic points. It proves the following theorem:

Theorem 12 (A triangle and 3 conics). *Let a, b, c be 3 points, let a', b', c' be 3 points with $a' \in bc, b' \in ac, c' \in ab$. Let I, J be 2 generic distinct points which do not lie on lines ab, ac, bc. Let A be the conic curve through 5 points a, b', c', I, J, let B' be the conic curve through 5 points b, a', c', I, J, let C' be the conic curve through 5 points c, a', b', I, J. Then the 3 conics share another intersection point ω.*

8   The 4 Circles Theorem

Theorem 13 (The 4 circles theorem.). *Let a, b, c, d be 4 points in generic position. Let $f = ab \cap cd$, and $f' = ac \cap bd$. Let C_{ab} be the circle through a, b, f; let C_{cd} be the circle through a, b, f; let C_{bc} be the circle through b, c, f'; let C_{ad} be the circle through a, d, f'. Then the 4 circles $C_{ab}, C_{cd}, C_{bc}, C_{ad}$ share another common point, which is not a cyclic point.*

Proof (by Chasles). See Fig.6. Let C'_{ab} be the cubic curve which is the union of C_{ab} and line cdf'; let C'_{cd} be the cubic curve which is the union of C_{cd} and line abf'; let C'_{bc} be the cubic curve which is the union of C_{bc} and line adf; let C'_{ad} be the cubic curve which is the union of C_{ad} and line bcf; then the 2 cubics C'_{ab} and C'_{cd} intersect in 9 distinct points $a, b, c, d, f, f', I, J, \omega$, where I, J are the two cyclic points common to all circles, and ω is the other intersection point of circles C_{ab} and C_{cd} (the 3 other intersection points are f and I, J). The cubic curve

C'_{bc} passes through the 8 first of these 9 points, so after Chasles' theorem, it also passes through the 9th point, ω. Since ω does not lie on the line adf, component of the cubic curve C'_{bc} (a figure or witness is a sufficient visual proof), it means that ω lies on the other component of C'_{bc}, the circle C_{bc}. Similarly for the cubic C'_{ad}, which is left to the reader (A symmetry argument, in fact a permutation, can also be used).

Remark 7. Again, Chasles' proof proves more: I and J can be generalized to any (generic) points.

Theorem 14. *Let a, b, c, d, i, j be any generic points and $f = ad \cap bc$, $f' = ab \cap cd$. Points i and j generalize previous cyclic points I and J, they are any point (i generic position). The 4 conics $C_{ab}, C_{cd}, C_{bc}, C_{ad}$ share points i and j. Moreover the conic C_{ab} passes through a, b, f, the conic C_{cd} passes through c, d, f, the conic C_{bc} passes through b, c, f', the conic C_{ad} passes through a, d, f'. Then the 4 conics share another common point ω.*

Proof. In the previous proof by Chasles, replace I with i, and replace J with j.

9   Automatization

All previous proofs share the same combinatorial flavor and resort to the same lemmas arguments (Pascal', Chasles', Brianchon's theorems), which suggests that the search of such proofs can be automatized with search methods [11]. It will extend the naive algorithm in [11]: it also considers brianchons, and it relies on a witness, *i.e.* it considers a typical figure to prune the combinatorial search space and discard irrelevant degenerate cases.

In a nutshell, users provide (possibly interactively) the hypothesis and the conclusion of a conjecture. Hypothesis and conclusion involve only incidences. All incidences (point-circle incidences, point-conic incidences, point cubic incidences) are internally reducible to point-line incidences: a circle is just a conic passing through two constant points (the cyclic points), 6 points on the same conic are an hexamys, and a cubic is the union of 3 distinct lines, or of a line and a proper conic.

Users also provide a witness. A witness is a figure, which illustrates the conjecture to be proved, and where vertices (and possibly lines and conics) have numerical coordinates (either rational, floating-point, interval), and names. H. Gelernter is the first to rely on a witness to discover and prove geometric theorems in 1959 [7,8,6,5]. More recently, witnesses are used to detect dependences in systems of geometric constraints, and to decompose and solve systems of geometric constraints [10,12].

The witness first permits to check that the user makes no mistake when specifying the hypothesis and the conclusion: the witness must satisfy the conjecture (otherwise the conjecture has a counterexample, or more likely, the user makes some mistake when specifying the problem). Also, when completing the figure with (typically) intersection points between lines, the witness is used to check that created intersection points are indeed new (different from the vertices) and all distinct. Note that when two intersecting points, or a vertex and an inter-section point are equal in the witness, it provides a conjecture, which the user may try to prove, but not a fact. Conversely, when two intersecting points, or a vertex and an intersection point, are numerically different[1], this is considered as a fact, and the witness is considered as a proof. In passing, we tried to prove non colinearities and non concurrences with logic and some matroid rules, but the computations are slow and the obtained proofs are long, tedious and boring; the visual proof provided by the witness is the best in all aspects.

The proof searcher ("proof assistant" would be confusing) provides several tools. One tool is a combinatorial and straightforward search of hexamys (as in [11]) and brianchons which prove the conjecture. It is also possible to search to apply Pappus' or Desargue's theorems.

When this search fails, users have two non exclusive possibilities: first, they can ask the proof searcher to complete the figure with intersection points between two lines (or conics) of the figure, or with lines joining two vertices; we already under-lined the essential role played by the witness during the completion (the previous method [11] used no witness, which is its main weakness: degeneracies could not be handled). Second they can ask the prover to search for other conjectures, *i.e.* other colinearities of 3 points or concurrences of 3 lines, which are not specified in the hypothesis, but which are (numerically, and approximately) fulfiled in the wit-ness. The proof searcher and users then interactively try to recursively prove these conjectures. Proved conjectures are added to the hypothesis. The proof searcher then checks if these enriched hypothesis contain an hexamys or a brianchon which proves the initial conjecture. Of course, many tactics (backward chaining / for-ward chaining) can be imagined and implemented, in the wake of search methods [5]. Also, several classes of inner representations can be considered; for instance one may imagine to rely on matroids [11,9], or a combination of several matroids (rank 3 for lines, rank 6 for conics, rank 10 for cubics, plus some transition rules). These questions deserve further study.

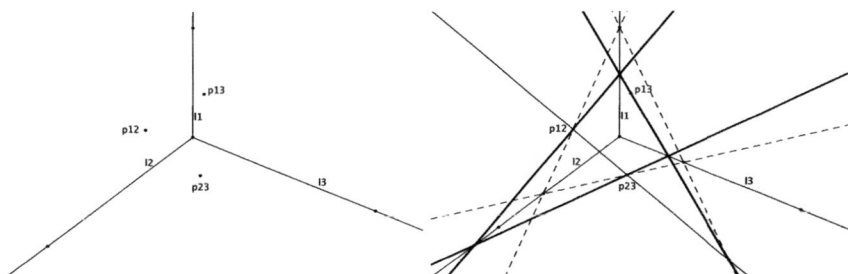

Fig. 7. Left: let l_1, l_2, l_3 be three given concurrent lines; let p_1, p_2, p_3 be 3 given points. Find 3 points $x_1 \in l_1$, $x_2 \in l2$, $x_3 \in l_3$ such that the line x_1x_2 passes through p_{12}, the line x_2x_3 passes through p_{23}, and the line x_1x_3 passes through p_{13}. Right: a construction with ruler only, which relies on Desargue' theorem.

---

[1] Far enough from each other, say one pixel, to account for the numerical inaccuracy; this heuristic is used in Cabri, Cindarella, and other dynamic geometry softwares.

10 Conclusion

In the hope to enable current search methods [5] to find short and human readable proofs of incidence theorems in projective geometry, this article proposes some rules, *i.e.* lemmas: Chasles, Pappus, Pascal and Brianchon's theorems which may be powerful enough. For conciseness, some relevant concepts could not be mentioned: projectivities, perspectivities, colineations, homographies, involutions, cross ratios, etc, though Coxeter [4] relies only on them to prove our lemmas, *i.e.* basic theorems of projective geometry: Pappus', Pascal', Brianchon's theorems, etc. Proofs à la Coxeter should also be considered and computed, and compared with proofs proposed in this article.

Finally, short and human readable proofs should permit to automatically extract, and prove geometric constructions with ruler and compass, or with ruler alone, for incidence problems like: solving the problem in Fig. 7, constructing the intersection points of a given line and a conic given by 5 points, constructing with the ruler only the second intersection point when the first is known, etc (see Cabri web pages for solutions).

Acknowledgements. I thank the anonymous reviewers, for helpful comments and the reference [15].

References

1. Apel, S., Richter-Gebert, J.: Cancellation patterns in automatic geometric theorem proving. Talk at ADG 2010, Munich, Germany (July 2010)
2. Chou, S.C., Gao, X.S., Zhang, J.Z.: Automated generation of readable proofs with geometric invariants, ii. theorem proving with full-angles. J. Automated Reasoning 17, 325–347 (1996)
3. Chou, S.-C., Gao, X.-S., Zhang, J.-Z.: A deductive database approach to automated geometry theorem proving and discovering. J. Autom. Reason. 25, 219–246 (2000)
4. Coxeter, H.: Projective Geometry. Springer, Heidelberg (1987)
5. Gao, X.-S.: Chapter 10: Search methods revisited. In: Gao, X.-S., Wang, D. (eds.) Mathematics Mechanization and Application, pp. 253–272. Academic Press (2000)
6. Gelernter, H.: Realization of a geometry theorem proving machine. In: IFIP Congress, pp. 273–281 (1959)
7. Gelernter, H.: Realization of a geometry-theorem proving machine. In: Siekmann, J., Wrightson, G. (eds.) Automation of Reasoning 1: Classical Papers on Computational Logic 1957-1966, pp. 99–122. Springer, Heidelberg (1983)
8. Gelernter, H., Hansen, J.R., Loveland, D.W.: Empirical explorations of the geometry-theorem proving machine. In: Siekmann, J., Wrightson, G. (eds.) Automation of Reasoning 1: Classical Papers on Computational Logic 1957-1966, pp. 140–150. Springer, Heidelberg (1983)
9. Magaud, N., Narboux, J., Schreck, P.: Formalizing Desargues' theorem in Coq using ranks. In: Shin and Ossowski [16], pp. 1110–1115
10. Michelucci, D., Foufou, S.: Interrogating witnesses for geometric constraints solving. In: ACM Conf. Solid and Physical Modelling, San Francisco, pp. 343–348 (2009)

11. Michelucci, D., Schreck, P.: Incidence constraints: a combinatorial approach. Int. J. Comput. Geometry Appl. 16(5-6), 443–460 (2006)
12. Michelucci, D., Schreck, P., Thierry, S.E.B., Fünfzig, C., Génevaux, J.-D.: Using the witness method to detect rigid subsystems of geometric constraints in CAD. In: SPM 2010: Proceedings of the ACM Conference on Solid and Physical Modeling, Haïfa, Israël. ACM (September 2010)
13. Pouzergues, R.: Les hexamys, web document (2002) (in French)
14. Richter-Gebert, J.: Meditations on Ceva's theorem. In: Davis, C., Ellers, E. (eds.) The Coxeter Legacy: Reflections and Projections, pp. 227–254. Fields Institute American Mathematical Society (2006)
15. Richter-Gebert, J.: Perspectives on Projective Geometry: A Guided Tour Through Real and Complex Geometry. Springer, Heidelberg (2011)
16. Shin, S.Y., Ossowski, S. (eds.): Proceedings of the 2009 ACM Symposium on Applied Computing (SAC), Honolulu, Hawaii, USA, March 9-12. ACM (2009)
17. Wilson, S., Fleuriot, J.D.: Combining dynamic geometry, automated geometry theorem proving and diagrammatic proofs. In: Proceedings of the European Joint Conferences on Theory and Practice of Software (ETAPS) Satellite Workshop on User Interfaces for Theorem Provers (UITP), Edinburgh, UK (April 2005)

What Is a Line ?

Dominique Michelucci

Dijon University, LE2I, CNRS 5158, France
Dominique.Michelucci@u-bourgogne.fr

Abstract. The playground is the projective complex plane. The article shows that usual, naive, lines are not all lines. From naive lines (level 0), Pappus geometry creates new geometric objects (circles or conics) which can also be considered as (level 1) lines, in the sense that they fulfil Pappus axioms for lines. But Pappus theory also applies to these new lines. A formalization of Pappus geometry should enable to automatize these generalizations of lines.

1 Introduction: What Is a Line ?

There are several ways to automatize deduction in geometry. The one which is investigated here is to extend the basic objects: *i.e.* lines and points, of some geometric theory. The playground is the complex plane projective geometry [1,6]: only incidence properties are considered, two distinct lines always meet in one point, two distinct proper conics always meet in four points. Since Pappus theorem will be used as the main axiom, let us call it the Pappus geometry.

The main idea is to see the Pappus geometry as a functor:

- its input are two types, point and line, which fulfill axioms A_1, A_2, A_3 (given below) of the Pappus geometry; the most important axiom is Pappus property, A_3; at the first time, points and lines are the basic, naive, ones; they can be seen as symbols. It is well known that, due to the symmetry of axioms involving points and lines, points and lines can be exchanged; it is the principle of duality.

- its output is a *theory*. A theory is a set of lemmas or theorems (Desargue, Pascal, the 3-circle theorem, the 4-circle theorem, etc), their proofs, new objects (like circles and conics), and proved algorithms (drawing the conic defined by five points; computing with the ruler only the second intersection point of a line and a conic, knowing the first intersection point; etc).

It turns out that some of these new objects (*e.g.* pair of inverse points, or conics through three fixed points) generated by the theory can be considered as points and lines, actually *are* points and lines, in the sense that they comply with axioms for points and lines of the Pappus geometry.

Thus the Pappus functor can be applied a second time on these new points and lines, which are no more the naive points and lines. But the previous theory still holds, its proofs and algorithms are still valid: it will generate new theorems (or extend existing ones) and new objects. This time the generated "conics" will be cubics or quartics; in spite of their higher degree, they are still defined with five points.

P. Schreck, J. Narboux, and J. Richter-Gebert (Eds.): ADG 2010, LNAI 6877, pp. 132–151, 2011.
© Springer-Verlag Berlin Heidelberg 2011

Again, some of the new objects can be considered as points and lines, because it is (or it should be, see below) a theorem in the Pappus theory. Thus we can apply the Pappus functor a third time. And so on.

In passing, note the similarity with compilers bootstrapping, *i.e.* compilers able to compile themself. The latter is an evidence of correctness and power of compilers.

If this approach can be formalized, say in Coq[1], it would give a way to automatically generate an infinity of non-trivial theorems. Up to now, Coq only proves already known theorems, it does not produce new ones. Also, if a dynamic geometry program can be automatically extracted from this Coq software, this dynamic geometry program would account for extended points and lines (contrarily to current dynamic geometry softwares).

Howewer, this approach imposes constraints on the Pappus theory: its proofs must rely only on explicit axioms of the theory, and not on implicit axioms like properties of naive points and lines, which should not be shared by non naive points and lines. In principle, axiomatic geometry should satisfy this constraint, by definition of the axiomatic approach... However, some theorems in projective geometry may have no such proof for the moment: it is often easier to find algebraic proofs (with Gröbner bases, Chou's method, etc) and these methods assume properties (*e.g.* that conics are second degree algebraic curves) or coordinates which no more hold for generalized points and lines. Second, Wu remarked in his pioneering book [9] that classical proofs often neglect degeneracies. Also, maybe some axioms are missing in the Pappus theory summarized in §2, but only a formal implantation of Pappus theory, in Coq or another proof assistant, will permit to detect the gaps. To give an idea, a possible missing axiom could be: if a, b, c, d are four distinct points, not three on the same line, then the three intersection points $ab \cap cd, ac \cap bd, ad \cap bc$ are distinct and not on the same line. Or it could be some "trivial" matroid axiom which is missing.

Other predictable difficulties for an implementation in Coq are subtleties or degeneracies which are neglected in this article: it focuses on the big picture.

Plane. §2 summarizes Pappus theory. Pappus theory considers only combinatorial properties, *i.e.* incidence theorems, like Pappus, Desargue, Pascal, etc. §3 defines three times constrained conics (TTCC), and show that they can be considered as lines. However, this proof does not lie in the Pappus theory: it does not rely only on axioms A_1, A_2, A_3 of the Pappus theory. An hexamys proof (see [4] for examples of such proofs) would; but I have no such proof for the moment. §4 give some standard constraints for a conic to be a circle, a parabola, etc. §5 presents several examples of TTCC. §6 illustrates how the Pappus functor may extend theorems on non naive lines or conics. §7 sketches the generalization of points. §8 presents several variants of planes, each of which manages degeneracies (the issues of parallel lines, points at infinity, non intersecting conics, etc) in its own way. §9 concludes.

Some TTCC, and the fact they satisfy Pappus, Pascal, or Desargue' theorems, are illustrated in GeoGebra files available on internet[2].

---

[1] http://coq.inria.fr/

[2] http://math.u-bourgogne.fr/michelucci/OCAML/GEOGEBRA/

2   Pappus Geometry: A Summary

Pappus geometry is seen as a functor which takes two arguments, a type for lines and a type for points. We do not know what are really lines and points, we only know that they fulfil three axioms:

A_1. Two distinct points define one line.

A_2. Two distinct lines meet in one point.

A_3. If three distinct points p_i, $i = 1, 2, 3$, lie on a common line P, and three distinct points q_1, q_2, q_3 lie on a common line Q, with P and Q distinct, then the three intersection points $p_i \cap q_j$, $i \neq j$, lie on a common line.

A_3 could be called Pappus axiom.

Remark about A_2: the complex projective plane is considered; it is the set of 3D complex lines incident to a given point, say the origin: this model does not require points at infinity, so axioms do not have to consider or distinguish them. It is only for the visualization of the (real part of the) projective plane that this set of 3D lines is cut with any (affine) plane not passing through the origin; points at infinity are introduced for the 3D lines which are parallel to the cutting plane.

Pappus theory can now unfold from these three axioms.

Pappus axiom permits first to define projectivities between lines; a projectivity γ from l to l' is defined by three pairs $(p_i \in l, p_i' = \gamma(p_i) \in l')$, where $i \in 1, 2, 3$. The axis of the projectivity γ is the line through the three intersection points $p_i p_j' \cap p_i' p_j$, $i \neq j$, which are aligned after Pappus' theorem. Let x, y be two points on l and $x' = \gamma(x), y' = \gamma(y)$; then $xy' \cap x'y$ lies on the axis of the projectivity. It permits to construct the image by γ of any point x on l, assuming three pairs $(p_i, p_i' = \gamma(p_i))$.

Coxeter's book [1] provides combinatorial proofs of classical projective geometry theorems, which rely only on properties of projectivities. His book also provides algebraic proofs, using computations on cartesian or homogeneous coordinates or cross ratios.

By duality, it is possible to define a projectivity between two bundles L and L' of lines; a bundle of lines is the set of all lines passing through a common point. The projectivity is defined by three pairs of lines $(l_i \in L, l_i' \in L')$. A dual construction permits to draw with the ruler only the image of any line of L.

One of the first theorems involves the harmonic conjugate.

Harmonic Conjugate Theorem. *Let O, A, B be three aligned points. The harmonic conjugate M of O, relatively to A and B, may be constructed in many ways, using an auxilliary point S not on the line OAB, and a second auxilliary point T on SA (T, S, A are distinct). Whatever S and $T \in SA$, M is fixed, and depends only on O, A, B. If O is a point at infinity, M is the middle of A and B. This theorem is illustrated in Fig. 5 and 9.*

Projectivities can be generalized to homographies. An homography is defined by four pairs of non aligned points and their images (p_i, p_i'), with $i = 0, 1, 2, 3$. Homography of a line is a line, and the restriction of the homography to a line and its image is a projectivity. Define $l_{ij} = p_i p_j$, $l'_{ij} = p'_i p'_j$ for $i, j \in 0, 1, 2, 3$. Then the image of l_{ij} is l'_{ij}, the image of $l_{ij} \cap l_{rs}$ is $l'_{ij} \cap l'_{rs}$, etc. It is possible to draw with the ruler only the image of any point of the plane by the homography.

Another result in Pappus theory is due to Hessenberg, who proved that Desargue' theorem is a consequence of A_1, A_2, A_3:

Desargue's Theorem $(o, p_1, p_2, p_3, q_1, q_2, q_3)$. *Three lines l_i, $i = 1, 2, 3$, concur at o, and points o, p_i, q_i lie on l_i. Triangles $p_1 p_2 p_3$ and $q_1 q_2 q_3$ are said to be perspective (viewed from o). Then the three intersection points between homologous sides $p_i p_j \cap q_i q_j$ (with $i \neq j$) lie on a common line.*

Other theorems of Pappus theory involves conics. Of course we have first to define conics. A possible definition uses Pascal's theorem:

Here is a first definition of conic. Let p_0, p_1, p_2, p_3, p_4 be five points, no four on a common line. Then p_0, p_1, p_2, p_3, p_4 define a unique conic, which is the set of points p_5 such that the three points $p_0 p_1 \cap p_3 p_4$, $p_1 p_2 \cap p_4 p_5$, and $p_2 p_3 \cap p_5 p_0$ lie on a common line.

Raymond Pouzergues reformulates Pascal's theorem eliminating any reference to conics. He calls this the hexamys theorem (a shortcut for Pascal's "mystical hexagram").

Hexamys Theorem $(p_0, p_1, p_2, p_3, p_4, p_5)$. *Six points $p_0, p_1, p_2, p_3, p_4, p_5$ (no four colinear) are an hexamys if, by definition, opposite sides cut in three points along a common line. The hexamys theorem states that all permutations of an hexamys are hexamys as well.*

Hexamys theorem can be derived from Pappus [4].

Remark: when points $p_0 p_2 p_4$ lie on a common line, and points $p_1 p_3 p_5$ lie on another common line, then $p_0, p_1, p_2, p_3, p_4, p_5$ is an hexamys: the three intersection points of opposite sides $p_0 p_1 \cap p_3 p_4$, $p_1 p_2 \cap p_4 p_5$, and $p_2 p_3 \cap p_5 p_0$ lie on a common line after Pappus property. Thus pairs of distinct lines are conics.

The hexamys theorem enables Pouzergues to prove a bunch of incidence theorems: from collinearities of a given geometric configuration, the hexamys theorem deduces new collinearities. Proofs are very short [4]. Moreover, these proofs lie in the Pappus theory, *i.e.* they remain valid when naive points and lines are replaced by non naive ones : the hexamys proofs only use Pascal theorem, which is provable with Pappus theorem. For example, hexamys prove Desargue, and the harmonic conjugate theorems.

Pouzergues gives another definition of conics. Define an involution α on a line l: this involution is defined by four colinear points a, a', b, b' on l such that $\alpha(a) = a'$, $\alpha(b) = b'$. Define two distinct points u, v not on l. The set of points p such that $\alpha(up \cap l) = vp \cap l$ is a conic. Intuitively, l can be seen as the vanishing line of the plane (or the line at infinity), thus points on l are directions, and $x' = \alpha(x \in l)$ is a the direction "orthogonal" to x.

A third definition of conics can be useful. If L and L' are two bundles of lines (a bundle of lines is a set of lines all passing through a common point) in homographic bijection β: $\beta(l \in L) = l' \in L'$, then the set of intersection points $l \cap \beta(l)$ is a conic.

Hexamys permit to prove the equivalence of all these definitions of conics.

Some special conics are circles. It turns out that circles are just conics which pass through 2 special points. Classically, these 2 points are called the cyclic points, and they are often represented with homogeneous coordinates (x, y, h) equal to, for instance, $(1, \sqrt{-1}, 0)$ and $(-1, \sqrt{-1}, 0)$. Circles with center $(x_c, y_c, 1)$ and radius r have equations $x^2 + y^2 + 2x_c x h + 2y_c y h + h^2(x_c^2 + y_c^2 - r^2) = 0$, which are satisfied by cyclic points, whatever x_c, y_c, r. However, cyclic points may be replaced by any pair of distinct points, and all combinatorial theorems (which do not mention metric properties, like angles or distances) still hold. For instance this theorem.

Three Circles Theorem. *(Fig. 10, 11, 12). Let a, b, c be three points, not on a common line. a' is a point on line (bc), b' is a point on line (ac), c' is a point on line (ab). Let C_a be the circle circumscribed (CC) to a, b', c', C_b the CC to b, a', c', and C_c the CC to c, a', b'. Then C_a, C_b, C_c have a common point (other than the 2 cyclic points).*

A short proof is given in §6.1, but this proof does not lie in Pappus theory, *i.e.* this proof is not precise enough to guarantee that it follows strictly from the axioms of Pappus theory. A proof inside Pappus theory would apply to generalized lines and circles.

A theory also provides algorithms.

An algorithm to draw a conic point by point relies on Pascal theorem. Let a, b, c, d, e five points defining a conic. Let $k = ab \cap ed$. Let D a line through k. Define $i = bc \cap D$ and $j = cd \cap D$. Then $x = aj \cap ie$ lies on the conic. When D rotates around k, x draws the conic. To prove the correctness of this method, just remark that $abcdex$ is an hexamys.

Pascal's theorem also gives an algorithm to find the second intersection point between a line and a conic, passing through five points a, b, c, d, e. We want the second intersection point between az and the conic. Define $k = ab \cap ed$, $j = az \cap cd$, $D = (jk)$, $i = D \cap bc$. Then the second intersection point is $az \cap ei$.

Pascal's theorem gives an algorithm (not detailed here) to find the fourth intersection point between two conics, when the three others are known. This algorithm is useful for computing the intersection point between two non naive lines, like TTCC.

3   Three Times Constrained Conics

For convenience, 2D points are represented with homogeneous complex coordinates (x, y, h). Define

$$\phi(x, y, h) = (x^2, y^2, h^2, xy, xh, yh)$$

A conic equation is $\phi(x, y, h) \cdot Q = 0$ where Q is a non zero vector in a $\mathbb{C}$ vector space with dimension 6 (the Hermitian scalar product is noted $\cdot$). Each time a conic Q is constrained to pass through a point $p = (x, y, h)$, it imposes a constraint on the vector Q (the same name is used for the conic and its representing vector): $\phi(x, y, h) \cdot Q = 0$, *i.e.* the vector Q must be orthogonal to

$\phi(p)$. Of course, the vector Q is determined, up to its norm, by five independent orthogonality conditions, thus by five points. It is consistent with the fact that conics are determined by five independent points.

But there are other constraints than passing through a specified point, which make sense, and which give the same kind of orthogonality condition on the vector Q representing a conic.

For instance, to specify that the conic Q is a circle, the vector Q must be orthogonal to $C_1 = (1, -1, 0, 0, 0, 0)$ and to the vector $C_2 = (0, 0, 0, 1, 0, 0)$; the orthogonality with C_1 imposes that the coefficients of x^2 and of y^2 in the equation of the conic Q are equal; the orthogonality with C_2 imposes that the coefficient of xy in this equation is 0. It is also possible to specify that the conic is a parabola, or a circle orthogonal to a specified circle, or a circle with its center on a specified line. The corresponding vectors are given below, §4.

Now, let C_1, C_2, C_3 be three independent such constraints. Call a conic constrained with these three constraints a three times constrained conic, a TTCC for short. These TTCC lies in a vector space with rank three : thus TTCC are 2D *lines* (or 2D points with the duality argument). 2D lines fulfil Pappus property, thus TTCC also. QED.

Unfortunately, the previous proof does not lie in the Pappus theory (it does not use only axioms A_1, A_2, A_3, it uses properties of vector spaces). A proof in the Pappus theory (for instance, an hexamys proof) would permit to apply the Pappus functor on TTCC considered as lines.

A last remark. The previous proof suggests that cubic curves constrained with 7 independent constraints, *e.g.* to pass through 7 specified points, could also be considered as lines. Since a non constrained cubic is defined by 9 (independent) points, a constrained cubic will be completely defined by two points, as naive lines; this condition is needed in order for constrained cubics to be considered as generalized lines. However:

- as for the conics, we need a definition of cubics which lie inside the Pappus theory; I think it is possible.

- two cubics must intersect in one point (the 7 constrained point do not count); this last constraint can not be satisfied: non constrained cubics cut in 9 points, after Bézout theorem; subtracting the 7 constraints, constrained cubics cut in 2 points, not 1.

More generally, which degree d algebraic curves can be considered as extended lines ?

The equation vector of an algebraic curve with degree d has $e = (d+1)(d+2)/2$ coordinates; it is a vector in a vectorial space of dimension (and rank) e. It is defined by $e-1$ constraints, *e.g.* $e-1$ points lying on the curve. Assuming the corresponding generalized line exists, it is defined by $e - 3$ fixed points (or other constraints); moreover two generalized lines must cut in just one point, ignoring the $e - 3$ fixed points; it means the two generalized lines meet in total at $e - 2$ points; but, after Bézout theorem, two degree d curves meet in d^2 points. Thus the degree d must fulfil: $d^2 = e - 2 \Leftrightarrow d^2 - 3d + 2 = 0 \Leftrightarrow d = 1$ or $d = 2$. So only algebraic curves with degree one or degre two can be considered as generalized lines.

There is here an apparent paradox, which may confuse the reader. The Pappus functor, when applied to TTCC lines, will generate new "conics", which will be cubics or quartics, and, if constrained three times, these curves can be considered as lines... The solution to this apparent paradox is that lines which feed the Pappus functor: naive lines, then TTCC, etc always lie in a vector space with *rank* three, even when the dimension is greater than three (*e.g.* six for TTCC).

4   Conditions, or Vector-Based Constraints for Conics

For short, vector-based constraints for conics are called conditions.

	x^2	y^2	h^2	xy	xh	yh
C_1	1	-1	0	0	0	0
C_2	0	0	0	1	0	0
$\phi(+1, i, 0)$	1	-1	0	i	0	0
$\phi(-1, i, 0)$	1	-1	0	$-i$	0	0
C_3	0	0	0	0	0	1
C_4	1	0	-1	0	0	1
C_5	0	1	0	0	0	0
C_6	1	0	0	0	0	1
$\phi(p) = C_p$	x_p^2	y_p^2	h_p^2	$x_p y_p$	$x_p h_p$	$y_p h_p$

Fig. 1. Possible constraints on a conic vector Q. i is $\sqrt{-1}$.

Let $Q = (a, b, c, d, e, f)$ be the vector representing a conic. The equation of the conic is $ax^2 + by^2 + ch^2 + dxy + exh + fyh = 0$. This section gives possible constraints on the conic, they are summarized in table 1.

The conic passes through a point $p = (x_p, y_p, h_p)$ if Q is orthogonal to the vector $C_p = \phi(p)$.

The conic is a circle if Q is orthogonal to C_1 and C_2. Orthogonality to C_1 implies that $a = b$, orthogonality to C_2 means coefficient of monomial xy is zero. Equivalent conditions are that Q passes through cyclic points $(\pm 1, i, 0)$ (with $i^2 = -1$), thus Q is orthogonal to both $\phi(\pm 1, i, 0)$.

The circle has its center on the line $y = 0$ if Q is orthogonal to C_3.

The circle is orthogonal to the unit circle with equation $x^2 + y^2 - 1 = 0$ if Q is orthogonal to C_4 (proof: see Fig. 2).

The circle cuts the unit circle (*i.e.* $x^2 + y^2 - 1 = 0$) in two points symmetric relatively to the origin $(0, 0)$ if Q is orthogonal to C_6. These circles have equations $x^2 + y^2 - 2ux - 2vy - 1 = 0$, the center is (u, v) and the radius is R such that $R^2 = 1 + u^2 + v^2$.

The conic is a parabola with axis Oy if Q is orthogonal to C_5 and C_2, *i.e.* the coefficients for y^2 and xy are 0.

Some constraints do not give orthogonality conditions, for instance the tangence of a circle Q to a prescribed line, say $y = 0$.

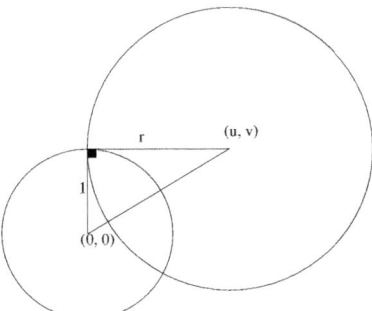

Fig. 2. The circle with center (u, v) and radius r is orthogonal to the unit circle. Thus $u^2 + v^2 = 1 + r^2$, after Pythagora. Its equation (in affine coordinates) is $x^2 + y^2 - 2ux - 2vy + (u^2 + v^2 - r^2) = 0$, *i.e.* $x^2 + y^2 - 2ux - 2vy + 1 = 0$. Thus the coefficient for the constant must equal the coefficient for x^2, and for y^2 in the homogeneous equation.

5   Examples of Non Naive Lines

§5.1 shows that circles through a given fixed point can be considered as lines. §5.2 shows that circles orthogonal to a given fixed circle and passing through a given fixed point can be considered as lines. §5.3 shows that circles (or half circles) with their centers lying on a given fixed line can be considered as lines. §5.4 shows that circles which cut the unit circle in two points symmetric relatively to the origin can be considered as lines. §5.5 shows that parabolas with axis parallel to a given fixed direction and passing through a given fixed point can be considered as lines. §5.6 shows that conics passing through three given fixed points can be considered as lines.

5.1   Circles through One Fixed Point

Let Ω be a fixed, arbitrary, point. Then circles (in the classical sense) through Ω can be considered as lines. For convenience, such circles are called clines in this section. Two distinct clines cut in one point (ignoring Ω and the two cyclic points); it can happen that Ω is a double intersection point; in this case, one may say that the two clines are parallel, and that they meet at a point at infinity, which is Ω. Two distinct points (and distinct of Ω) define an unique cline. Clines satisfy the Pappus property, as illustrated in Fig. 3.

Clines satisfy Pappus property: *i.e.* if p_0, p_1, p_2 lie on a common cline, and q_0, q_1, q_2 lie on another common cline, then the three intersection points r_{ij} between the cline $p_i q_j$ and $p_j q_i$, $i \neq j$, lie on a common cline.

It has already been proved, but this new proof may be instructive. An inversion relatively to any circle (say with radius 1) with center Ω maps points p_i to point p_i', points q_j to points q_j', and points r_{ij} to points r_{ij}', and it maps clines to naive lines not passing through Ω. Thus the points p_i', q_j', and r_{ij}' satisfy the Pappus property, *i.e.* the intersection points r_{ij}' lie on the same line, call it R'. The preimage of R' by the inversion is a cline R; in the peculiar case where R'

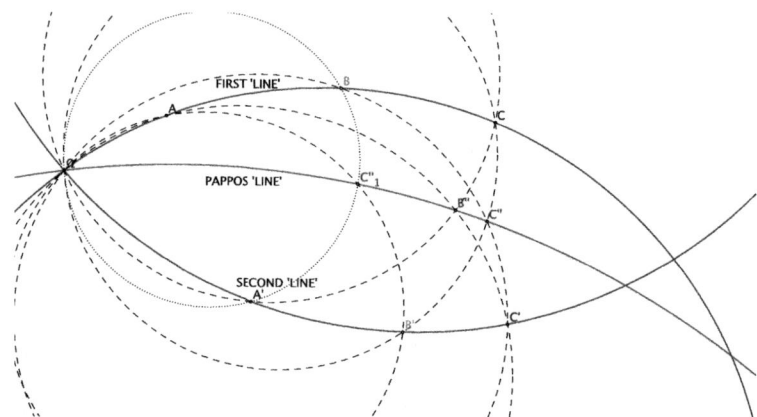

Fig. 3. Clines fulfil Pappus property. They can be considered as lines.

pass through Ω, its preimage is $R = R'$, so it is a (degenerate) cline. In all cases, the preimage R of R' is a cline, thus the r_{ij} lie on a common cline, R. QED.

Thus all theorems of Pappus theory still hold when the word "line" is replaced by the word "cline". For instance the hexamys theorem holds. Define a C-hexamys as a set of six points, no four on the same cline, such that opposite clines meet in three points lying on a common cline. Then any permutation of the six points is also a C-hexamys.

Fig. 4 illustrates Pascal's theorem with clines. For simplicity, the six points lie on a common circle (which does not pass through Ω). The three pairs of opposite clines indeed lie on a common cline, *i.e.* they are cocyclic with Ω.

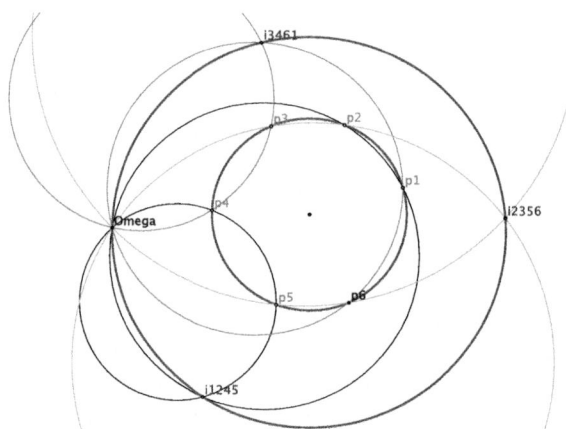

Fig. 4. Pascal theorem. Points p_i lie on the magenta circle. The lines $p_i p_j$ are replaced with clines (circles through Omega). The intersection points lie on a common cline (red circle).

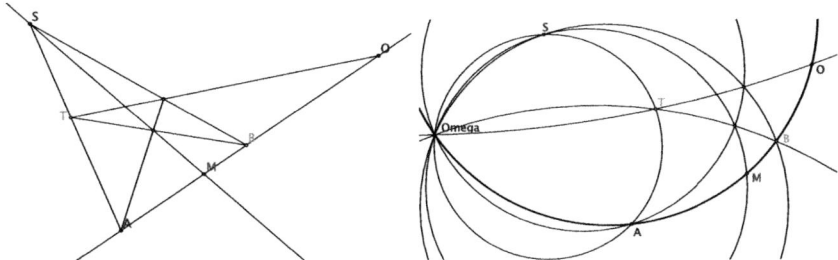

Fig. 5. The harmonic conjugate theorem. Left: for given points O, A, B on a common line, for any point S, for any point T on the line SA, the point M is invariant (hint: M is the harmonic conjugate of O relatively to A, B; if O is a point at infinity, M is the middle of AB. Right: all lines are replaced with clines, M is still invariant.

Fig. 5, Right, illustrates the harmonic conjugate theorem with clines.

What are conics in the Pappus of clines ? They are images of a naive conic by an inversion, thus they are quartic curves, or cubic curves in degenerate cases (the inversion center lies on the conic).

Remark. In the projective complex plane, the inversion is not defined on Ω. It can be defined for other planes (§8). These details are predictable sources of complications for a Coq implementation.

5.2   Orthogonal Circles

Circles orthogonal to a given fixed circle can be considered as lines. A difficulty is due to the fact that such circles cut in two points. These two points are inverse of each other and always come in pairs. Thus it is sufficient to consider these pairs as generalized points. Another solution is to consider only one side (either the inside, or the outside) of the given fixed circle.

5.3   Poincaré Half Circles Are Lines

Circles the centers of which lie on a given line, for example $y = 0$, can be considered as lines. Fig. 6 illustrates the Pappus property for these generalized lines. Points come in pairs, with a symmetry relatively to the line $y = 0$. To define related generalized points, either only points and half circles above the line $y = 0$ are considered, or pairs of symmetric points are considered.

In passing, the Poincaré model for the hyperbolic plane uses these half circles, it is called the Poincaré half plane [8] (curiously, this book does not mention the Pappusian feature of the Poincaré plane).

5.4   Other Circles

Circles which cuts the unit circle (having equation: $x^2 + y^2 - 1 = 0$) in two points symmetric relatively to the origin $(0, 0)$ can also be considered as generalized

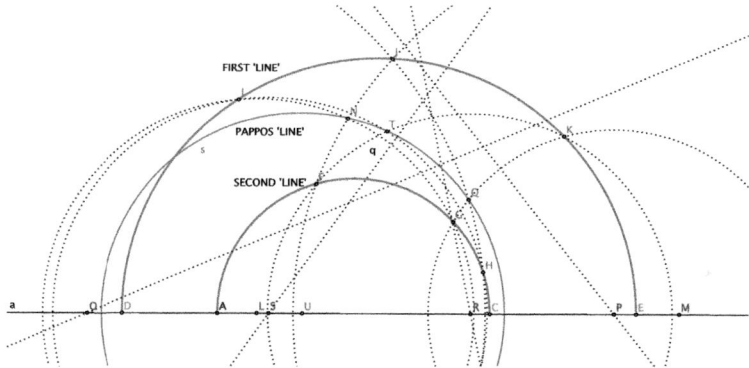

Fig. 6. Circles with centers on a common line (*e.g.* $y = 0$) fulfil Pappus property. Points come in pairs.

lines. Like all circles, their vector Q is orthogonal to C_1 and C_2; moreover their vector Q is orthogonal to C_6. They have equations $x^2+y^2-2ux-2vy-1 = 0$, their center is (u, v), their radius is $\sqrt{1 + u^2 + v^2}$. Two distinct circles in this family always meet in two antipodal points of the unit circle. Fig. 7 shows a bundle of such circles. It illustrates the fact that all these circles belong to a bundle generated by the unit circle $x^2+y^2-1 = 0$ and a line with equation $ux+vy = 0$. Thus all circles of this bundle pass through points $(v/\sqrt{u^2 + v^2}, -u/\sqrt{u^2 + v^2})$ and $(-v/\sqrt{u^2 + v^2}, u/\sqrt{u^2 + v^2})$.

Points for these generalized lines are pair of naive points, which are symmetric w.r.t. the origin.

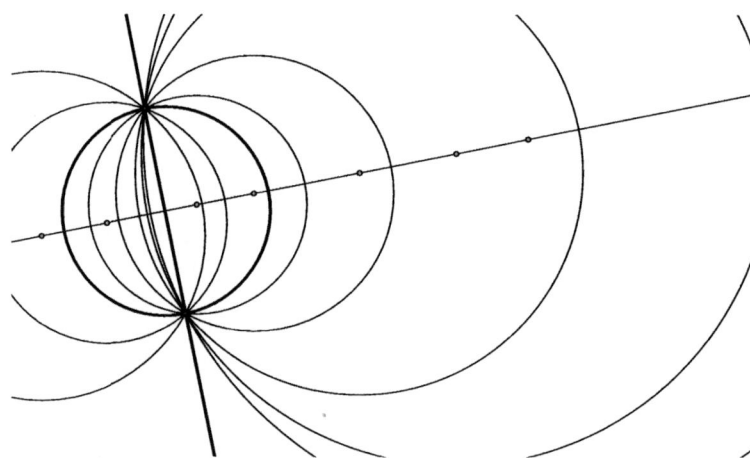

Fig. 7. A bundle of circles. The thick circle and the thick line generate the bundle. The full class of these circles is obtained when rotating the line.

Another way to generate circles in this class is to compose two projections; it also gives another proof of the fact that these circles are generalized lines; first project naive lines in the plane to great circle on a 3D sphere, with the center of the sphere as the center of projection. This projection maps each point of the plane to two antipodal points on the sphere, which are equivalent. Then apply a stereographic projection from the sphere to the (say, equatorial) plane, *i.e.* the center of the projection is a pole of the sphere. The proof relies on easy but tedious computations which are omitted for conciseness. Both projections preserve incidences, thus the Pappus property holds for great circles on the sphere, and for the final circles.

These circles are lines in the Beltrami model of the hyperbolic plane [8,2].

5.5   Some Parabolas Are Lines

Parabolas with a prescribed axis direction (say Oy) and passing through a given fixed point can be considered as lines. They are completely defined with two other points, like naive lines. These parabolas cut in at most one point (ignoring the fixed common point, and the double point at infinity: $(0, 1, 0)$). As usual, two parabolas non intersecting in the affine real plane do meet in the projective complex plane.

5.6   Conics through Three Fixed Points Are Lines

The GeoGebra figure 8 illustrates that conics through three given distinct points (non colinear) can be considered as lines: they fulfil A_3, the Pappus axiom. They also fulfil A_1 and A_2. Fig. 9 illustrates the harmonic conjugate theorem.

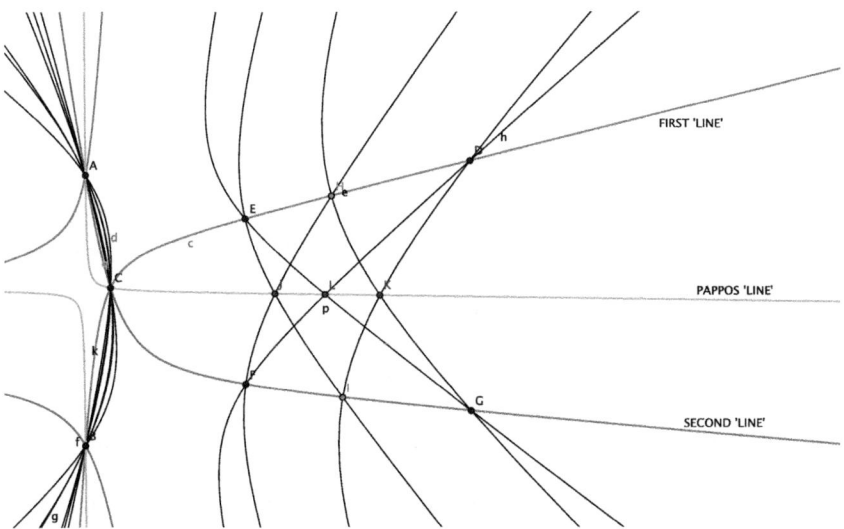

Fig. 8. Conics passing through three given distinct points $(A, , B, C$ on the figure) fulfil Pappus axioms. Thus they can be considered as lines.

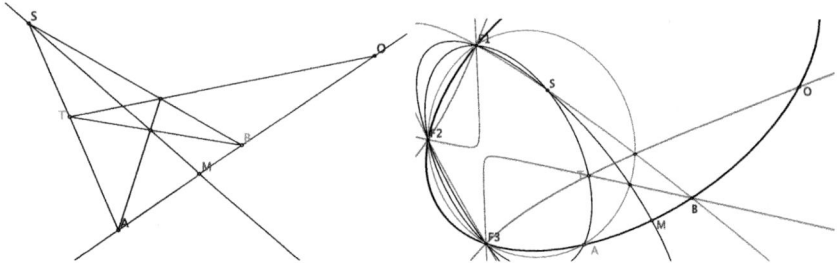

Fig. 9. The harmonic conjugate theorem. Left: the harmonic conjugate theorem for naive lines. Right: the harmonic conjugate theorem for conics passing through three fixed points F_1, F_2, F_3.

6   Playing with Some Theorems

This section illustrates how a Pappus functor may extend theorems, on three examples.

6.1   Proof of the Three Circles Theorem

The three circles theorem is used as an example of a theorem, for which I know no proof lying in the Pappus theory for the moment.

Three Circles Theorem. *Let a, b, c be three points, not on a common line. a' is a point on line (bc), b' is a point on line (ac), c' is a point on line (ab). Let C_a be the circle circumscribed (CC) to a, b', c'. C_b is the CC to b, a', c', and C_c is the CC to c, a', b'. Then C_a, C_b, C_c have a common point (other than the two cyclic points).*

A short proof is given here, but it does not lie inside Pappus theory. A proof inside Pappus theory would permit to extend this theorem to generalized lines.

The proof considers lines. The lines of the triangle are indexed 1, 2, 3, see Fig. 10 for the definition of lines 5,6,7. By hypothesis, the points $1\cap2, 2\cap4, 4\cap5, 5\cap1$ are cocyclic, as well as the points $5\cap6, 6\cap3, 3\cap1, 1\cap5$. We need to prove that the points $2\cap3, 3\cap6, 6\cap4, 4\cap2$ are cocyclic too. Note $1, 2,\ldots 6$ the orthogonal symmetry relatively to line $1, 2, \ldots 6$.

We first need the lemma: the transform 5124 is a translation. I use the convention that in the transform 5124, the symmetry 5 is performed first, but anyway it does not matter: the reader can uses the opposite convention when reading the proof. In the transform $5124 = (51)(24)$, the transforms 51 and 24 are rotations; 51 is a rotation around $5\cap 1$, with angle twice the angle between lines 5 and 1. Similarly, 24 is a rotation around $2\cap 4$, with angle twice the angle between lines 2 and 4. But opposite angles in a cocyclic quadrilateral are either opposite, or their sum equals π. In both cases, the effect of rotations 51 and 24 on vectors annihilate each other, so 5124 is just a translation. QED. The converse also holds.

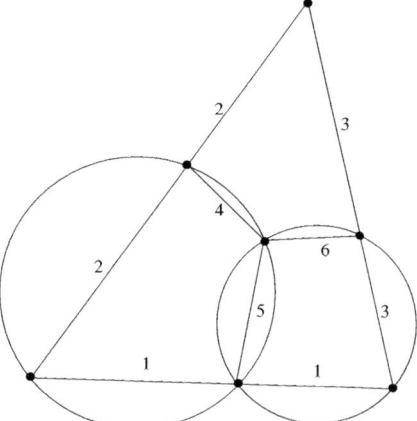

Fig. 10. The three circles theorem. The three circles have a common point.

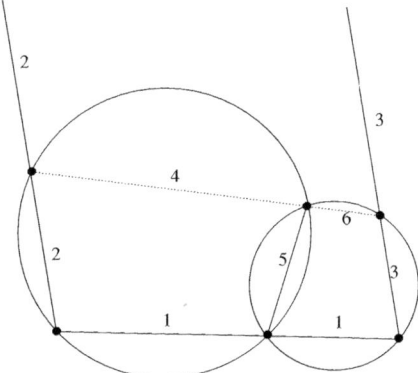

Fig. 11. Here we do not know that lines 4 and 6 are equal, we have to prove it. As in the generic case, 6315 and 5124 are translations. Thus their composition $(6315)(5124) = 6324$ is a translation too. Circular permutations of a translation are translations too [3], thus 3246 is a translation too. Moreover 32 and its inverse 23 are translations because lines 2 and 3 are parallel in this special case. Thus $(23)(3246) = 46$ is a translation. Thus lines 4 and 6 are parallel. But they have a common point ($6 \cap 5$ and $4 \cap 5$), thus they are equal. QED.

Similarly, 6315 is a translation.

Thus the composition $(6315)(5124) = 63(1(55)1)24 = 6324$ is a translation, thus the four points $6 \cap 3, 3 \cap 2, 2 \cap 4, 4 \cap 6$ are cocyclic. It is worth to mention that this proof works also when the triangle 1, 2, 3 is degenerate, *e.g.* when lines 2 and 3 are parallel, as in Fig. 11.

Actually the three circles theorem still holds when circle are replaced with conics passing through two distinct arbitrary points. See Fig. 12.

Another correct generalization of the three circles theorem is illustrated Fig.13. It replaces Euclidean lines with conics passing through three distinct fixed (non

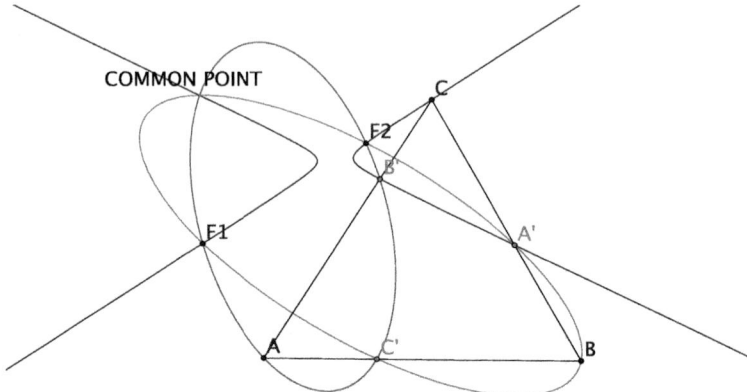

Fig. 12. A generalization of the three circles theorem. Circles are replaced with conics passing through 2 distinct arbitrary points F_1, F_2. These three conics have a common point (other than the two arbitrary points).

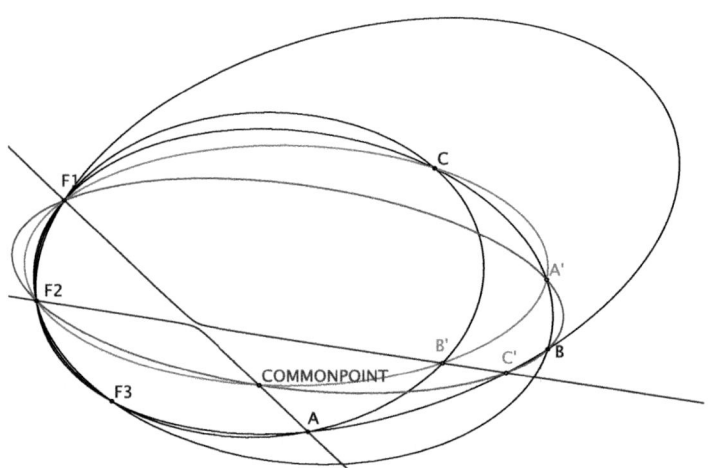

Fig. 13. An extension of the three circles theorem. Lines (AB, AC, BC) are replaced with conics passing through three fixed points F_1, F_2, F_3, and circles are replaced with conics through two of the fixed three points, for instance F_1 and F_2. The three generalized circles have a common point, different of F_1 and F_2.

aligned) points F_1, F_2 and F_3, and replaces circles with conics through F_1 and F_2. Then the three "circles" have a common pont.

A Pappus functor should be able to automatically produce such non trivial generalizations of the three circles theorem and the corresponding proofs.

6.2   The Four Circles Theorem

Four Circles Theorem. *It is also called Miquel's four circles theorem. It states that the four circles circumscribed to three points of a complete quadrilateral have a common point, see Fig.14.*

I know no proof in Pappus theory up to now (a combinatorial search for hexamys by computer should find one). Anyway, the theorem can be proved easily from Chasles theorem: each circle union the "opposite" line defines a cubic curve; the four cubic curves meet in 8 common points: the two cyclic points and the six points of the complete quadrilateral. Thus after Chasles theorem, these cubics meet in another nineth point.

Another short and nice proof relies on orthogonal symmetries relatively to lines of the complete quadrilateral, see Fig.14 for the names of the lines. By hypothesis, $ACUH$ is cocyclic, thus the transform $ACUH$ is a translation. Idem for $HVDA$. Thus the composition $(ACUH)(HVDA) = ACUVDA$ is a translation as well. Thus $A(ACUVDA)A = CUVD$ is a translation too. Thus $CUVD$ is cocyclic. QED. Unfortunately this last proof can not be generalized to generalized lines.

A Pappus functor should be able to produce this non trivial generalization (Fig. 15) of the four circles theorem. Let F_1, F_2, F_3 be three distinct non aligned points. The six points (which were the vertices of the complete quadrilateral in the initial four circles theorem) are called $Q_i, i = 1, \ldots 6$, and there are four conics. The conic K_{134} passes through points $F_1, F_2, F_3, Q_1, Q_3, Q_4$; the conic K_{156} passes through points $F_1, F_2, F_3, Q_1, Q_5, Q_6$; the conic K_{235} passes through points $F_1, F_2, F_3, Q_2, Q_3, Q_5$; the conic K_{246} passes through points $F_1, F_2, F_3, Q_2, Q_4, Q_6$. Replace circles in the initial four circles theorems with conics passing through points F_1 and F_2. Then the four conics: $K_{134}, K_{156}, K_{235}, K_{246}$ all pass through another common point, Z in Fig. 15.

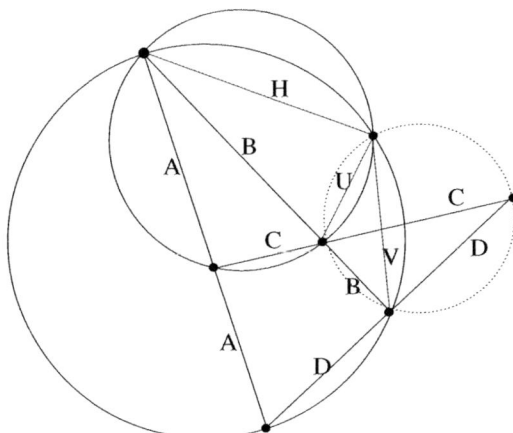

Fig. 14. Miquel's four circles theorem: the four circles have a common point (distinct of the two cyclic points)

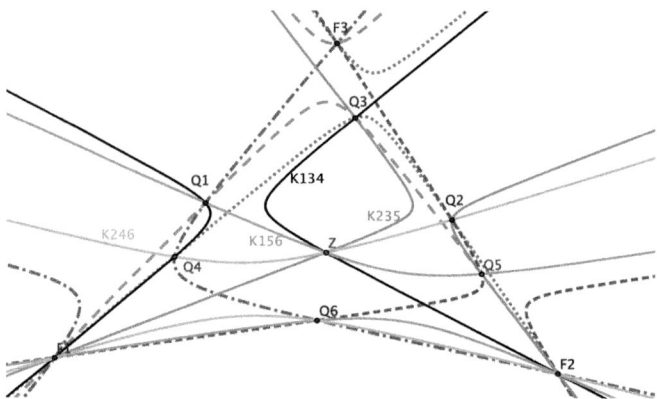

Fig. 15. An extension of Miquel's four circles theorem: lines are replaced with conics through three fixed points F_1, F_2, F_3, circles are replaced with conics through two of the fixed points, namely F_1 and F_2. Then the four generalized circles have a common point, Z in the figure.

6.3 A Butterfly Theorem

We conclude with this last theorem, Fig. 16. Let C be a fixed circle, and E a point not on C. The symmetric to a point $M \in C$ is by definition $M' = (EM) \cap C$. It is clearly an involution. The symmetry is extended to all points in the plane with a Butterfly theorem which states that for all chords (A_1, A_2) through M (where $A_1 \in C, A_2 \in C$), the symmetric chords (A_1', A_2') passes through a common point, which is M'. Any conic can be used in place of the circle C (for instance two lines, which gives a variant of Pappus theorem), and the theorem still holds. For conciseness, no proof is provided. A Pappus functor should be able to generalize (Fig. 16) this theorem and its proof (if it lies in Pappus theory). A first generalization replaces linear chords with clines, *i.e.* circles through a fixed point. Since this generalization reduces to applying some inversion, it may be considered trivial. A second generalization is less obvious; it replaces lines with conics through three fixed points F_1, F_2, F_3, and the circle C is replaced with a circle (or any conic) passing through F_1 and F_2.

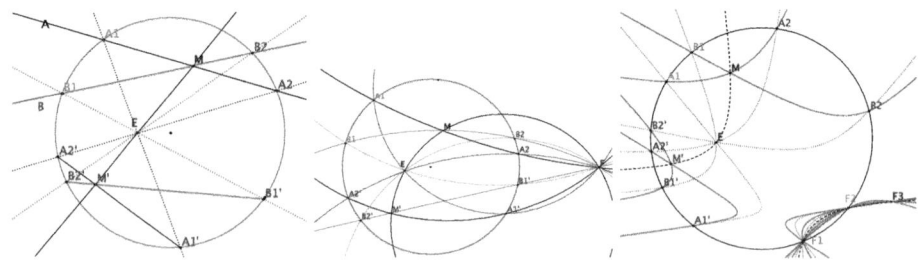

Fig. 16. From left to right: a butterfly theorem, a first generalization, and a second one

7   What Is a Point ?

This article mainly generalized lines. Another way to extend Pappus theory is to generalize points. It is well known that, due to duality, lines and points can be exchanged. It is conics which I will consider as points, in some sense.

A conic is represented with a non zero vector in a six dimensions vector space. So it can be seen as a point in a projective space in five dimensions. We can say that three conics Q_1, Q_2, Q_3 are aligned iff there are non all zeros numbers a_1, a_2, a_3 such that $a_1 Q_1 + a_2 Q_2 + a_3 Q_3 = 0$. Two distinct conics Q_1 and Q_2 generates a "line of conics", *i.e.* the set of conics equal to $a_1 Q_1 + a_2 Q_2$ for some numbers a_1, a_2. To avoid ambiguity, call it a 2-bundle of conics.

Similarly, four conics are coplanar iff there are non all zeros numbers a_1, a_2, a_3, a_4 such that $a_1 Q_1 + a_2 Q_2 + a_3 Q_3 + a_4 Q_4 = 0$. Three non aligned conics generate a plane of conics, called a 3-bundle of conics.

A 3-bundle of conics is a Pappus plane, its points are conics. Thus all theorems of the Pappus plane apply: Pappus, Desargue, Pascal, three-circle theorems, etc. We can apply the Pappus functor.

8   Variants: A Zoo of Planes

For simplicity, we considered only the strongest axioms, so two distinct lines always meet in one point, and two distinct proper conics always meet in four points. It is the complex projective plane.

Weaker axioms, and other planes, are possible. For instance, we can accept that two distinct lines meet in at most one point, and that two proper conics meet in at most four points. The essential constraint is that no configuration contradicts Pappus axiom, which can be rephrased as follows: if three distinct points p_i are aligned, and if three distinct points q_j are aligned on another line, and if the three intersection points $r_{ij} = p_i \cap q_j, i < j$ exist, then they are aligned.

This freedom of choice for axioms is related to the fact that the plane can be coordinalized in several ways [6]. A point can be represented with two real cartesian coordinates (x, y): it is the affine real plane, $\mathbb{R}^2$; it contains parallel lines which do not meet. A point can be represented with homogenous real coordinates $(x, y, h) \in \mathbb{R}^3 \setminus (0, 0, 0)$, two colinear vectors representing the same point; this representation can be made canonic, using only values zero and one for the homogeneous coordinate h; points $(x, y, 0)$ are points at infinity; this is the real projective plane $P^2(\mathbb{R})$; all pair of distinct lines meet in one point; but two distinct proper conics can meet in less than four points because $\mathbb{R}$ is not algebraically closed. Geometrically, $P^2(\mathbb{R})$ is the set of 3D lines incident to a given point, say the origin; to visualize the plane, the set of lines is cut with an arbitrary plane not passing through the origin. To get more regularity, a solution is to use complex coordinates, either cartesian coordinates $(x, y) \in \mathbb{C}^2$, or homogeneous coordinates $(x, y, h) \in \mathbb{C}^3 \setminus (0, 0, 0)$, *i.e.* $P^2(\mathbb{C})$. Another classical representation represents each point $(x \in \mathbb{R}, y \in \mathbb{R})$ of the plane with a complex number $c = x + iy \in \mathbb{C}$: this is the complex line $\mathbb{C}$; if $\mathbb{C}$ is augmented with $1/0$ for convenience, the complex projective line $P^1(\mathbb{C})$ is obtained: this plane has only one point at infinity; through stereographic projection, this plane is mapped to the sphere $\mathcal{S}^2$ ($\mathcal{S}^2 = \{(x, y, z) \in \mathbb{R}^3 \mid x^2 + y^2 + z^2 = 1\}$) and its point at infinity $1/0$ is mapped to the North pole $(0, 0, 1)$ of the sphere. All these planes are locally equivalent for their "visible" part, but they are no more when points at infinity or imaginary points are involved; also they have not the same topology.

Example. Consider the circle with center $(-3, 0)$ and radius 1, and the circle with center $(3, 0)$ and radius 1. In the affine real plane $\mathbb{R}^2$, in the projective real plane $P^2(\mathbb{R})$, and in the complex projective line $P^1(\mathbb{C})$, they do not intersect. In the complex affine plane $\mathbb{C}^2$ (points are represented with $(x \in \mathbb{C}, y \in \mathbb{C})$), they intersect in two points: $(0, \pm i\sqrt{3})$. In the complex projective plane $P^2(\mathbb{C})$, they intersect in four points, the two previous ones, and the cyclic points $(\pm i, 1, 0)$.

Remark. In the complex projective line $P^1(\mathbb{C})$, inversions, for example: $T(z) = 1/\bar{z}$ and $T'(z) = 1/z$, can be extended to their pole 0: the pole and the point at infinity are inverse of each other. In the complex projective plane $P^2(\mathbb{C})$, the inversion: $T(x, y) = (x/(x^2 + y^2), y/(x^2 + y^2))$ (using cartesian coordinates for short) can not be consistently extended to the point $(0, 0)$: it is because $P^2(\mathbb{C})$ has a line at infinity, and not one point at infinity like $P^1(\mathbb{C})$.

Remark. It is convenient to map $P^1(\mathbb{C})$ to the sphere $\{(x, y, z) \in \mathbb{R}^3 \mid x^2 + y^2 + z^2 - 1 = 0\}$ with the stereographic projection s. For convenience, place the plane $P^1(\mathbb{C})$: $c = x + iy$ horizontally at altitude $z = 0$; then the stereographic projection maps $c = (x, y, 0)$ to $s(c) = (2x/(x^2 + y^2 + 1), 2y/(x^2 + y^2 + 1), (x^2 + y^2 - 1)/(x^2 + y^2 + 1))$. $s(c)$ is the intersection point of the sphere and the line (Nc), where $N = (0, 0, 1)$ is the North pole of the sphere. Naive lines in $P^1(\mathbb{C})$ are mapped to circles on the sphere, all passing through N. The point at infinity of $P^1(\mathbb{C})$ is mapped to N. Some properties of $P^1(\mathbb{C})$ are more easily seen on the sphere, e.g. in $P^1(\mathbb{C})$, the point at infinity $1/0$ belongs to all (naive) lines; thus non parallel lines (in the usual, naive sense) in $P^1(\mathbb{C})$ cut in two points.

The "Pappus tower" can likely be built with these planes. However, each of them manages degeneracies (parallel lines, non intersecting conics, points at infinity) in its own way, which may complicate implementations.

9   Conclusion

Two remarks before concluding:

- From another viewpoint, the content of this article is sometimes trivial. We just apply many homographies and inversions to the whole naive Pappus plane, so naive lines and naive conics are mapped to curves with arbitrary high degree. What is essential is that all these transforms (homographies and inversions) preserve incidences. More general non linear diffeomorphisms could be used as well.

- Jürgens Richter-Gebert et al [5] show that tropical lines do not always fulfil Pappus property.

In conclusion, this article considers the Pappus theory as a functor: its inputs are points and lines which must fulfil axioms of Pappus geometry. The output is a set of proved theorems and methods, and new geometric objects, some of which fulfil axioms of Pappus geometry. Theorems are incidence theorems, and have a combinatorial flavor.

For this approach to work in practice, *e.g.* to be programmed in Coq, all proofs must lie inside the Pappus theory, *i.e.* all proofs must use only axioms of the Pappus theory. A computer combinatorial search inspired by the area method or the full-angle method [7], but through the set of Hexamys (or their duals, Brianchons) as in [4], and relying on some numerical example (a figure, or a witness) like the area method to help prune the search space, may help find such proofs in an automatic way.

This article was written with in mind a geometric formalization, *i.e.* theorems and algorithms are proved applying the Pappus axiom, or the hexamys theorems, or relying on properties of projectivities or homographies, like in Coxeter's book [1]. However a more algebraic approach can also be considered; for instance, lines can be seen algebraically as vectors in some rank three vector space.

Acknowledgements. I thank the anonymous referees: their remarks and comments helped me to improve the clarity of this article.

References

1. Coxeter, H.: Projective geometry. Springer, Heidelberg (1987)
2. Henle, M.: Modern Geometries: Non-Euclidean, Projective, and Discrete, 2nd edn. Prentice Hall (2001)
3. Michelucci, D.: Isometry group, words and proofs of geometric theorems. In: SAC 2008: Proceedings of the 2008 ACM Symposium on Applied Computing, pp. 1821–1825. ACM, New York (2008)
4. Michelucci, D., Schreck, P.: Incidence constraints: a combinatorial approach. Int. J. Comput. Geometry Appl. 16(5-6), 443–460 (2006)
5. Richter-Gebert, J., Sturmfels, B., Theobald, T.: First steps in tropical geometry (2003)
6. Richter-Gebert, J.: Perspectives on Projective Geometry: A Guided Tour Through Real and Complex Geometry. Springer, Heidelberg (2011)
7. Gao, X.s.: Search methods revisited. In: Mathematics Mechanization and Application, ch. 10, pp. 253–272. Academic Press (2000)
8. Stahl, S.: The Poincaré Half-Plane. Jones and Bartlett Books in Mathematics (1993)
9. Wen-Tsün, W.: Mechanical Theorem Proving in Geometries - Basic Principles. Texts and monographs in symbolic computation. Springer, Heidelberg (1994)

On One Method of Proving Inequalities in Automated Way

Pavel Pech

Faculty of Education, University of South Bohemia,
Jeronýmova 10, 371 15 České Budějovice, Czech Republic
pech@pf.jcu.cz

Abstract. The paper describes proving geometric inequalities in automated way without cell decomposition. Firstly an overview of known methods of proving inequalities is given including the method which is based on reduction of a conclusion polynomial to the canonical form modulo a hypotheses ideal. Then a parametrization method of proving geometric inequalities is introduced. Further a method of proving geometric inequalities which introduces an auxiliary polynomial is described.

Keywords: geometric inequalities, automated geometry theorem proving.

1 Introduction

Theory of automated theorem proving usually requires to translate a geometric statement into an algebraic form in some coordinate system. Then one of proving methods is applied (Gröbner bases, Wu–Ritt, Collins CAD,...). Besides this, also coordinate-free methods of proving theorems are used. Many papers are dealing with equality type statements, only few articles are devoted to proving inequalities. The reason seems to be clear — proving inequalities is much more intricate.

In this paper we will be concerned with geometric inequalities in polygons, especially in triangles and quadrilaterals. Many such inequalities are described in the well–known book by O. Bottema et al: Geometric inequalities [1] which appeared in 1969 and in the book by Mitrinovič et al: Advances in geometric inequalities [11] which appeared twenty years later. All inequalities in these two books are proved classically. Since then many attempts to prove inequalities in automated way were done. One of the most promising method is that coming from L. Yang [22]. It is based on cell decomposition of a parametric space similarly as Collins CAD method. Many inequalities from [1] were proved by this method. Another two techniques — reduction of a polynomial to canonical form modulo ideal and the Rabinowitsch/Seidenberg device which converts the inequalities to equations by introducing new variables — are given in [4]. See also [21].

There are also other automated methods of proving (algebraic) inequalities as *sos* (sum of squares) method [16], [13], *sds* (successive difference substitution) method [23], [24], [9] etc.

P. Schreck, J. Narboux, and J. Richter-Gebert (Eds.): ADG 2010, LNAI 6877, pp. 152–168, 2011.

To prove geometric inequality automatically we usually need to carry out *two basic steps:*

- To translate a geometric inequality into an algebraic inequality.
- To prove an algebraic inequality.

In the paper we will concentrate on the first step — the translation of a geometric inequality into an algebraic inequality — and problems which may occur during the translation. The second step — the proof of algebraic inequality — is beyond our consideration.

The paper is organized as follows:

First we recall the method which is based on expressing a conclusion polynomial modulo a hypotheses ideal together with a few examples, to realize problems we encounter.

Then a parametrization of n-gons is described. It is shown how a parametrization of the side lengths of an n-gon can help by proving geometric inequalities.

Next a new method of proving geometric inequalities which uses an auxiliary polynomial is introduced.

The methods we describe in this paper do not use cell decomposition of a parametric space.

Throughout the paper we will use Gröbner bases (GB) and Wu–Ritt (WR) approach, see [2], [3], [6], [7], [10], [14], [15], [17], [18], [19], [21].

2   Basic Method

We will deal with automated proving of geometric statements which are of the form $H \Rightarrow c$, where H is a set of hypotheses of equality type and c is a conclusion of inequality type. Let the given inequality be in the form

$$\forall x \ (h_1(x) = 0, h_2(x) = 0, \ldots, h_r(x) = 0) \Rightarrow (N(x) \geq 0), \tag{1}$$

where $x = (x_1, x_2, \ldots, x_n)$, h_i are polynomials with rational coefficients, and x_i are real numbers.

When proving geometric inequalities we can use the method which is based on reduction of a conclusion polynomial to canonical form modulo the hypotheses ideal [4].

In the method we distinguish two steps:

In the first step we carry out reduction of the conclusion polynomial N to canonical form modulo the hypotheses ideal $I = (h_1, h_2, \ldots, h_r)$. To do this we introduce a slack variable t such that

$$N - t = 0. \tag{2}$$

Considering the ideal $J = (h_1, h_2, \ldots, h_r, N - t)$ we usually eliminate dependent variables in the ideal J to express t in terms of independent variables (parameters).

In the second step we try to express t in such a form from which its non-negativity follows. Here we can use e.g. the theory of *sos* or *sds*.

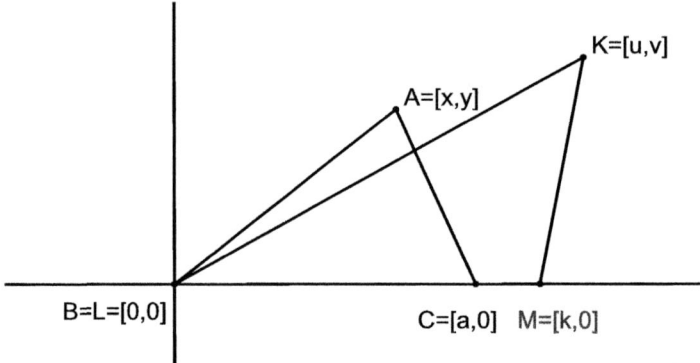

Fig. 1. Proof of the Weitzeböck inequality

To see how this method works we start with the following example.

Example 1. Given a triangle ABC with side lengths a, b, c and the area P and a triangle KLM with side lengths k, l, m and the area Q. Prove that

$$k^2(-a^2 + b^2 + c^2) + l^2(a^2 - b^2 + c^2) + m^2(a^2 + b^2 - c^2) \geq 16\,PQ. \quad (3)$$

When the equality is attained?

The inequality (3) is known as the Neuberg–Pedoe inequality [11].

Proof. Let $A = [x, y]$, $B = [0, 0]$, $C = [a, 0]$, $K = [u, v]$, $L = [0, 0]$, $M = [k, 0]$, Fig. 1. We express the side lengths a, b, c, k, l, m and the areas P, Q in algebraic equations:

$b = |CA| \Rightarrow h_1 := (x - a)^2 + y^2 - b^2 = 0,$

$c = |AB| \Rightarrow h_2 := x^2 + y^2 - c^2 = 0,$

$l = |MK| \Rightarrow h_3 := (u - k)^2 + v^2 - l^2 = 0,$

$m = |KL| \Rightarrow h_4 := u^2 + v^2 - m^2 = 0,$

$P = \text{area } ABC \Rightarrow h_5 := 2P - ay = 0,$

$Q = \text{area } KLM \Rightarrow h_6 := 2Q - kv = 0.$

Introduce a slack variable t such that

$h_7 := k^2(-a^2 + b^2 + c^2) + l^2(a^2 - b^2 + c^2) + m^2(a^2 + b^2 - c^2) - 16PQ - t = 0.$

We will proceed in the following manner:

1. We express the variable t in terms of *independent* variables x, y, a, u, v, k in the ideal $I = (h_1, h_2, \ldots, h_7)$. We also say that we reduce t to canonical form modulo ideal I [4].
2. We write t in such a form from which its non-negativity follows.

To do this we eliminate dependent variables b, c, l, m, p, q in the ideal $I = (h_1, h_2, \ldots, h_7)$. Using GB approach, in the program CoCoA[1] we get

```
Use R::=Q[x,y,a,u,v,k,b,c,l,m,p,q,t];
I:=Ideal((x-a)^2+y^2-b^2,x^2+y^2-c^2,(u-k)^2+v^2-l^2,
u^2+v^2-m^2,2p-ay,2q-kv,
k^2(-a^2+b^2+c^2)+l^2(a^2-b^2+c^2)+m^2(a^2+b^2-c^2)-16pq-t);
Elim(b..q,I);
```

the polynomial which leads to the equation

$$t = 2a^2u^2 + 2a^2v^2 - 4xauk - 4yavk + 2x^2k^2 + 2y^2k^2$$

which is equivalent to

$$t = 2(xk - ua)^2 + 2(yk - va)^2.$$

We expressed t in the form of the sum of squares, hence $t \geq 0$.

The equality in (3) occurs iff $t = 0$ which is equivalent to $xk - ua = 0$ and $yk - va = 0$ which means that triangles ABC and KLM are similar. The inequality (3) is proved.

Remark 1. We get the same result using WR approach with characteristic sets and the Epsilon library[2] [19], [18], using the same variable ordering as above.

Remark 2. If KLM is equilateral then $k^2 = 4Q/\sqrt{3}$ and (3) transforms into the Weitzenböck inequality [20]:

$$a^2 + b^2 + c^2 \geq 4\sqrt{3}P, \tag{4}$$

where equality occurs iff a triangle ABC is equilateral.

We see that (4) is equivalent to

$$\frac{a^2\sqrt{3}}{4} + \frac{b^2\sqrt{3}}{4} + \frac{c^2\sqrt{3}}{4} \geq 3P,$$

where on the left side is the sum of areas of three equilateral triangles with side lengths a, b, c. Then (4) can be demonstrated in the style *proof without words*, see [12], Fig. 2. This proof without words method can be considered as another powerful dynamic geometry tool of proving.

In the given example we could see the strength of the reduction of a conclusion polynomial to the canonical form modulo hypotheses ideal. But be careful, the same method can fail, see the next example which is from [1].

Example 2. Let a, b, c be the side lengths of a triangle. Then

$$2(a + b + c)(a^2 + b^2 + c^2) - 3(a^3 + b^3 + c^3 + 3abc) \geq 0. \tag{5}$$

---

[1] Program CoCoA is freely distributed at `http://cocoa.dima.unige.it`
[2] Program Geother is freely distributed at
   `http://www-calfor.lip6.fr/~wang/epsilon/`

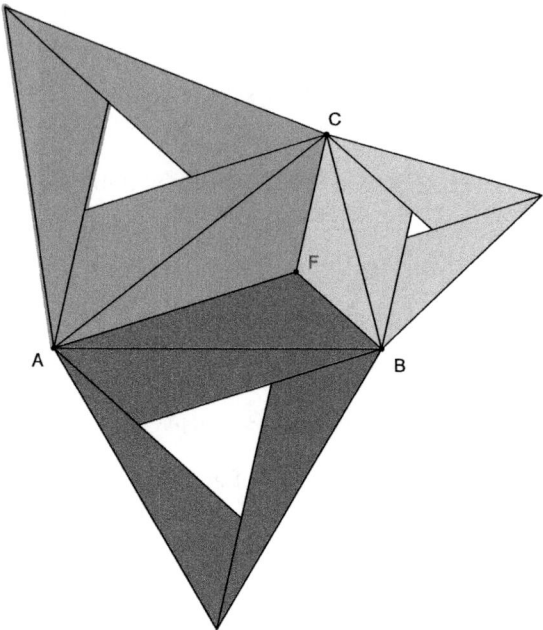

Fig. 2. Weitzenböck inequality: $\frac{a^2\sqrt{3}}{4} + \frac{b^2\sqrt{3}}{4} + \frac{c^2\sqrt{3}}{4} \geq 3P$

Proof. Let us introduce an orthogonal system of coordinates so that $A = [u, v]$, $B = [0, 0]$, $C = [a, 0]$, Fig. 3. To prove (5) we translate the geometric situation into the following algebraic equations:

$$b = |CA| \Rightarrow h_1 := (u - a)^2 + v^2 - b^2 = 0,$$

$$c = |AB| \Rightarrow h_2 := u^2 + v^2 - c^2 = 0.$$

Let

$$h_3 := 2(a + b + c)(a^2 + b^2 + c^2) - 3(a^3 + b^3 + c^3 + 3abc) - t = 0,$$

where t is a slack variable. We express the left side of (5) modulo ideal $I = (h_1, h_2, h_3)$ and get

```
Use R::=Q[u,v,p,q,a,b,c,t];
I:=Ideal((u-a)^2+v^2-b^2,u^2+v^2-c^2,2(a+b+c)(a^2+b^2+c^2)-
3(a^3+b^3+c^3+3abc)-t);
Elim(b..c,I);
```

the polynomial equation of the fourth degree in t

$576u^{10}a^2 + 2980u^8v^2a^2 + 6160u^6v^4a^2 + 6360u^4v^6a^2 + 3280u^2v^8a^2 + 676v^{10}a^2 - 2880u^9a^3 - 11920u^7v^2a^3 - 18480u^5v^4a^3 - 12720u^3v^6a^3 - 3280uv^8a^3 + 3600u^8a^4 + 11080u^6v^2a^4 + 11400u^4v^4a^4 + 3960u^2v^6a^4 + 40v^8a^4 + 2880u^7a^5 + 8480u^5v^2a^5 + 8000u^3v^4a^5 + 2400uv^6a^5 - 8352u^6a^6 - 16820u^4v^2a^6 - 9480u^2v^4a^6 - 948v^6a^6 +$

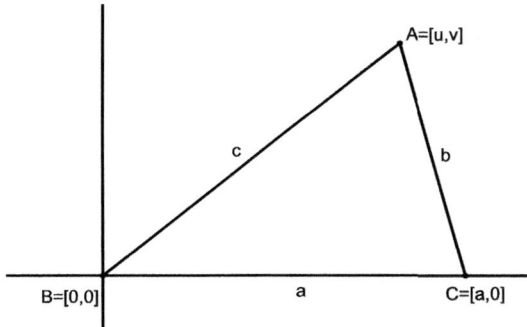

Fig. 3.

$2880u^5a^7+5600u^3v^2a^7+2400uv^4a^7+3600u^4a^8+3880u^2v^2a^8+40v^4a^8-2880u^3a^9-3280uv^2a^9+576u^2a^{10}+676v^2a^{10}-104u^8at-416u^6v^2at-624u^4v^4at-416u^2v^6at-104v^8at + 416u^7a^2t + 1248u^5v^2a^2t + 1248u^3v^4a^2t + 416uv^6a^2t - 1656u^6a^3t - 4968u^4v^2a^3t-4968u^2v^4a^3t-1656v^6a^3t+3512u^5a^4t+7856u^3v^2a^4t+4344uv^4a^4t-3656u^4a^5t - 5952u^2v^2a^5t - 2424v^4a^5t + 1944u^3a^6t + 2232uv^2a^6t - 416u^2a^7t-656v^2a^7t-40ua^8t+4u^6t^2+12u^4v^2t^2+12u^2v^4t^2+4v^6t^2-12u^5at^2-24u^3v^2at^2-12uv^4at^2 + 112u^4a^2t^2 + 200u^2v^2a^2t^2 + 88v^4a^2t^2 - 204u^3a^3t^2 - 188uv^2a^3t^2 + 48u^2a^4t^2 + 152v^2a^4t^2 + 52ua^5t^2 - 4a^6t^2 + 16u^2at^3 + 16v^2at^3 - 16ua^2t^3 + 4a^3t^3 - t^4 = 0$

with unclear solution.

What happened? The resulting equation is not linear in t as in the previous case. By the translation into algebra we obtain extraneous factors and the degree of the polynomial is higher. This is caused by the presence of radicals when expressing the side lengths b, c in the polynomial on the left in (5) in terms of independent variables a, u, v.

To solve the last example we can use the method based on *parametrization* of explored geometric objects, see the next section.

3 Parametrization of n-gons

A parametrization of a triangle is used by various authors, see [22], though they do not call it so. We can also call it *the auxiliary variables method* similarly as the auxiliary polynomial method, see the next section.

The fundamental problem of proving (in)equalities in a triangle given by its three side lengths a, b, c is that a, b, c must obey triangle inequalities $a+b-c \geq 0$, $b + c - a \geq 0$ and $c + a - b \geq 0$ (here we admit also degenerate cases $a + b - c = 0, \dots$).

Basic Outline of the Parametrization Method:

Parametrization is a kind of change of variables that simplifies:

a) some basic restrictions (traditional and more complicated constraints about the side lengths are replaced by the non-negativity of some other variables through bijective mapping from an octant (in affine space x, y, z) to the semialgebraic set in space a, b, c defined by the locus of (a, b, c) such that a, b, c are the side lengths of a triangle).

b) the elimination polynomial (complicated in a, b, c easier in x, y, z).

To avoid triangle inequalities we express the side lengths a, b, c by non-negative parameters x, y, z which do not have to obey triangle inequalities.

Theorem 1. *Given a triangle with side lengths a, b, c. Then the side lengths a, b, c can be expressed in terms of non-negative real numbers x, y, z which are not subject to triangle inequalities.*

Proof. Given a triangle with side lengths a, b, c we define $x = c+a-b$, $y = a+b-c$, $z = b+c-a$. From the triangle inequalities it follows that x, y, z are non-negative real numbers which are not subject to triangle inequalities. To see this let x, y, z be *arbitrary* non-negative real numbers. Then $a = (x + y)/2$, $b = (y + z)/2$, $c = (z + x)/2$.

Definition 1. *Non-negative real numbers x, y, z from the previous theorem we call parameters of a triangle with side lengths a, b, c.*

By the Theorem 1 there is one to one correspondence between side lengths a, b, c and parameters x, y, z. Namely given side lengths a, b, c we can determine parameters x, y, z and conversely. Let us look at the Example 2 and prove it using parametrization.

Proof of (5) — Example 2 continued. To prove the inequality

$$N := 2(a + b + c)(a^2 + b^2 + c^2) - 3(a^3 + b^3 + c^3 + 3abc) \geq 0$$

we will parametrize the side lengths a, b, c of a triangle. Put $c + a - b = x$, $a + b - c = y$, $b + c - a = z$.
Eliminating a, b, c in the ideal $I = (a+b-c-x, b+c-a-y, c+a-b-z, N-t)$ we get

$$t = 1/4(x^3 - x^2y - xy^2 + y^3 - x^2z + 3xyz - y^2z - xz^2 - yz^2 + z^3). \quad (6)$$

Now we are to show that $t \geq 0$, where $x \geq 0$, $y \geq 0$, $x \geq 0$. We will do it by *sds* method [23]. Let us briefly show how this method works.

First suppose that $x \geq y \geq z \geq 0$. We introduce new variables $t_1 \geq 0$, $t_2 \geq 0$, $t_3 \geq 0$ so that

$$
\begin{array}{ll}
t_1 = x - y & x = t_1 + t_2 + t_3 \\
t_2 = y - z \text{ or equivalently} & y = t_2 + t_3 \\
t_3 = z & z = t_3.
\end{array}
\quad (7)
$$

By elimination of x, y, z we get

```
Use R::=Q[x,y,z,t[1..3]];
K:=Ideal(1/4(x^3-x^2y-xy^2+y^3-x^2z+3xyz-y^2z-xz^2-yz^2+z^3),
x-(t[1]+t[2]+t[3]),y-(t[2]+t[3]),z-t[3]);
Elim(x..z,K);
```

$$t = 1/4(t_1^3 + 2t_1^2 t_2 + t_1^2 t_3 + t_1 t_2 t_3 + t_2^2 t_3) \tag{8}$$

from which non-negativity of t follows.

Another choice, for instance $x \geq z \geq y \geq 0$ leads to relations $x = t_1 + t_2 + t_3$, $z = t_2 + t_3$, $y = t_3$ which give the same result. Due to the symmetry of the polynomial (6) we always get the same expression (8).[3]

The sign of equality in (5) occurs iff $t = 0$ which is equivalent to $t_1 = t_2 = t_3 = 0$. This means that $a = b = c$ and the triangle is equilateral.

Remark 3. If a, b, c in (5) are *arbitrary* non-negative real numbers then the inequality does not hold! A counter-example $a = 1$, $b = 1$, $c = 6$.

Parametrization of a triangle enables to prove plenty of geometric inequalities, e.g. those from [1]. Sometimes the situation can be more complicated. Let us see the following example which is the well-known isoperimetric inequality for $n = 3$.

Example 3. Given a triangle ABC with side lengths a, b, c and the area p. Prove that

$$(a + b + c)^2 - 12\sqrt{3}p \geq 0. \tag{9}$$

When the equality is attained?

Proof. We will show two ways of automated proofs, from which the first one fails.

1st attempt - modulo ideal approach:

Let $A = [u, v]$, $B = [0, 0]$, $C = [a, 0]$, Fig. 3. Then

$b = |CA| \Rightarrow h_1 := (u - a)^2 + v^2 - b^2 = 0,$

$c = |AB| \Rightarrow h_2 := u^2 + v^2 - c^2 = 0,$

$p = \text{area } ABC \Rightarrow h_3 := 2p - av = 0,$

$q = \sqrt{3} \Rightarrow h_4 := q^2 - 3 = 0,$

and let

$h_5 := (a + b + c)^2 - 12\sqrt{3}p - t = 0,$

where t is a slack variable. Then the elimination of dependent variables b, c, p in the ideal $I = (h_1, h_2, h_3, h_4, h_5)$ gives

```
Use R::=Q[u,v,a,b,c,p,q,t];
```

---

[3] This step can be done using the software TSDS which was developed by L. Yang and Y. Yao.

```
I:=Ideal((u-a)^2+v^2-b^2,u^2+v^2-c^2,2p-av,q^2-3,
(a+b+c)^2-12qp-t);
Elim(b..p,I);
```

the polynomial which leads to the equation of fourth degree in t

$384u^4va^3q + 5760u^2v^3a^3q + 5376v^5a^3q - 768u^3va^4q - 5760uv^3a^4q + 384u^2va^5q + 5376v^3a^5q - 1728u^4v^2a^2 - 3456u^2v^4a^2 - 1728v^6a^2 + 3456u^3v^2a^3 + 3456uv^4a^3 - 5184u^2v^2a^4 - 14272v^4a^4 + 3456uv^2a^5 - 1728v^2a^6 - 192u^4vaqt - 384u^2v^3aqt - 192v^5aqt + 384u^3va^2qt + 384uv^3a^2qt - 576u^2va^3qt - 2880v^3a^3qt + 384uva^4qt - 192va^5qt + 64u^4a^2t + 2688u^2v^2a^2t + 2624v^4a^2t - 128u^3a^3t - 2688uv^2a^3t + 64u^2a^4t + 2624v^2a^4t + 144u^2vaqt^2 + 144v^3aqt^2 - 144uva^2qt^2 + 144va^3qt^2 - 16u^4t^2 - 32u^2v^2t^2 - 16v^4t^2 + 32u^3at^2 + 32uv^2at^2 - 48u^2a^2t^2 - 672v^2a^2t^2 + 32ua^3t^2 - 16a^4t^2 - 24vaqt^3 + 8u^2t^3 + 8v^2t^3 - 8uat^3 + 8a^2t^3 - t^4 = 0$

with unclear solution.

2nd attempt — parametric plus modulo ideal approach:

Now we will use a parametric approach to prove the inequality (9). Let the coordinates of the vertices A, B, C and respective algebraic relations be the same as in the previous attempt.

Put $c+a-b = x$, $a+b-c = y$, $b+c-a = z$. The elimination of u, v, a, b, c, p, q in the ideal J gives

```
Use R::=Q[x,y,z,u,v,a,b,c,p,q,t];
J:=Ideal((u-a)^2+v^2-b^2,u^2+v^2-c^2,2p-av,q^2-3,c+a-b-x,a+b-c-y,
b+c-a-z,(a+b+c)^2-12qp-t);
Elim(u..q,J);
```

a *quadratic* equation

$$t^2 + Bt + C = 0, \tag{10}$$

where

$$B = -2(x + y + z)^2,$$
$$C = (x^3 + 3x^2y + 3xy^2 + y^3 + 3x^2z - 21xyz + 3y^2z + 3xz^2 + 3yz^2 + z^3)$$
$$(x + y + z). \tag{11}$$

Unlike the Example 2 now the parameter t is not expressed linearly in (10).

In this case we can use formulas of Viète to investigate the signs of the roots of (10).

For the roots t_1, t_2 we get $t_1 + t_2 = -B$ and $t_1t_2 = C$, where both $-B$ and C are non-negative as we can verify by *sds* method. Thus we get $t_1 \geq 0$, $t_2 \geq 0$, and the inequality (9) is proved.

The equality is attained iff the triangle is equilateral as from *sds* decomposition of C follows.

In the next section we will show one more proof of this statement using a new method based on the introduction of an auxiliary polynomial.

Remark 4. In the previous example both roots of the quadratic equation were non-negative. In general we could use the method of Sturm to investigate signs of roots of an algebraic equation. But it could happen that the roots of the conclusion polynomial equation have different signs and we can not decide.

There are many parametrizations of a triangle. Expression of side lengths a, b, c of a triangle ABC by coordinates u, v, a of the vertices $A = [a, 0]$, $B = [u, v]$, $C = [0, 0]$ can be considered as another parametrization. This parametrization is as follows:

$$(u - a)^2 + v^2 = b^2, \ u^2 + v^2 = c^2.$$

In this case coordinates u, v, a are parameters, with $a \geq 0$ and arbitrary u, v. Parametrization of a quadrilateral is not known to me to date. It could help to prove many inequalities between sides and diagonals of quadrilaterals.

Similarly, parametrization of a simplex in E^n has not been known yet. We are able to parametrize only special cases of a simplex in E^n for instance an orthocentric simplex, whose heights are concurrent. The parametrization of an orthocentric tetrahedron in E^3 is as follows [8]:

Theorem 2. *Let $ABCD$ be an orthocentric tetrahedron with edges $a = |AB|$, $b = |BC|$, $c = |CD|$, $d = |DA|$, $e = |AC|$, $f = |BD|$. Then*

$$a^2 = x + y, \ b^2 = y + z, \ c^2 = z + u, \ d^2 = u + x, \ e^2 = x + z, \ f^2 = y + u,$$

where x, y, z, u are parameters, is a parametrization.

We verify whether the parameters x, y, z, u obey the following identities which characterize orthocentric tetrahedron:

$$a^2 + c^2 = b^2 + d^2 = e^2 + f^2.$$

It holds

$$a^2 + c^2 = b^2 + d^2 = e^2 + f^2 = x + y + z + u$$

and the identities are confirmed.

Conversely given arbitrary x, y, z, u we get

$$x = (d^2 + a^2 - f^2)/2, \ y = (a^2 + b^2 - e^2)/2, \ z = (b^2 + c^2 - f^2)/2, \ u = (c^2 + d^2 - e^2)/2.$$

Notice that parameters may also attain negative values.

4 Proving Inequalities by Introduction of Auxiliary Polynomials

In this section we will introduce a method which can help us in proving geometric inequalities. This method is based on the use of an auxiliary polynomial. We will call it briefly *the auxiliary polynomial method.*

For the determination of definiteness of polynomials by factorization we will need the important theorem [4]:

Theorem 3. *Let f, g, h be polynomials in real variables $x = (x_1, x_2, \ldots, x_n)$ such that $f = gh$, where g and h have no common factors. Then*

$$(\forall x \in R^n : f \geq 0) \Leftrightarrow [(\forall x \in R^n : g \geq 0 \wedge h \geq 0) \vee (\forall x \in R^n : g \leq 0 \wedge h \leq 0)].$$

In the next example we demonstrate the strength of the auxiliary polynomial method, whereas the method based on the reduction of a conclusion polynomial to the canonical form modulo ideal fails.

Example 4. Let $ABCD$ be a plane quadrilateral with side lengths a, b, c, d and diagonals e, f. Prove that

$$(bc + ad)^2 - e^2(a^2 + b^2 + c^2 + d^2 - e^2 - f^2) \geq 0. \tag{12}$$

When the equality is attained?

The inequality (12) seems to be new. I did not find it anywhere.

Proof. Let $A = [0, 0]$, $B = [a, 0]$, $C = [u, v]$, $D = [w, z]$ and denote $|AB| = a$, $|BC| = b$, $|CD| = c$, $|DA| = d$, $|AC| = e$ $|BD| = f$, Fig. 4. Algebraic translation of the side lengths a, b, c, d and diagonals e, f of $ABCD$ is as follows:

$b = |BC| \Rightarrow h_1 := (u - a)^2 + v^2 - b^2 = 0,$

$c = |CD| \Rightarrow h_2 := (w - u)^2 + (z - v)^2 - c^2 = 0,$

$d = |DA| \Rightarrow h_3 := w^2 + z^2 - d^2 = 0,$

$e = |AC| \Rightarrow h_4 := u^2 + v^2 - e^2 = 0,$

$f = |BD| \Rightarrow h_5 := (w - a)^2 + z^2 - f^2 = 0.$

Let us denote

$N := (bc + ad)^2 - e^2(a^2 + b^2 + c^2 + d^2 - e^2 - f^2).$

We introduce a slack variable t to express the left side of the inequality (12):

$h_6 := N - t = 0.$

We are to show that from $h_1 = 0$, $h_2 = 0$, $h_3 = 0$, $h_4 = 0$, $h_5 = 0$ the conclusion $t \geq 0$ follows.

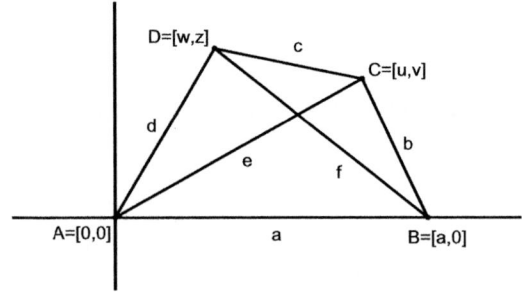

Fig. 4.

Consider the ideal $I = (h_1, h_2, h_3, h_4, h_5)$ which describes the quadrilateral $ABCD$.

1st attempt — modulo ideal approach:

By elimination of dependent variables b, c, d, e, f in the ideal $J = (h_1, h_2, h_3, h_4, h_5, N - t)$ we get

```
Use R::=Q[a,b,c,d,e,f,u,v,w,z,t];
J:=Ideal((u-a)^2+v^2-b^2,(w-u)^2+(z-v)^2-c^2,w^2+z^2-d^2,u^2+v^2
-e^2,(w-a)^2+z^2-f^2,N-t);
Elim(b..f,J);
```

the quadratic polynomial in t

$$-8a^3v^3wz+16a^2uv^3wz-8a^3uvw^2z+8a^2u^2vw^2z-8a^2v^3w^2z+4a^4u^2z^2-8a^3u^3z^2+ \\ 4a^2u^4z^2+8a^3uv^2z^2-8a^2u^2v^2z^2+4a^2v^4z^2+8a^3v^2wz^2-16a^2uv^2wz^2+8a^2v^2w^2z^2- \\ 8a^3uvz^3+8a^2u^2vz^3-8a^2v^3z^3+4a^2v^2z^4+4u^3vwz-4a^2uwt+4au^2wt-4av^2wt+ \\ 4a^2w^2t-4auw^2t-4a^2vzt+8auvzt+4a^2z^2t-4auz^2t-t^2$$

with both negative and positive roots, e.g. the choice $a = 3$, $u = 1$, $v = 2$, $w = 0$, $z = 1$ gives the roots $t_1 = 24$, $t_2 = -24$. Thus we cannot decide our statement and the method fails at the moment.

2nd attempt — auxiliary polynomial approach:

Eliminating variables b, c, d, e, f in the ideal $K = (h_1, h_2, h_3, h_4, h_5, N)$ we get

```
Use R::=Q[a,b,c,d,e,f,u,v,w,z];
K:=Ideal((u-a)^2+v^2-b^2,(w-u)^2+(z-v)^2-c^2,w^2+z^2-d^2,u^2+v^2
-e^2,(w-a)^2+z^2-f^2,N);
Elim(b..f,K);
```

a polynomial S

$$S = a(-avw + 2uvw - vw^2 + auz - u^2z + v^2z - vz^2)^2. \tag{13}$$

which is non-negative (we suppose that $a > 0$). Realize that $S = 0$ represents a circle k which is depicted in the Fig. 5, where circles k and k' are symmetric with respect to the line AC.

It is obvious that $S \in K$. On the other hand $N \notin L$, where $L = (h_1, h_2, h_3, h_4, h_5, S)$, as we easily verify. Therefore we search for an auxiliary polynomial M in variables a, b, c, d, e, f such that MN belongs to the ideal L. Saturation of L with respect to N gives

```
Use R::=Q[u,v,w,z,a,b,c,d,e,f];
L:=Ideal((u-a)^2+v^2-b^2,(w-u)^2+(z-v)^2-c^2,w^2+z^2-d^2,u^2+v^2
-e^2,(w-a)^2+z^2-f^2,S);
Elim(u..z,Saturation(L,Ideal(N)));
```

a polynomial M

$$M := (bc - ad)^2 - e^2(a^2 + b^2 + c^2 + d^2 - e^2 - f^2). \tag{14}$$

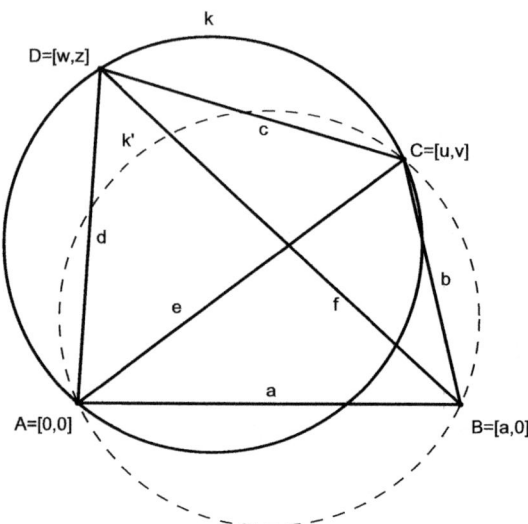

Fig. 5. The angles by B and D in a quadrilateral $ABCD$ are equal

We multiply N with M and search for an integer $q > 0$ such that MN^q belongs to the ideal L. The normal form NF(MN,L)

```
Use R::=Q[u,v,w,z,a,b,c,d,e,f];
L:=Ideal((u-a)^2+v^2-b^2,(w-u)^2+(z-v)^2-c^2,w^2+z^2-d^2,u^2+v^2
-e^2,(w-a)^2+z^2-f^2,S);
NF(M*N,L);
```

equals zero, hence $MN \in L$. Using the command GenRepr(MN,L) we find representation of MN with respect to the ideal L

$$MN = f_1 h_1 + f_2 h_2 + \cdots + f_5 h_5 - 4aS,$$

where $f_1, \ldots, f_5$ are some polynomials. As $h_1 = h_2 = \cdots = h_5 = 0$ we obtain the identity

$$MN = -4aS,$$

from which $MN \leq 0$ follows. Then from the Theorem and the fact that $N - M = 4abcd > 0$ we get $N \geq 0$ and $M \leq 0$. Realize that in addition we proved that $M \leq 0$.

The sign of equality in (12) is attained iff the sum of angles by the vertices B and D of a quadrilateral $ABCD$ equals $180°$ and B, D lie on the same side of AC, see Fig. 6, whereas the equality in $M \leq 0$ occurs iff the opposite angles by the vertices B and D are equal and B, D lie on either side of AC, see Fig. 5. This result follows from the decomposition of the variety of the ideal $L = (h_1, h_2, h_3, h_4, h_5, S)$ into *two* non-degenerate irreducible components which represent two circular arcs. The circle k is divided by the points A, C into two

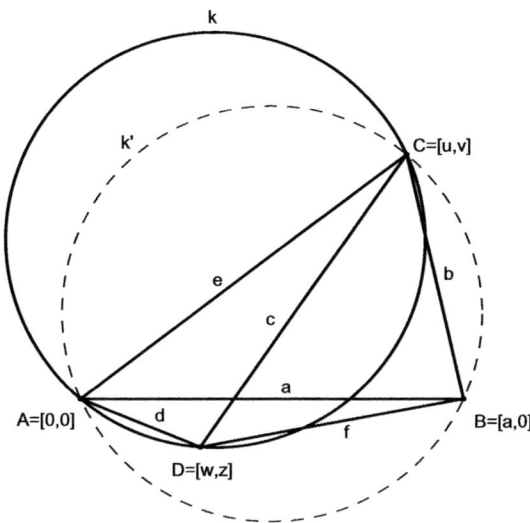

Fig. 6. The sum of angles by B and D in a quadrilateral $ABCD$ equals $180°$

arcs. The upper arc of the circle k belongs to the case $M = 0$, and the lower arc to the case $N = 0$. The decomposition was carried out using the software Epsilon based on WR approach.

The inequality (12) is proved.

Now we will summarize basic steps of the auxiliary polynomial method. Suppose that $I = (h_1, h_2, \ldots, h_r)$ is a hypotheses ideal and N is a conclusion polynomial. We are to prove that $N \geq 0$ (or $N \leq 0$.)

Basic Steps of the Auxiliary Polynomial Method:

1. Eliminate dependent variables in the ideal $(h_1, h_2, \ldots, h_r, N)$ to obtain a polynomial S.
2. If S is not positively (or negatively) semidefinite the procedure terminates. We can not apply it.
3. If S is positively (or negatively) semidefinite verify whether $N \in L$, where $L = (h_1, h_2, \ldots, h_r, S)$. If yes, then carry out a general representation of N with respect to L and get the relation between N and S.
4. If S is positively (or negatively) semidefinite and $N \notin L$ then carry out the saturation of L with respect to N. This should yield a positively (or negatively) semidefinite polynomial M.
5. Verify for which integer $q > 0$, $MN^q \in L$.
6. Express a general representation of MN^q with respect to the ideal L.
7. Determine the resulting relation between MN^q and S.
8. Decompose the variety of L into irreducible components in order to determine when the sign of equality is attained.

Remark 5. In the fifth step we usually get $MN \in L$, i.e., $q = 1$. To compute polynomials M with MN in L the use of syzygies or quotients comes into consideration.

Let as see the use of the auxiliary polynomial method in the Example 3.

Example 3 revisited

We are to prove that in a triangle the inequality

$$N := (a + b + c)^2 - 12\sqrt{3}p \geq 0$$

holds.

With the same notation as above consider the ideal $I = (h_1, h_2, h_3, h_4)$. We will eliminate variables x, y, p, q in the ideal $K = (h_1, h_2, h_3, h_4, N)$ to obtain

```
Use R ::= Q[x,y,p,q,a,b,c];
K:=Ideal((x-a)^2+y^2-b^2,x^2+y^2-c^2,2p-ay,q^2-3,(a+b+c)^2-12qp);
Elim(x..q,J);
```

a polynomial S

$$S := (a+b+c)(7a^3 - 6a^2b - 6ab^2 + 7b^3 - 6a^2c + 15abc - 6b^2c - 6ac^2 - 6bc^2 + 7c^3)$$

in *dependent* variables a, b, c.

By *sds* method we can prove that $S \geq 0$. Further we verify that N does not belong to $L = (h_1, h_2, h_3, h_4, S)$.

Saturation of the ideal L with respect to N gives an auxiliary

```
Use R ::= Q[x,y,p,q,a,b,c];
L:=Ideal((x-a)^2+y^2-b^2,x^2+y^2-c^2,2p-ay,q^2-3,S);
Elim(x..y,Saturation(L,Ideal(N)));
```

polynomial M

$$M := (a + b + c)^2 + 12qp$$

which is non-negative. Next we find out that MN belongs the ideal L. Finally we express the product MN of polynomials M and N with respect to the ideal L using the command `GenRepr(MN,L)` and obtain

$$MN = 4S.$$

Thus

$$((a + b + c)^2 - 12pq)((a + b + c)^2 + 12pq) \geq 0$$

from which the inequality (9) follows.

The equality is attained iff $S = 0$ which by *sds* occurs iff $a = b = c$ and the triangle is equilateral.

5   Concluding Remarks

The described method which takes advantage of the saturation of the ideal with respect to a polynomial to obtain an auxiliary polynomial is not complete. It can be useful when other methods fail.

Method of parametrization of a triangle is successful in proving many geometric inequalities. We need to find similar parametrizations of n-gons and simplices in E^n for $n > 3$.

In the examples above we presented geometric inequalities which were proved mostly in a coordinate system. Perhaps in future we should concentrate on coordinate-free methods.

Acknowledgments. The author wish to thank the referees for their valuable and helpful suggestions.

The research is partially supported by the University of South Bohemia grant GAJU 089/2010/S.

References

1. Bottema, O., et al.: Geometric inequalities, Groningen (1969)
2. Buchberger, B.: Groebner bases: an algorithmic method in polynomial ideal theory. In: Bose, N.-K. (ed.) Multidimensional Systems Theory, pp. 184–232. Reidel, Dordrecht (1985)
3. Chou, S.C.: Mechanical Geometry Theorem Proving. D. Reidel Publishing Company, Dordrecht (1987)
4. Chou, S.C., Gao, X.S., Arnon, D.S.: On the mechanical proof of geometry theorems involving inequalities. Advances in Computing Research 6, 139–181 (1992)
5. Collins, G.E.: Quantifier elimination for the elementary theory of real closed fields by cylindrical algebraic decomposition. LNCS, vol. 33, pp. 134–183. Springer, Berlin (1975)
6. Cox, D., Little, J., O'Shea, D.: Ideals, Varieties and Algorithms. Springer, Berlin (1997)
7. Dalzotto, G., Recio, T.: On protocols for the automated discovery of theorems in elementary geometry. Journal of Automated Reasoning 43, 203–236 (2009)
8. Gerber, L.: The orthocentric simplex as an extreme simplex. Pacific J. of Math. 56, 97–111 (1975)
9. Hou, X., Xu, S., Shao, J.: Some geometric properties of successive difference substitutions. Science China, Information Sciences 54, 778–786 (2011)
10. Kapur, D.: A Refutational Approach to Geometry Theorem Proving. Artificial Intelligence Journal 37, 61–93 (1988)
11. Mitrinovic, D.S., Pecaric, J.E., Volenec, V.: Recent Advances in Geometric Inequalities. Kluwer Acad. Publ., Dordrecht (1989)
12. Nelsen, R.: Proofs Without Words. MAA (1993)
13. Parrilo, P.A.: Structured Semidefinite Programs and Semialgebraic Geometry Methods in Robustness and Optimization. PhD. thesis. California Institute of Technology, Pasadena, California (2000)
14. Pech, P.: Selected topics in geometry with classical vs. computer proving. World Scientific Publishing, New Jersey (2007)

15. Recio, T., Sterk, H., Vélez, M.P.: Project 1. Automatic Geometry Theorem Proving. In: Cohen, A., Cuipers, H., Sterk, H. (eds.) Some Tapas of Computer Algebra, Algorithms and Computations in Mathematics, vol. 4, pp. 276–296. Springer, Heidelberg (1998)
16. Reznick, B.: Some concrete aspects of Hilbert's 17th Problem. Contemp. Math. 253, 257–272 (2000)
17. Wang, D.: Gröbner Bases Applied to Geometric Theorem Proving and Discovering. In: Buchberger, B., Winkler, F. (eds.) Gröbner Bases and Applications. Lecture Notes of Computer Algebra, pp. 281–301. Cambridge Univ. Press, Cambridge (1998)
18. Wang, D.: Elimination Methods. Springer Wien, New York (2001)
19. Wang, D.: Elimination Practice. Software Tools and Applications. Imperial College Press, London (2004)
20. Weitzenböck, R.: Math. Zeitschrift 5, 137–146 (1919)
21. Wu, W.-T.: Mathematics Mechanization. Science Press, Kluwer Academic Publishers, Beijing, Dordrecht (2000)
22. Yang, L., Zhang, J.: A Practical Program of Automated Proving for a Class of Geometric Inequalities. In: Richter-Gebert, J., Wang, D. (eds.) ADG 2000. LNCS (LNAI), vol. 2061, pp. 41–57. Springer, Heidelberg (2001)
23. Yang, L.: Difference substitution and automated inequality proving. Journal of Guangzhou Univ., Natural Science Edition 5(2), 1–7 (2006)
24. Yao, Y.: Termination od the sequence of SDS and machine decision for positive semi-definite forms. arXiv:0904.4030v1 (2009)

Thousands of Geometric Problems for Geometric Theorem Provers (TGTP)

Pedro Quaresma

CISUC, Department of Mathematics, University of Coimbra
3001-454 Coimbra, Portugal
pedro@mat.uc.pt

Abstract. Thousands of Geometric problems for geometric Theorem Provers (*TGTP*) is a Web-based library of problems in geometry.

The principal motivation in building *TGTP* is to create an appropriate context for testing and evaluating geometric automated theorem proving systems (GATP). For that purpose *TGTP* provides a centralised common library of geometric problems with an already significant size but aiming to became large enough to ensure meaningful system evaluations and comparisons. *TGTP* provides also a workbench were it is possible to test any given geometric conjecture.

TGTP is independent of any given GATP. For each problem the code for each GATP (whenever available) is kept in the library. A common format for geometric conjectures, extending the i2g format, is being developed. This common format, plus a list of converters, one for each GATP, will allow to test all the GATPs with all the problems in the library.

TGTP is well structured, documented and with a powerful querying mechanism, allowing an easy access to the information. All information in the library, and also the supporting formats and tools are freely available.

TGTP aims, in a similar spirit of *TPTP* and other libraries, to provide the automated reasoning in geometry community with a comprehensive and easily accessible library of GATP test problems. The development of *TGTP* problem library is an ongoing project.

Keywords: Library of problems in geometry, Geometric Automated Theorem Proving.

1   Introduction

Automated theorem provers, applications, and libraries of problems are often developed separately. In some cases, joint efforts of many of researchers led to standards such as *DIMACS* (for propositional logic) [6] and *SMT* (for satisfiability modulo theory) [1] and libraries of problems such as *SATLIB* (for propositional logic) [10], *TPTP*[1] (for predicate logic) [21], *SMT-lib* (for satisfiability modulo theory) [1] etc. Such efforts, standards, and libraries are fruitful for easier exchange of problems, ideas, and even program code. However, this is often

---

[1] http://www.cs.miami.edu/~tptp

P. Schreck, J. Narboux, and J. Richter-Gebert (Eds.): ADG 2010, LNAI 6877, pp. 169–181, 2011.

very demanding and there are not many systems smoothly integrating libraries of problems and theorem provers.

There are also several systems integrating dynamic geometry software (DGS), GATPs, and a set of examples. For example: *Java Geometry Expert*[2] (*JGEX*) is a system that combines dynamic geometry, automated geometry theorem proving and visual dynamic presentation of proofs. It contains a large set of examples; *GEOTHER*[3] is an environment for manipulating and proving geometric theorems implemented in Maple. It contains a collection of theorems in both elementary and differential geometry [22]; *Ludi Geometrici*[4] has a vast library of problems in the area of classical constructive (ruler and compass only) Euclidean geometry. It does not provide a GATP so no formal proofs are provided; *GeoThms*[5] is a Web workbench in the field of constructive problems in Euclidean geometry. It links DGSs and GATPs and contains a large library of geometry problems [19]. Many of the DGSs, e.g. *GeoGebra*[6] [8], *Cabri*[7], *Cinderella*[8][15,20], etc, DGSs/GATPs, e.g. *GCLC* [12], *GeoView* [2], *GeoProof* [16], *Geometry Explorer* [23], *MMP/Geometer*[9] [9], *GEX* [9], *Discover* [3], and also GATPs like *Theorema* [4] come with a (some times large) set of examples. However none of these systems try to provide a common platform for meaningful system evaluations and comparisons.

In the rest of this paper we present Thousands of Geometric problems for geometric Theorem Provers (*TGTP*[10]) which is a Web-based library of GATP test problems. It aims to become a comprehensive common library of problems with a significant size and unambiguous reference mechanism, easily accessible to all researchers in the automated reasoning in geometry community. *TGTP* tries to address all relevant issues. In particular:

- is Web-based and is thus easily available to the research community;
- is easy to use;
- aims to provide a common format to conjectures in geometry;
- tries to cover the different forms of automated proving in geometry, e.g. synthetic proofs and algebraic proofs;
- aims to become large enough for statistically significant testing. In its current version it contains already over 180 problems;
- aims to become a comprehensive, up-to-date library;
- is independent of any particular GATP system;
- is well structured and documented. This allows effective and efficient use of the library;

---

[2] http://www.cs.wichita.edu/~ye/;
[3] http://www-calfor.lip6.fr/~wang/GEOTHER/
[4] http://www.polarprof.org/geometriagon/
[5] http://hilbert.mat.uc.pt/GeoThms/
[6] http://www.geogebra.org/cms/
[7] http://www.cabri.com/
[8] http://www.cinderella.de
[9] http://www.mmrc.iss.ac.cn/~xgao/software.html
[10] http://hilbert.mat.uc.pt/TGTP

- documents each problem. This contributes to the unambiguous identification of each problem;
- provides query mechanisms;
- provides a mechanism for adding new problems;
- provides a workbench for an easy testing of any given conjecture.

Paper Overview. Section 2 describes the *TGTP* system, its realm, the Web-interface, the structure of the information, the queries, the performance information. Section 3 talks about a common format for geometric conjectures. Section 4 discusses further work, and in Section 5 some final conclusions are drawn.

2 TGTP

Thousands of Geometric problems for geometric Theorem Provers (TGTP) is a Web-based library of geometric problems for testing and evaluating geometric automated theorem proving systems.

2.1 Realm

TGTP is a library of problems (conjectures) in geometry for GATP systems evaluation. *TGTP* aims to supply the automatic reasoning in geometry community with a comprehensive library of GATPs problems.

The *TGTP* library is independent of any GATP system, for each problem generic information is kept (see Section 2.3 for details) and, connected to this, the code for the different GATPs that are already associated with the problem.

A common XML-format is being developed based in the author's previous experience [18] and in the i2g common file format [7], extending this last format, allowing it to cope with conjectures. From this common format converters to GATP specific formats will be written, which can be used to provide the GATPs code whenever a specific realisation was not provided.

As said above *TGTP* stores, for each problem, some generic information, namely the name of the problem, a short textual information, a formal statement of the conjecture, a set of keywords and bibliographic references (some of this fields are optional), this linked with powerful query mechanisms allows keeping the list of problems coherent, avoiding duplications (see Section 2.3 for details).

The goal for building *TGTP* is, in a similar spirit of *TPTP* and other libraries, to provide the GATP community with a centralised problem collection with an easy access to all researchers. The *TGTP* aims to become a comprehensive up-to-date library of problems for the GATPs testing and evaluation.

2.2 The Web Interface

The *TGTP*'s Web interface aims to fulfil the goal of an easy availability of all the information to the GATP community. It is structured in only three levels (see Fig. 1), two, if we do not consider the entry level: a first level for login and also to browse some generic info about the system (HELP) and a second level (after the login) divided in four sections plus a LOGOUT option.

There are three different type of *TGTP*'s users: anonymous/regular users, contributors and the administrator. The administrator has access to a simple interface that allows to see logging information and to do some administrative duties.

The anonymous/regular user has access to the "public" interface. All the access is given in terms of "see but do not touch" mode. Exception to this is the WORKBENCH ,where any user can test the problems with the already installed provers. A personal scrapbook, i.e. a list of problems, is available. The anonymous users will share a common list, the other (registered) users will have his/her own list. This type of user has full access to the information and to the downloadable materials.

Contributors will have, in addition to all the regular users' features, the ability to add new problems, i.e., in the section "Problems List" the contributors will have the possibility of submit new problems and/or update the existing ones (see markers (r)egular and (c)ontributer in Fig. 1)

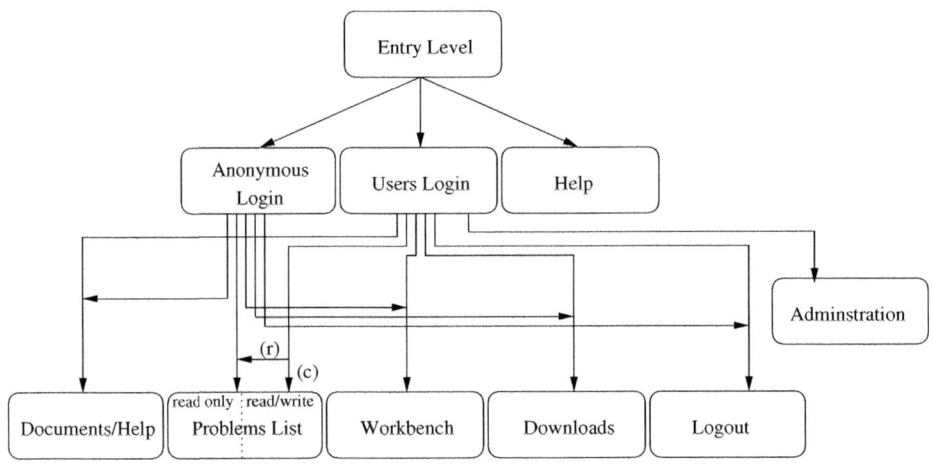

Fig. 1. Structure of the Web-page

The contributors can also produce a new set of evaluation data, i.e. a new set of performance values for the different GATPs when run over the *TGTP* set of problems. For instance, after the introduction of a new set of problems.

The *TGTP* shares with the *GeoThms* system the list of users.

The interface is divided in six main sections (see Fig. 1). The *Administration* is the section reserved for administrative duties. The *Logout* section is for a well-behaved exit, closing the Web-session and registering some information about the time spent by the user in the system.

The *Documents/Help* section contains documents, for instance, a list of bibliographic references about GATPs (in BibT_EX format) cited, or not, in the problems; a list of provers and a list of authors with information about GATPs and its authors. It will also contain information regarding the use of *TGTP*: manuals, frequently asked question list, *How-Tos*.

This section contains also the performance information regarding the GATPs and the list of problems: number of attempts, number of proof attempts succeeded, i.e. the GATP has reached a conclusion within the time limit of 600s; the percentage of success, and information of the CPU time spent in the proofs, the minimum time, the maximum time, and the average time. The information of each individual proof attempt is also displayed (see Section 2.5).

Also in this section is a link to the TGTP's Forum, a place where TGTP's users can freely exchange information.

The *Problem List* section contains the list of all problems introduced up to the present day. It is presented in a concise form: a list of 10 (or 20, or 50, or all) lines with the unique identification problem, the name of the problem, a short description (if present), and the number of proofs succeeded and the number of proof attempts. Each line contains also a link to another Web-page where all the info about the problem is presented.

For each problem it is possible to get all the details about it, and its proofs. From this page it is possible to download the information about the problem in textual form for an easy reading: its identification name, the submission date, its name, a short description and a formal statement (in LaTeX format), a list of keywords and for each proof attempt its status, the GATP used, and the GATP code.

The contributors can update/change the info on every existing problem in the database. They have also the capability to add new problems, the insertion of new problems is safeguarded with a validation step where a search for similar problems, already in the database, is done.

It will also be possible to submit a list of new problems for a bulk insertion into the database. The automatic processing of the list will be done with the help of a given XML-format (see Appendix A).

It is also possible to query the database to look for a problem or a set of related problems (see Section 2.4).

In the *Workbench* section it is possible to test conjectures with the "in-house" GATPs. A user (of any type) will have a simple Web-editor to write the conjecture he/she wants to submit to one of the GATPs that are available in the server, for now GCLCprover [13] and CoqAM [16]. The GATPs are called with a 600s time limit and after a successful run, or after 600s, the results of the proof are made available.

The problems to be submit can be: new problems, written by the user; existing problems, selected from *TGTP* list of problems, or from the personal scrapbook. The scrapbook is unique to every user.

The *Downloads* section is the place where it is possible to download documents related to the *TGTP* database itself and to GATP's codes listing.

The *TGTP* database can be, with the exception of the tables with the information of the *TGTP* users, downloaded in full, i.e. it is possible to download a file with the result of a "mysqldump" command [17]. It is also possible to download the entity-relationship diagram that describes the database (see Fig.2 for a condensed version of the ERD).

From this section the GATP's codes listings are also available, i.e. a text file with all the codes in the database related to any given GATP. This file is a simple text file with a simple separator between each problem's code. This lists of problems is also available in a compressed file containing the list of problems in XML format (see Appendix A) for an easy automatic parsing.

2.3   The List of Problems

The information is organised in five different aggregations (see Fig. 2). The aggregations *Conjectures* and *Proofs* are the core of *TGTP*. In *Conjectures* we have the list of all problems and in *Proofs* we have, for each problem, all its proofs attempts.

The *Users* aggregation is used to control the access of registered (and anonymous) users to *TGTP* and to keep information about the user's login history. The workbench is connected to this section by the `CodeTmpProver` table.

The aggregations *Measures of Efficiency* and *Statistics* (a more correct name should be *Performance Information*) have all the details about performance information. In *Statistics* a snapshot of all the measures of efficiency, at a given time, is kept. The purpose of this information is to keep an historical record of the *TGTP* status allowing an evaluation of the problems/GATPs/TGTP development along the years.

The `TGTP` table is used to keep track of different (majors) versions of *TGTP*.

Since *TGTP* shares with the *GeoThms* system the database of problems we can also have, for many of the conjectures but not necessarily for all, the DGS's code for the geometric construction. The DGS constructions are only available within the *GeoThms* system.

2.4   Queries

The list of problems can be queried in two ways: a simple query using MySQL's regular expressions and a more powerful using the full-text search of MySQL [17]. The first one is done over the `name` attribute of the table `Conjectures` after the user has provided a word to be searched. This word will be matched against any of the words in the list of words that constitute the conjectures names. The second one is done over the attributes `name`, `description`, `shortDescription` and `keywords` of the table `Conjectures` and allows, for a given input sentence, to get the list of most similar sentences in either of the these attributes.

2.5   Performance Information

The *TGTP* database contains now (2011/06/17) 180 problems and contains results of proof attempts from two GATP: CoqAM [16], and GCLCprover [13],

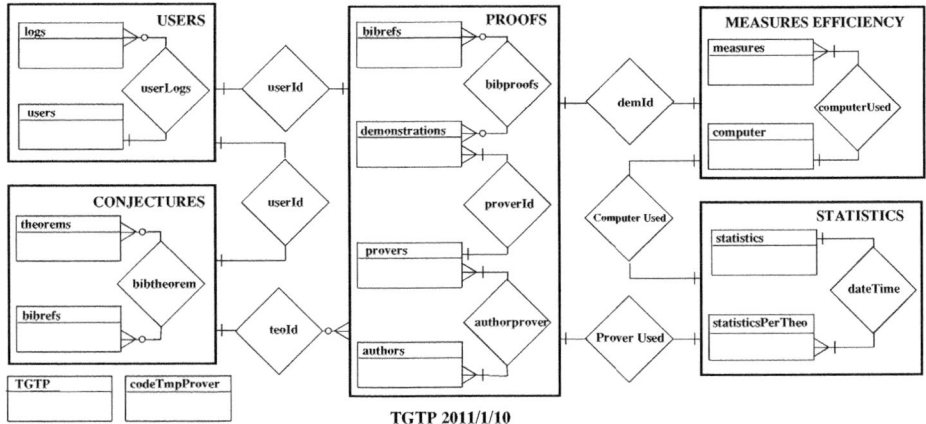

Fig. 2. Structure of the data base (E-R diagram)

covering the methods: Wu's method [24], Gröbner basis method [14] and area method [11].

A new set of performance values is taken whenever a major change in TGTP database occurs: a increase in the number of problems; a change in the computer that is used to run the GATPs; a inclusion of a new GATP or a change in version of an existing GATP. That is, all the data in the `measures` table is collected and tagged with the current date and saved in the table `statistics`. If needed, new measures of efficiency are taken, e.g. a new computer will imply a new set of measures for all the problems times all the GATPs; a new GATP, or a new version of an existing one, will imply a set of measures for all the problems with that particular GATP. For new problems all the GATPs (whenever applicable) are executed and the values added. A new version of *TGTP* is only relevant for this issue if the change would imply new, hopefully improved, codes for the problems, e.g. new converters, or a new version of the common format.

These snapshots of the information contained in *TGTP* and the fact that any kind of change: a new problem, a new proof to an existing problem; a new GATP, or a change in version of an existing GATP, etc, will add to the existing information (not update the information but to add new one) will allow to trace the evolution of a given GATP (through its changes of versions), or of a given problem, or the *TGTP* system itself.

The values are taken per proof attempt (see Table 1), that is, for each pair of problem and GATP's code, the performance values of that attempt are saved in the `measures` table. All the proofs attempts have a time limit of 600s after which the process is killed by the operating system.

The proof status are : "Proved"; "Disproved"; "Failed to prove the conjecture"; "Time-out: failed to prove the conjecture", when the process is killed before it reaches and end; "Maximal number of proof steps reached: failed to prove the conjecture", a limit that some GATPs (for example GCLC AM), have themselves; "The conjecture out of scope of the prover", whenever the GATP could not deal with the problem, e.g. the provers based in the area method have a limited range of problems that they can deal with (see [11] for details). The correspondent numeric codes range from 1 to 6.

Table 1. Results of Proof Attempts (fragment)

TheoId	Coq (AM)		GCLC (AM)		GCLC (WM)		GCLC (GBM)	
	status	time	status	time	status	time	status	time
GEO0230			4	600.021	4	1.468	4	605.362
GEO0231	1	17.89	3	0	2	0.004	3	0.224
GEO0232			3	0.024	2	0	3	0.004
GEO0233			3	0.252	1	0.044	1	1.392
GEO0234	1	1.07	1	0	1	0	1	0
GEO0235	4	600.44	1	1.4	2	0.008	3	0.004
GEO0236	4	600.29	4	600.17	2	1.668	1	5.22
GEO0237	4	600.6	3	0.788	1	0.048	4	599.169
GEO0238	4	601.27	1	0.032	1	0.024	1	0.092
GEO0239			1	0.004	1	0.008	4	609.362

Apart from this, per problem results, some overall values are also collected (see Table 2). For each GATP the following measures are taken: the number of proof attempts, i.e. the number of code entries contained in the database; the number of times the GATP succeeded in proving (or disproving); the percentage of success; and some measures of CPU times: the minimum time needed, the maximum time needed, and the average time. This last values are taken only for those cases where the time-out limit was not reached.

The script used to run the GATPs, imposing a time limit, and getting the CPU time used by the GATPs, is this *bash script*:

```
#!/bin/bash
ulimit -t $1
/usr/bin/time --output=$2 -f "CPU time in seconds: %e" $3 $4 > $5
```

where `ulimit` and `time` are Linux tools to impose a time limit and to get the CPU time spent by a given process respectively. The arguments of the script are: the time limit (600s); the name of the file where the CPU time will be written; the name of the GATP; the argument (code) to the GATP; and the file that will receive (by a Linux redirection) all the output of the GATP.

After each run, a set of other scripts will parse the resulting files getting the desired results.

3   Common File Format for Conjectures

In [18] an XML-suite for constructive descriptions of geometrical figures and geometrical proofs is described. This format is used in the *GeoThms* system to

Table 2. Overall Results

	attempts	succeeded	%of success	min	max	avg
Coq (AM)	76	68	0.89	0.73	213.71	17.698
GCLC (AM)	123	62	0.5	0	360.235	9.194
GCLC (WM)	96	88	0.92	0	6.404	0.422
GCLC (GBM)	96	56	0.58	0	112.319	5.393

provide a common format for its list of problems, and where the conversion of this format to the DGSs/GATPs format is done via XSLT files.

Since then, the i2g common file format of the Intergeo consortium was specified, which is a file format designed to describe any construction made with the help a DGS [7].

Having this in mind we decided to adopt the i2g format and to extend it with an XML-based format for geometrical proofs (from our previous work). As said in [7] the Content Dictionaries [5] of OpenMath[11] can be used to define a new set of symbols, to describe geometric conjectures, and in this way to enrich the expressive power of the i2g common file format (see Appendix B for details).

We intend to support the automatic conversion from this new extended common format to all the GATPs formats available in the *TGTP* system.

4   Future Work

The *TGTP* project is, and it will always be, an ongoing project. New problems should be added to the existing list of problems, new GATPs, or new versions of existing GATPs should be considered.

Apart from this long term commitment, there are short/medium term improvements to be done: the common format for GATPs and the corresponding converters; direct conversions between the different GATPs (e.g. the Coq AM and GCLCprover) for an optimized comparison between GATPs; improvements in the performance information, namely the inclusion of graphical outputs for a better reading; improvements in the documentation and the Web-page.

5   Conclusions

In the *GeoThms* system the author of this article and Predrag Janičić already addressed some of the issues that are now being laid down for *TGTP*, namely the XML common format, and the list of problems. Where the *GeoThms* goal is to have a publicly accessible and widely used Internet based framework for constructive geometry with a strong integration of DGSs, GATPs and a library of problems, the *TGTP* goal is to provide the GATP community with a centralised collection of problems, independent of any particular GATP system.

The development of *TGTP* problem library is an ongoing project, aiming to provide all of the desired properties described above.

---

[11] http://www.openmath.org/

A   List of Problems XML Format

The lists of problems (for each GATP) are available in files written in a simple XML format for an easy automatic parsing. This format is used for the bulk automatic insert of a given list of problems in the database, but it is also used to assemble a file with all the conjectures in *TGTP*. This file is accessible to download in the Web-page.

The XML format has the necessary tags to describe any given problem (an load it into the database). The tags are self-explanatory, the example below describes the format. The author of this article is open to any suggestions/improvements to this format that the readers might be willing to suggest.

```
<results>
 <gatpid>
  GATP id
 </gatpid>
 <result>
  <userid>
   Contributor Id (mandatory)
  </userid>
  <theoid>
    Theorem Id (output file, in the input file it will be ignored)
  </theoid>
  <theoname>
   Theorem Name (mandatory)
  </theoname>
  <description>
   Theorem statement in LaTeX format (optional)
  </description>
  <shortDescription>
   Theorem statement in text format (optional, but highly desirable)
  </shortDescription>
  <keywords>
   <keyword>
    keyword (list of keywords, optional, but highly desirable)
   </keyword>
   ...
  </keywords>
  <biblist>
   <bibitem>
    Bibliographic entry, in BibTeX format (optional)
   </bibitem>
   ...
  </biblist>
  <DGSid>
   <n> - the DGS id number (optional)
  </DGSid>
  <figcode>
   DGS code for the rendering of the Geometric Construction (optional)
```

```
  </figcode>
  <GATPid>
   <n> - the GATP id
  </GATPid>
  <proofscode>
   GATPs code
  </proofscode>
 </result>
 ...
</results>
```

The `biblist` and `keywords` lists may be empty.

The `theoid` tag is only meaningful when the XML file was generated by the *TGTP* system. If provided in the input file it will be ignored, the system will provide a unique identifier for each new problem disregarding any given value.

The `DGSids` are: (0,i2g),(2, Eukleides - 1.0.2), (3,GCLC - 9.00), (4,GeoGebra - 3.2.0.0). The `GATPids` are: (0,i2gGATP), (2,GCLC Area Method - 9.00), (4,COQ Area Method - 1.0), (5,GCLC Wu's Method - 9.00) and (6,GCLC Gröbner Basis Method - 9.00).

B   The Common Format for GATPs

The proof methods considered in *TGTP* for now are: the area method, the Wu's method and the Gröbner Basis method. Having that in mind, we begin to introduce the symbols needed to support those methods. We have to consider algebraic polynomials, the area method quantities, and the geometric conjectures:

Algebraic Polynomials. The symbols needed for this can be imported from Open-Math Polynomial CD Group "polygrp"[12] which in turn use the symbols for arithmetic operators from other CDs (e.g. the "arith1" CD).

Area Method Symbols. The Area Method introduce the *ratio of directed segments*, the *signed area* and the *Pythagoras difference* of triangles and rectangles. Given that we will need to introduce:

```
sratio, signed_area3, signed_area4, pythagoras_difference3,
pythagoras_difference4
```

these symbols will be applied to points, the axiomatic elements of the area method, which are in the i2g CD.

The area method needs also the symbol for equality and the operators of a field $(F, +, \cdot, 0, 1)$ of characteristic different from 2, these symbols can be found in the CDs for arithmetic operators.

Symbols to express the *non-degeneracy conditions* [11] are also required.

---

[12] http://www.openmath.org/cd/

Geometric Conjectures. Introducing geometric conjectures we need the symbols for expressing conjectures, e.g. identical, collinear, perpendicular, parallel, midpoint, etc.

And also the symbols for the proof itself: conjecture, prove, lemmas.

An example of a file in this format is given below:

```
<conjecture>
  <equality>
   <expression>
    <signed_area3>
     <point>P</point>
     <point>Q</point>
     <point>R</point>
    </signed_area3>
   </expression>
   <expression>
    <number>0.000000</number>
   </expression>
  </equality>
 </prove>
</conjecture>
```

This is an ongoing project, any help will be welcome.

References

1. Barrett, C., Stump, A., Tinelli, C.: The SMT-LIB Standard: Version 2.0. In: Gupta, A., Kroening, D. (eds.) Proceedings of the 8th International Workshop on Satisfiability Modulo Theories, Edinburgh, England (2010)
2. Bertot, Y., Guilhot, F., Pottier, L.: Visualizing geometrical statements with GeoView. Electronic Notes in Theoretical Computer Science 103, 49–65 (2004)
3. Botana, F., Valcarce, J.L.: A dynamic-symbolic interface for geometric theorem discovery. Computers and Education 38, 21–35 (2002)
4. Buchberger, B., Craciun, A., Jebelean, T., Kovacs, L., Kutsia, T., Nakagawa, K., Piroi, F., Popov, N., Robu, J., Rosenkranz, M., Windsteiger, W.: Theorema: Towards computer-aided mathematical theory exploration. Journal of Applied Logic 4(4), 470–504 (2006)
5. Davenport, J.H.: On writing OpenMath content dictionaries. SIGSAM Bulletin 34(2), 12–15 (2000)
6. DIMACS: Satisfiability suggested format (May 1993), ftp://dimacs.rutgers.edu/pub/challenge/satisfiability/doc/
7. Egido, S., Hendriks, M., Kreis, Y., Kortenkamp, U., Marquès, D.: i2g Common File Format Final Version. Tech. Rep. D3.10, The Intergeo Consortium (2010)
8. Fuchs, K., Hohenwarter, M.: Combination of dynamic geometry, algebra and calculus in the software system geogebra. In: Computer Algebra Systems and Dynamic Geometry Systems in Mathematics Teaching Conference 2004, pp. 128–133. Pécs, Hungary (2004)

9. Gao, X.-S., Lin, Q.: MMP/Geometer - a Software Package for Automated Geometric Reasoning. In: Winkler, F. (ed.) ADG 2002. LNCS (LNAI), vol. 2930, pp. 44–66. Springer, Heidelberg (2004)
10. Hoos, H., Stützle, T.: SATLIB: An online resource for research on SAT. In: Gent, I.P., Maaren, H.V., Walsh, T. (eds.) SAT 2000, pp. 283–292. IOS Press, Amsterdam (2000)
11. Janičić, P., Narboux, J., Quaresma, P.: The Area Method: a recapitulation. Journal of Automated Reasoning 7 (to appear, 2011), doi:10.1007/s10817-010-9209-7
12. Janičić, P.: GCLC — A Tool for Constructive Euclidean Geometry and More Than That. In: Iglesias, A., Takayama, N. (eds.) ICMS 2006. LNCS, vol. 4151, pp. 58–73. Springer, Heidelberg (2006)
13. Janičić, P., Quaresma, P.: System Description: GCLCprover + Geothms. In: Furbach, U., Shankar, N. (eds.) IJCAR 2006. LNCS (LNAI), vol. 4130, pp. 145–150. Springer, Heidelberg (2006)
14. Kapur, D.: Using Gröbner bases to reason about geometry problems. Journal of Symbolic Computation 2(4), 399–408 (1986)
15. Kortenkamp, U., Richter-Gebert, J.: Using automatic theorem proving to improve the usability of geometry software. In: Procedings of the Mathematical User-Interfaces Workshop 2004 (2004)
16. Narboux, J.: A graphical user interface for formal proofs in geometry. Journal of Automated Reasoning 39, 161–180 (2007)
17. Oracle: MySQL 5.5 Reference Manual, 5.5 edn. (January 2011), revision: 24956
18. Quaresma, P., Janičić, P., Tomašević, J., Vujošević-Janičić, M., Tošić, D.: XML-Bases Format for Descriptions of Geometric Constructions and Proofs. In: Communicating Mathematics in The Digital Era, pp. 183–197. A. K. Peters, Ltd. (2008)
19. Quaresma, P., Janičić, P.: Geothms – a Web System for Euclidean Constructive Geometry. Electronic Notes in Theoretical Computer Science 174(2), 35–48 (2007)
20. Richter-Gebert, J., Kortenkamp, U.: The Interactive Geometry Software Cinderella. Springer, Heidelberg (1999)
21. Sutcliffe, G.: The TPTP problem library and associated infrastructure. Jounal of Automated Reasoning 43(4), 337–362 (2009)
22. Wang, D.: GEOTHER 1.1: Handling and Proving Geometric Theorems Automatically. In: Winkler, F. (ed.) ADG 2002. LNCS (LNAI), vol. 2930, pp. 194–215. Springer, Heidelberg (2004)
23. Wilson, S., Fleuriot, J.: Combining dynamic geometry, automated geometry theorem proving and diagrammatic proofs. In: Proceedings of the European Joint Conferences on Theory and Practice of Software (ETAPS) Satellite Workshop on User Interfaces for Theorem Provers (UITP). Springer, Heidelberg (2005)
24. Wu, W.T.: The characteristic set method and its application. In: Gao, X.S., Wang, D. (eds.) Mathematics Mechanization and Applications, pp. 3–41. Academic Press, San Diego (2000)

An Investigation of Hilbert's Implicit Reasoning through Proof Discovery in Idle-Time

Phil Scott and Jacques Fleuriot

Centre for Intelligent Systems and their Applications, Informatics Forum, University of Edinburgh, 10 Crichton Street, Edinburgh, UK, EH8 9AB
phil.scott@ed.ac.uk, jdf@inf.ed.ac.uk

Abstract. In this paper, we describe how we captured and investigated incidence reasoning in Hilbert's *Foundations of Geometry* by using a new discovery tool integrated into an interactive proof assistant. Our tool exploits concurrency, inferring facts independently of the user with the incomplete proof as a guide. It explores the proof space, contributes tedious lemmas and discovers alternative proofs. We show how this tool allowed us to write readable formalised proof-scripts that correspond very closely to Hilbert's prose arguments.

1 Introduction

The *Foundations of Geometry* [9] is the successor to the most influential mathematical text in history [3], Euclid's *Elements* [5], and is claimed to be the most influential book on geometry written in the 20th century [1]. Hilbert sought to close all the logical gaps of the *Elements* and

> [...] to establish for geometry a *complete*, and *as simple as possible*, set of axioms and to deduce from them the most important geometric theorems.[9]

The logical rigour was supposedly ensured by stripping all interpretation from the basic concepts, so that the words "point", "line" and "plane" could be replaced by "table", "chair" and "mug". Weyl has since claimed that the deductions which follow have no gaps [18], but in their attempt to formalise the axiomatics and its elementary theorems, Meikle and Fleuriot found many missing lemmas [11]. Indeed, it is sufficient to note that Hilbert followed Euclid in one pervasive omission: they both give proofs on an ambient plane when the axioms characterise *solid* geometry [8].

Our aim is to completely formalise Hilbert's axiomatics in an interactive proof assistant, and investigate the space of missing lemmas. Our strategy has been to integrate concurrent discovery tools which can systematically and collaboratively explore that space and contribute the necessary lemmas automatically. This should give us formalised proofs whose structure corresponds closely to the prose, or else give us evidence that the prose arguments should be finer grained.

P. Schreck, J. Narboux, and J. Richter-Gebert (Eds.): ADG 2010, LNAI 6877, pp. 182–200, 2011.
© Springer-Verlag Berlin Heidelberg 2011

2   Declarative Formalisation

Interactive proof assistants have been used extensively to verify theorems in geometry [6,10,12], and are generally needed for verification tasks where theorems are so challenging that they cannot be solved in a timely fashion by unguided automation, and instead require human assistance. They are also useful when we wish to analyse axiomatic systems for which no sophisticated automated tools are yet available, and to analyse the structure of informal proofs which such tools fail to replicate.

The two main ways to approach interactive proof roughly divide into *procedural* and *declarative*[1]. In a procedural setting, automated tools, known as *tactics*, are invoked on demand to simplify the current goal and advance the proof. By carefully composing these tactics, the user can simplify a theorem all the way back to its premises, thereby proving it.

The procedural approach involves explicitly stating the procedures needed to transform an implicit proof state. In contrast, the *declarative* approach leaves the procedures *implicit*, so that the user instead states the formulas which connect premises to the desired conclusion. The challenge of declarative proof is to make this connection fine-grained enough that the implicit and usually very generic procedures are effective.

The resulting declarative proofs are therefore more readable, generally speaking, than their procedural counterparts. Firstly, it is relatively easy to determine the proof state at any point in the proof without knowing the details of the automation. Secondly, proof commands are named after the mathematician's ordinary logical vocabulary. So declarative proofs can potentially resemble ordinary textbook mathematics both structurally and syntactically. We can therefore hope that they are potentially accessible to ordinary mathematicians, and not just expert users of the respective systems.

It is for these reasons that we opted to formalise Hilbert's text in the declarative style. We want a readable, formally verified version of Hilbert's text, and we want to have a proof script whose logical structure can be *analysed* and compared with the prose.

However, we found that declarative proof leaves little scope for proof *exploration*. The user typically has to work out the correct inferential paths by hand, and hope that the granularity is adequate for the proof tools. We hope to rectify this with the discovery tool described in this paper.

3   HOL Light

Our chosen proof assistant, HOL Light [7], is an LCF-style prover [13], in which proofs are carried out by invoking functions directly at the ML top-level. The correctness of proofs is guaranteed via encapsulation. Theorem objects are exposed by an abstract data type whose signature corresponds to the primitive inference rules of a higher-order logic.

---

[1] See Wiedijk [19] for some analysis of this distinction.

The upshot is that we can express our proofs with the full support of ML, and easily develop new tactics and tools. These tools are typically written in embedded combinator languages [14], making them almost trivially interoperable with each other and with the rest of ML.

The declarative proof language we are using is inspired by the purely declarative proof system Mizar [4]. It is known as *Mizar Light* [19], and was developed by Wiedijk and embedded in HOL Light as a set of further ML combinators. We give an overview of its primitives in Figure 1. We have emphasised a declarative semantics; that is, rather than describing *how* each primitive affects the state of the prover, we describe *what* each primitive asserts at a given point in a script.

Primitive	Meaning
theorem *term*	Begins a proof of *term*.
assume *term*	Asserts *term* as a justified assumption at this point.
so	Refers to the previous step as justifying the current step.
have *term*	Asserts *term* as derivable at this point.
consider *vars* **st** *term*	Introduces *vars* witnessing *term*.
by *thms*	Refers to previously established theorems *thms* as justifications for the current step.
qed	Asserts that the (sub)theorem is justified at this point.

Fig. 1. An overview of Mizar Light

Inspired by HOL Light's design, Wiedijk made the data structures used by this language public, so that the Mizar Light system is highly customisable and extensible: adding a new primitive is often as simple as defining a new function. Indeed, we have added our own combinator `obviously`, which we describe in §6.

4   Incidence Reasoning

Hilbert's axiomatic system is divided into five groups. After splitting conjunctions, there are a total of twenty-three axioms. The first group, concerned with incidence relations between the primitives *point*, *line* and *plane*, requires ten of them, and as such, one would expect its axioms to feature significantly in proofs. This was indeed the case in Meikle and Fleuriot's formalisation [11], and the matter is especially clear in our own formalisation [16]. But in Hilbert's prose, the axioms are rarely cited. When we fill in the detail, we find the complex incidence lemmas are appealed to sporadically, with some prose steps being justified by overlapping sets of lemmas. In other words, the formalisation does not preserve the structure of Hilbert's original arguments.

Our philosophy has been to justify Hilbert's omission, and explain away the complicated incidence reasoning as fussy detail which does nothing to enlighten the reader of the core ideas behind his proofs. We wish, then, to eliminate this reasoning from our proof scripts and leave it to automation. However, we do not want to add *ad-hoc* proof automation to HOL Light, nor do we want algorithms which cover both Hilbert's implicit and explicit proof steps equally, since we wish to analyse the latter. Instead, we need algorithms which *systematically* target *only* the missing incidence reasoning.

4.1   Incidence Rules

We recovered the incidence reasoning using a forward-chaining discovery algorithm, which can be parameterised by any given set of rules. Our rules, shown in Figure 2, were derived directly from Hilbert's first group of axioms, and govern finite sets of collinear and planar points, rather than the primitives of points, lines and planes. We favoured set-theoretic rules to drive the algorithm, since the incidence reasoning in the *Foundations of Geometry* has a distinctly *combinatorial* flavour.

While these rules are given for arbitrary sets, they are only used by our tool against finite sets built from the empty set and adjoin. Thus, we are only using sets as a convenient *representation*. They do not give us additional reasoning power above the elementary theory, and are effectively short-hand for a recursively enumerable set of conjunctive and disjunctive rules governing Hilbert's primitive incidence relations. For example,

$$\text{collinear}\{A, B, C\} \leftrightarrow \exists a. A \text{ on } a \wedge B \text{ on } a \wedge C \text{ on } a.$$

Note that under certain conditions, the rules (1), (4), (6), (10) and (11) are *interderivable* with axioms I,1, I,2, I,4, I,5, I,6, while rules (4), (7), (9) and (11) yield Hilbert's two theorems from the first group (the axioms and theorems are given in Appendix A).

To see an example of the correspondence, we first need to recover the concept of a line and a plane in the following way: given two distinct points P and Q, we will say that the line PQ is defined as a maximal collinear set of points containing P and Q. Given three points P, Q and R which form a non-collinear set, we say that the plane PQR is defined as a maximal planar set of points containing P, Q and R.

In this way, rules (1) and (4) correspond to Hilbert's first two axioms, which assert that two points determine a line. Indeed, given distinct points P and Q, we know from rule (1) that $\{P, Q\}$ is collinear, and thus by rule (4), that the (necessarily finite) union of all collinear sets containing P and Q must be collinear and maximal. Moreover, given any two maximal collinear sets S and T sharing P and Q, we know by rule (4) that $S \cup T$ is collinear and thus by the maximality of each, that $S = T$. Thus, the points P and Q determine a unique maximal collinear set: the line PQ.

$$\forall P\ Q.\ \text{collinear}\ \{P,Q\} \tag{1}$$

$$\forall S\ T.S \subseteq T \land \text{collinear}\ T \longrightarrow \text{collinear}\ S \tag{2}$$

$$\forall P\ Q\ R.\ \neg \text{collinear}\ \{P,Q,R\} \longrightarrow P \neq Q \land P \neq R \land Q \neq R \tag{3}$$

$$\forall S\ T\ P\ Q.\ \text{collinear}\ S \land \text{collinear}\ T$$
$$\land\ Q \in S \land P \in S \land Q \in T \land P \in T \land P \neq Q$$
$$\longrightarrow \text{collinear}(S \cup T) \tag{4}$$

$$\forall S\ X\ Y\ P\ Q\ R.\ \text{collinear}\ S \land \neg \text{collinear}\{P,Q,R\}$$
$$\land\ X \in S \land Y \in S \land P \in S \land Q \in S \land X \neq Y$$
$$\longrightarrow \neg \text{collinear}\{X,Y,R\} \tag{5}$$

$$\forall P\ Q\ R.\ \text{planar}\ \{P,Q,R\} \tag{6}$$

$$\forall S\ T.S \subseteq T \land \text{planar}\ T \longrightarrow \text{planar}\ S \tag{7}$$

$$\forall S.\ \text{collinear}\ S \longrightarrow \text{planar}\ S \tag{8}$$

$$\forall S\ T\ P.\ \text{collinear}\ S \land \text{collinear}\ T \land P \in S \land P \in T$$
$$\longrightarrow \text{planar}(S \cup T) \tag{9}$$

$$\forall S\ T\ P\ Q.\ \text{collinear}\ S \land \text{planar}\ T$$
$$\land\ Q \in S \land P \in S \land Q \in T \land P \in T \land P \neq Q$$
$$\longrightarrow \text{planar}(S \cup T) \tag{10}$$

$$\forall S\ T\ U.\ \text{planar}\ S \land \text{planar}\ T \land \neg \text{collinear}\ U \land U \subseteq (S \cap T)$$
$$\longrightarrow \text{planar}(S \cup T) \tag{11}$$

$$\forall S\ T\ U\ P\ Q.\ \text{collinear}\ S \land \text{collinear}\ T \land \neg \text{collinear}\ U$$
$$\land\ U \subseteq (S \cup T) \land P \in S \land Q \in S \land P \in T \land Q \in T$$
$$\longrightarrow P = Q \tag{12}$$

Fig. 2. Incidence Rules

4.2   Incidence and Pasch's Axiom

Hilbert's first *proofs* appear in Group II, and they make extensive use of an axiom due to Pasch. This axiom asserts that any line a which enters a triangle ABC on one side and does not meet any of the vertices, must leave by one of the other two sides, such as in the case depicted in Figure 3.

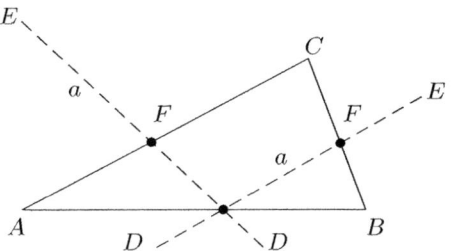

Fig. 3. Axiom II,4

This is a complex axiom to apply. In terms of our point-set predicates, the claim that a line cannot pass through any of the triangle's vertices becomes three claims of *non-collinearity* or the assertion that five points form three triangles. Together with the triangle that is being intersected, this leaves us to prove the existence of four triangles every time we apply Pasch's Axiom. The matter is formalised by deriving the following version of the axiom:

$\forall A\,B\,C\,D\,E.$

 between $ADB \wedge \neg\,\mathrm{collinear}\{A,B,C\}$

 $\wedge\,\neg\,\mathrm{collinear}\{A,D,E\} \wedge \neg\,\mathrm{collinear}\{B,D,E\} \wedge \neg\,\mathrm{collinear}\{C,D,E\}$

 $\wedge\,\mathrm{planar}\{A,B,C,D,E\}$

  $\longrightarrow \exists F.\ \mathrm{collinear}\{D,E,F\} \wedge (\text{between } AFC \vee \text{between } BFC)$ (13)

The axiom is further complicated by its existential and disjunctive conclusion. Every time this axiom is applied in Hilbert's proofs, at least one of the disjuncts is eliminated, by deriving a falsehood on its assumption. The existential is typically eliminated by showing that the point in question is already witnessed. In all but one case, Hilbert elides the case-splitting and the existential elimination, but in our original Isabelle formalisation [16], explicit and complex incidence arguments based on point sets were needed.

 Finally note that it is assumed in Pasch's Axiom that all points lie in a plane. This fact needs to be derived whenever we apply Pasch's Axiom, but as we mentioned in §1, Hilbert almost never mentions planes in his proofs.

5   Idle-Time Discovery

Most of the logic involved in reproducing the first three proofs of *Foundations of Geometry* involved finding the triangles which justify the applications of

Pasch's Axiom. This can be challenging, since the relevant axioms are elementary beyond geometric intuition and do not always fit with the constraints apparent in diagrams. It is not always obvious when the assumptions are strong enough to derive the conditions on Pasch's Axiom, or to determine which other triangles could be shown to exist which might yield alternative ways to apply it. In fact, the challenge had left us unnecessarily deviating from Hilbert's prose in our earlier formalisation. To alleviate the difficulties, we decided to leave the exhaustive search to a computer, and implement automated *discovery* algorithms.

Now the automation available in most interactive theorem provers is invoked on demand by the user when they evaluate individual proof steps. But when the user writes the formal proof for the first time, or comes to edit it later, they will spend most of their time *thinking*, consulting texts, backtracking and typing in individual proof commands. The CPU is mostly idling during this process, and we can exploit this idle time to run automated tools concurrently.

We wanted this tool to then complement the user's own interactive and declarative development of the proof. The automation thus identifies and derives implicit facts which might interest the user, or even solves the goal outright, while they investigate their own chains of deduction independently. They can perhaps focus on high-level strategies that require human insight, while the automation explores lower level mechanical details.

One interesting way we can allow these two independent systems — the human user and the machine automation — to *cooperate* is to focus on forward derivations. When a user writes a declarative proof script in an interactive setting, they invoke functions which add facts to a growing proof context. As these facts are added, an automated tool can inspect them and choose whether to use them as part of its own independent derivations.

The user, in turn, can inspect the facts derived by the automated tool, and choose whether to use them as part of their derivations. The symmetry leads to feedback. The user assists the automated tool by deriving new facts into the proof context, and the automated tool assists the user by outputting its own facts. These new facts are used by the user to produce more facts, and so on. The two systems work continuously in tandem, assisting each other as they drive forward towards the goal theorem.

6   Overview of the System

In a forthcoming paper [17], we define a generic discovery algebra and explain how our incidence discovery tool is just one of its applications. The algebra includes combinators which allow the user to combine forward-chaining, data-driven search, filtering, customisable data-flow and term-rewriting, and allows the user to lift arbitrary inference rules into the discovery mechanism.

The tools integrate with declarative proofs via the proof-context. The proof-context is a data-structure containing, firstly, hypotheses, which represent intermediate facts explicitly derived during a declarative proof, and secondly, the term which the user is trying to prove. Thus, one primitive in the algebra pulls in the hypotheses from the context, thereby allowing the discovery to be driven by the user's explicitly inferred facts, while a filtering function halts all further inference once the goal is found.

For incidence discovery, we compose these primitives with the incidence rules from Figure 2, and use the term-rewriter to unfold conclusions involving finite sets. We thereby derive all three-element non-collinear sets, the largest collinear sets and the largest planar sets that can be inferred from the goal context. A fixpoint is always reached, since our rules only work against a finite number of points. The fixpoint is announced to the user, and the tool sleeps until the proof context changes.

Since we are using an LCF style prover [13], we can ensure that the tool's derivations are *fully-expansive* [2]. This means that to carry out its derivations, the tool applies the same ML functions available in the core system to generate fully machine verified lemmas. These lemmas can then be seamlessly integrated into the user's proof script to produce a fully machine-checked proof.

The algorithm exploits *laziness* pervasively so that the front-end can output facts to a separate terminal as they are generated, though it should be trivial to write new front-ends and new mechanisms to filter out uninteresting facts. We have also extended the Mizar Light proof language, as mentioned in §3, by adding an `obviously` primitive. This combinator transforms an ordinary declarative proof step into one which picks up the discovered theorems, and adds them as justification[2].

7   Analysing Hilbert's Proofs

In this section, we discuss what we have learned about incidence reasoning with point sets based on our discovery tool. We then discuss two proofs which required significant effort to formalise in Isabelle, and show what their HOL Light formalisation looks like when written in tandem with our discovery tool. We compare these new proofs with their prose counterparts, suggest weaknesses of the prose, and discuss an alternative proof which the automation allowed us to explore.

7.1   Theorem 4

We give the prose version of the proof of Theorem 4 as it appears in the tenth edition of *Foundations of Geometry*:

> THEOREM 4. Of any three points A, B, C on a line there always is one that lies between the other two.

---

[2] The tool will be made freely available once we are satisfied it is suitable for end-users.

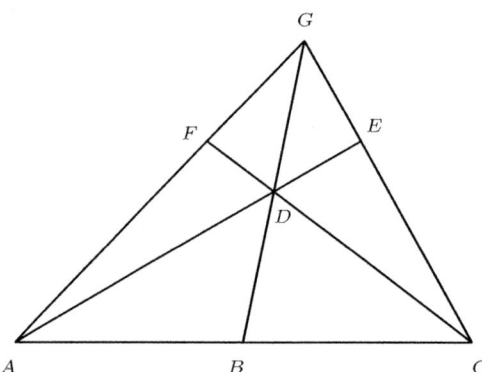

Fig. 4. Proof of Theorem 4

PROOF. Let A not lie between B and C and let also C not lie between A and B. Join a point D that does not lie on the line AC with B and choose by Axiom II,2 a point G on the connecting line such that D lies between B and G. By an application of Axiom II,4 [(Pasch's Axiom)] to the triangle BCG and to the line AD it follows that the lines AD and CG intersect at a point E that lies between C and G. In the same way, it follows that the lines CD and AG meet at a point F that lies between A and G.

   If Axiom II,4 is applied now to the triangle AEG and to the line CF it becomes evident that D lies between A and E, and by an application of the same axiom to the triangle AEC and to the line BG one realizes that B lies between A and C.

Four applications of Pasch's Axiom are needed to prove Theorem 4. In three of them, Hilbert names the triangle and the line on which the axiom is applied (the remaining application being a symmetry of the first). We have represented this reasoning in our HOL Light formalisation by implementing a function of two arguments: a triple of points defining a triangle and a pair of endpoints of a line. The function uses these to specialise the first five quantified variables in Pasch's Axiom (rule (13) in §4.2), runs the discovery tool to a fixpoint, and then returns the discovered theorems to be used as justification for the current step. Since the tool automatically eliminates existentials and splits cases, one of the disjuncts in the conclusion of Pasch's Axiom can be eliminated and the remaining disjunct fed as a justifying theorem.

   Before we turn to our HOL Light formalisation, we consider our earlier Isabelle formalisation. In Figure 5, we give an extract where we first apply Pasch's Axiom, and then, in the block introduced by **moreover**, we eliminate a disjunct from its conclusion. Hilbert never mentions this elimination, and as with the rest of the formalisation, most of the missing steps are unilluminating, consisting of a complex combination of tactics and picky variable instantiations that are not reflected in the prose. Indeed, we had to use comments judiciously to show any correspondence between the prose and the formal proof steps [16], which made it difficult to justify the proofs as *readable*, even though they were written in a declarative style.

```
with AxiomII4_col[of B G C D A]
    '¬collinear {B, C, G}' '¬collinear {A, B, D}'
    '¬collinear {A, C, D}' '¬collinear {A, D, G}'
    and 'between B D G' obtain E
  where "collinear {A, D, E}" and "(between B E C ∨ between G E C)"
  by auto

moreover
{
  assume   "between B E C"
  with '¬between B A C' have "A ≠ E" by auto
  from 'between B E C' and AxiomII1b[of B E C]
      have "collinear {B, C, E}" by simp
  with 'collinear {A, B, C}' and 'B ≠ C'
      have "collinear {A, B, C, E}"
  by (blast intro: collinear_subset[where T = "{A, B, C} ∪ {B, C, E}"]
          collinear_union)
  with 'collinear {A, D, E}' and 'A ≠ E'
      have "collinear {A, C, D}"
  by (blast intro: collinear_subset[where T = "{A,B,C,E} ∪ {A,D,E}"]
          collinear_union)
}
```

Fig. 5. Some Isabelle formalisation

While the Isabelle proof ran to a total of sixty-nine complicated proof steps, our new proof in Figure 6 has just nine lines, each readily understandable, and matching the prose in an almost one-to-one fashion (the formalisation of theorems exists_triangle and g22 are given in Appendix B.3).

Alternative Proof. We should remark that in the first edition of the *Foundations of Geometry*, Theorem 4 is only given as an axiom. This, despite Hilbert's claim:

> [...] so far as the particular axioms of groups I, II, and IV are concerned,
> it is easy to show that the axioms of these groups are each independent
> of the other of the same group.

This was fixed by the tenth edition, which contains the above proof of Theorem 4, attributed to Wald. This should be evidence enough that the theorem is not trivially obvious, and so we decided to investigate further. Starting from the

```
theorem collinear {A, B, C} ∧ A ≠ B ∧ A ≠ C ∧ B ≠ C
      ∧ ¬ between A C B ∧ ¬between B A C ⟹ between A B C
assume collinear {A, B, C} ∧ A ≠ B ∧ A ≠ C ∧ B ≠ C
      ∧ ¬ between A C B ∧ ¬between B A C
so consider D such that ¬collinear {A, B, D} by exists_triangle
obviously³consider G such that between B D G by g22
consider E such that collinear {A, D, E} ∧ between C E G
      by pasch_on B,C,G and A,D
consider F such that collinear {C, D, F} ∧ between A F G
      by pasch_on A,B,G and C,D
have between A D E by pasch_on A,E,G and C,F
have between A B C by pasch_on A,C,E and B,G
qed
```

Fig. 6. Theorem 4 in HOL Light

same basic diagram, we added Pasch's Axiom to our automated tool and used the tool purely for proof *discovery*, allowing it to fill in the last four applications of Pasch's Axiom in Figure 6 autonomously.

We had to limit the search, since after adding Pasch's Axiom, it is possible to derive an infinity of points. Moreover, we were not interested in proofs that were substantially *longer* than the original. We therefore allowed the discovery tool to non-deterministically make four applications of Pasch's Axiom against all possible triangles and lines in the diagram, filtering for those applications which were formally verified to derive the goal theorem. By incorporating some basic proof-recording, we could later recover the triangles and lines against which Pasch's Axiom is applied.

From the recorded data, we extracted many "alternative" proofs, but, in fact, most of these yield symmetries of the original. In some cases, two independent applications of Pasch's Axiom were applied in reverse order. In other cases, the proof was identical to the original up to a relabelling of points. Only one new proof was revealed up to symmetry. We give it now in a prose formulation with an accompanying diagram (Figure 7).

DISCOVERED PROOF OF THEOREM 4.

Assume A, B and C are collinear, with A not between B or C and C not between A or B. We find a point D off the line AC and extend it to G using Axiom II,2. We then use Axiom II,4 on the triangle BCG and the line AD to find the point E between C and G. We use Axiom II,4 on the triangle BEG and the line CD to find the point F between B and E. We use the axiom again on the triangle ABE and the line CF to show that D lies between A and E. Finally, we can use the axiom on the triangle ACE and the line BG to find B between A and C.

---

[3] Here we use our combinator (see §6) to pick up inequalities needed for axiom g22.

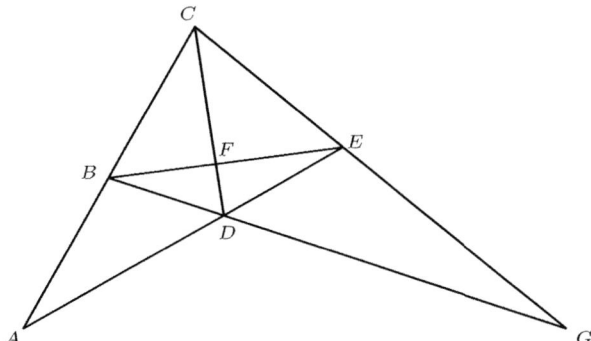

Fig. 7. Alternative Proof of Theorem 4

In both proofs, we first identify a point D off the line AC and extend the line BD to G. This tells us that D lies between B and G, which gives us our first opportunities to use Pasch's Axiom. In both cases, our goal is to use this axiom in order to place the point D between A and E, so that a final application of Pasch's Axiom to $\triangle ACE$ and the line BG will place B between A and C. The two proofs only differ in how they find the point F, and how they use F to place D between A and E.

Wald's original proof has more symmetry than the new proof: E could be replaced with F; and it is clear that the third application of Pasch's Axiom could be made on $\triangle CFG$ and the line AE, instead of $\triangle AEG$ and the line CF. Our proof makes it clear that, while E and F can be found symmetrically and independently, only one of these points is distinguished in the final few steps.

It is worth drawing some attention to the subtlety of the incidence reasoning here. We could have applied Pasch's Axiom differently to find the point F, using $\triangle CDG$ and the line BE. This would tell us that F lies on the line BE between C and D (before, it told us that F lies on the line CD between B and E). Now it might seem that we can use a symmetrical application of Pasch's Axiom on the same line BE and $\triangle ACD$, which would solve the goal putting B between A and C. But at this stage in the proof, we must consider the possibility that BF exits $\triangle ACD$ between A and D. This possibility is not yet eliminable by incidence reasoning alone.

This fact is not apparent in the proof. In his eleven uses of Pasch's Axiom across Theorems 3, 4 and 5, Hilbert only considers the case-split implied by the axiom twice. And yet our formalisations show that it takes up a significant amount of combinatorial reasoning about incidence. It is difficult to justify leaving this complexity implicit, when it has consequences on the shape of the proof which we find difficult to argue as *obvious*. The reasoning may be laborious, but we demand rigour when the proofs supposedly contain no gaps. At the very least, with machine-checked automation, we can be confident in the correctness of the implicit steps.

7.2   Theorem 5

The proof of Theorem 5 is split into three parts and is the most involved proof given in Group II, taking up almost an entire page of the English translation of *Foundations of Geometry*. Effectively, the theorem gives a transitivity property for point ordering. As with Theorem 4, it was originally provided as an axiom of Group II, but by the tenth edition, it had a proof based on one given by Moore [15].

> THEOREM 5. Given any four points on a line, it is always possible to label them A, B, C, D in such a way that the point labeled B lies between A and C and also between A and D, and furthermore, that the point labeled C lies between A ad D and also between B and D.

> PROOF. Let A, B, C, D be four points on a line g. The following will now be shown:
> 1. If B lies on the segment AC and C lies on the segment BD then the points B and C also lie on the segment AD. By Axioms I,3 and II,2 choose a point E that does not lie on g, [and] a point F such that E lies between C and F. By repeated applications of Axioms II,3 and II,4 it follows that the segments AE and BF meet at a point G, and moreover, that the line CF meets the segment GD at a point H. Since H thus lies on the segment GD and since, however, by Axiom II,3, E does not lie on the segment AG, the line EH by Axiom II,4 meets the segment AD, i.e. C lies on the segment AD. In exactly the same way one shows analogously that B also lies on this segment.

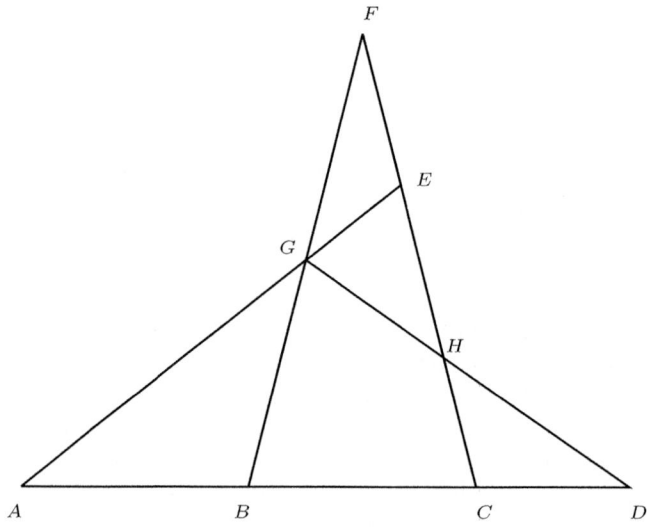

Fig. 8. Proof of Theorem 5

We have only given the first part of Hilbert's proof. We give its formalisation in Figure 9. Again, the formalisations of named theorems are given in Appendix B.3.

```
theorem between A B C ∧ between B C D ⟹ between A C D
assume between A B C ∧ between B C D
obviously consider E such that ¬collinear {A, B, E} by
        exists_triangle
obviously consider F such that between C E F by g22
consider G such that between A G E by pasch_on A,C,E and B,F
have between B G F by pasch_on B,C,F and A,E
consider H such that collinear {C, E, H} ∧ between D H G
        by pasch_on B,D,G and C,F
have between A C D by pasch_on A,D,G and E,H
qed
```

Fig. 9. Theorem 5 in HOL Light

Some Idle-Time Discoveries. We now describe some of the details of the incidence reasoning needed to prove this theorem. We implemented our tool with optional book-keeping, so that as it derived all the discovered facts, it retained information about which rules had been applied when and in what order. The results revealed fairly complicated chains of inference.

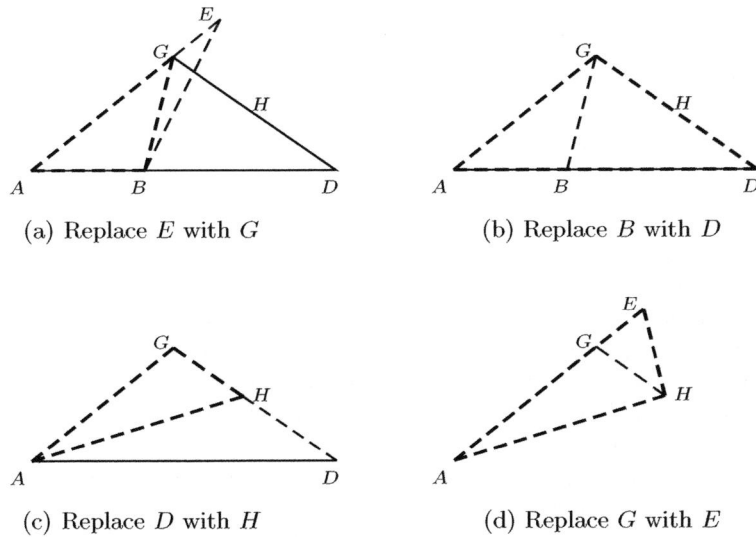

(a) Replace E with G   (b) Replace B with D

(c) Replace D with H   (d) Replace G with E

Fig. 10. Finding △AEH from △ABE

In the example of Figure 10, we must infer that three points A, E and H form a triangle. Each of the steps shown corresponds to an application of rule (5) from Figure 2 and can be understood as *substituting* points of a non-collinear triple one-at-a-time, until we have rewritten the initial triangle $\triangle ABE$ to $\triangle AEH$.

It might seem that we can just replace the point B with the point H to rewrite $\triangle ABE$ to $\triangle AEH$, but the triangle introduction rule requires a hypothesis about an appropriate collinear set and an appropriate point inequality. So the inference is indirect. At first, we can substitute G for E using the line AGE, producing $\triangle ABG$ from $\triangle ABE$. We then substitute D for B using the line ABD, and then H for D using the line DGH. Finally, E and H are shown distinct on the basis of $\triangle AGH$ and the line AGE, after which we can substitute H for G using the line AGE.

The task of manually reproducing these steps in our original Isabelle formalisation had got the better of us, and we had not been able to realise this particular chain of argument, though we had written many like it. At that time, we had always found ourselves experimenting with different starting triangles to get the right conclusions, and manually determining the instantiations for the point variables in our rules. Now with our automated tool, the inferences are automatically discovered, relieving us of the complex combinatorial steps on which they depend and the error-prone task of finding them.

A Comparison with Hilbert's Prose. Our new formalisation gives an almost one-to-one correspondence between our formal proof and Hilbert's prose, but there are still differences which draw our attention.

Firstly, notice that in the prose, Hilbert compresses his applications of Pasch's Axiom (Axiom II,4):

> By repeated applications of Axioms II,3 and II,4 it follows that the segments AE and BF meet at a point G, and moreover, that the line CF meets the segment GD at a point H.

We can see from our formal version that this becomes exactly three applications of Pasch's Axiom. While Hilbert does not give this number explicitly, it is implied by a subtle use of language in the English translation: "the segments AE and BF meet at a point G" while "the *line* CF meets the segment GD at a point H" (our emphasis). Now if we are to show that two *segments* intersect, we must derive *two* facts of betweenness, and therefore we need *two* applications of Pasch's Axiom. But if we are to show that a *line* and a segment intersect, we only need only one fact of betweeness. It seems that Hilbert was aware of this detail, and chose his words carefully, and as with Theorem 4, it indicates a subtlety of the logic which is not immediately apparent from the diagram.

We were able to omit the final step, where Hilbert proves that B lies between A and D. Hilbert implies this can be achieved by an analogous proof, but it is actually a corollary of the above. Hilbert's first axiom in the group tells us that `between A B C` implies `between C B A`. Therefore, we take Theorem 5, swap A and D, and swap B and C, and obtain the required result.

Finally, we briefly mention the second part of the proof, which derives the following:

> 2. If B lies on the segment AC and C lies on the segment AD then C also lies on the segment BD and B also lies on the segment AD.

We omit the details of its formalisation. Suffice to say, we noticed the same neat correspondence between our formal proof steps and Hilbert's prose.

8   Conclusion and Further Work

We have described a way to combine concurrent proof discovery with declarative interactive geometry theorem proving in a way which allows user and machine to cooperate, continually supplying each other with facts that bring the proof-state closer towards the goal.

We have shown how this tool can be used to capture incidence reasoning in Hilbert's *Foundations of Geometry*. It allows us firstly to investigate the proofs by helping us understand the complexity involved in applying the axioms. Secondly, we were able to use the tool to automatically *discover* fully-verified alternative proofs. And finally, by integrating the tool's results into the proof language of HOL Light, we have shown how to create formally verified *readable* versions of Hilbert's proofs whose steps correspond almost one-to-one with the prose. From this, we can suggest that Hilbert was justified in leaving gaps in his proofs, since they serve no pedagogical purpose and do not capture the essential strategy. Instead, they are just trivial combinatorial details to be left to automated verification, and thus rightly left implicit.

As we make progress formalising Hilbert's *Foundations of Geometry*, we expect to see more gaps in his prose, and these will yield case-studies for our tool. As we find new ways to plug these gaps, we will be in a position to gather evidence of the tool's usefulness, and find ways to improve it. Furthermore, we will be in a better position to analyse the logical rigour of Hilbert's proofs, understanding his gaps either as more implicit and fussy combinatioral reasoning that is justifiably omitted, or as genuine oversights. We also plan to use our tool in a non-geometrical domain, trying to determine if our approach has broader use to the theorem-proving community.

Our system is currently not designed for *proof replay*, where complete and correct proof scripts are rerun in batch fashion. Here, we might find that without the abundant idle time available in proof development, the discovery tool is too inefficient to be worthwhile. Future work will improve this situation. In §7.2, when describing the idle-time discoveries, we mentioned how the discovery can run with book-keeping, so that we can recover the direct paths to the discovered facts. Later, we shall investigate ways to automatically yet *cleanly* embed this information in the final proof script.

Finally, we should note that our system currently focuses on sets of ground facts initialised by the user, and most of our rules assume no complicated first-order or higher-order reasoning. We want to consider cases where rules must be applied with universal hypotheses, such as inductive rules, and we want to consider general ways that facts can be *speculated*, allowing us to infer implications, carry out proofs-by-contradiction and perform boolean case-splits. It is not clear whether Hilbert's text will be a useful case-study for these extensions, so we may need to move to a different setting.

Acknowledgements. We would like to thank our reviewers for their helpful and detailed comments.

References

1. Birkhoff, G., Bennett, M.: Hilbert's Grundlagen der Geometrie. Rendiconti del Circolo Matematico di Palermo 36, 343–389 (1987)
2. Boulton, R.: Efficiency in a Fully-Expansive Theorem Prover. Ph.D. thesis. Cambridge University (1993)
3. Boyer, C.B.: A History of Mathematics. John Wiley & Sons (1991)
4. de Bruijn, N.G.: The Mathematical Vernacular, a language for Mathematics with typed sets. In: Dybjer, P., et al. (eds.) Proceedings from the Workshop on Programming Logic, vol. 37 (1987)
5. Euclid: Elements (1998),
   http://aleph0.clarku.edu/~djoyce/java/elements/elements.html
6. Hales, T.: Introduction to the Flyspeck Project,
   http://drops.dagstuhl.de/opus/newlinevolltexte/2006/432/
   pdf/05021.HalesThomas.Paper.432.pdf
7. Harrison, J.: HOL Light: a Tutorial Introduction. In: Srivas, M., Camilleri, A. (eds.) FMCAD 1996. LNCS, vol. 1166, pp. 265–269. Springer, Heidelberg (1996)
8. Heath, T.L.: Euclid: The Thirteen Books of The Elements, vol. 1. Dover Publications (1956)
9. Hilbert, D.: Foundations of Geometry. Open Court Classics, 10th edn. (1971)
10. Magaud, N., Narboux, J., Schreck, P.: Formalizing Desargues' theorem in Coq using ranks. In: Symposium on Applied Computing, pp. 1110–1115 (2009)
11. Meikle, L.I., Fleuriot, J.D.: Formalizing Hilbert's Grundlagen in Isabelle/Isar. In: Basin, D., Wolff, B. (eds.) TPHOLs 2003. LNCS, vol. 2758, pp. 319–334. Springer, Heidelberg (2003)
12. Meikle, L.I., Fleuriot, J.D.: Mechanical Theorem Proving in Computational Geometry. In: Hong, H., Wang, D. (eds.) ADG 2004. LNCS (LNAI), vol. 3763, pp. 1–18. Springer, Heidelberg (2006)
13. Gordon, M., Wadsworth, C.P., Milner, R.: Edinburgh LCF. LNCS, vol. 78. Springer, Heidelberg (1979)
14. Milner, R., Bird, R.S.: The Use of Machines to Assist in Rigorous Proof [and Discussion]. Philosophical Transactions of the Royal Society of London. Series A, Mathematical and Physical Sciences 312(1522), 411–422 (1984)
15. Moore, E.H.: On the projective axioms of geometry. Transactions of the American Mathematical Society 3, 142–158 (1902)
16. Scott, P.: Mechanising Hilbert's Foundations of Geometry in Isabelle. Master's thesis. University of Edinburgh (2008)
17. Scott, P., Fleuriot, J.: Composable Discovery Engines for Interactive Theorem Proving. In: van Eekelen, M., Geuvers, H., Schmaltz, J., Wiedijk, F. (eds.) ITP 2011. LNCS, vol. 6898, pp. 370–375. Springer, Heidelberg (2011)

18. Weyl, H.: David Hilbert and his mathematical work. Bulletin of the American Mathematical Society 50, 635 (1944)
19. Wiedijk, F.: Mizar Light for HOL Light. In: Boulton, R.J., Jackson, P.B. (eds.) TPHOLs 2001. LNCS, vol. 2152, pp. 378–394. Springer, Heidelberg (2001)

A   Summary of Group I

A.1   Axioms

I,1 For every two points A, B there exists a line a that contains each of the points A, B.

I,2 For every two points A, B there exits [sic] no more than one line that contains each of the points A, B.

I,3 There exist at least two points on a line. There exist at least three points that do not lie on a line.

I,4 For any three points A, B, C that do not lie on the same line there exits [sic] a plane α that contains each of the points A, B, C. For every plane there exists a point which it contains.

I,5 For any three points A, B, C that do not lie on one and the same line there exists no more than one plane that contains each of the three points A, B, C.

I,6 If two points A, B of a line a lie in a plane α then every point of a lies in the plane α.

I,7 If two planes α, β have a point A in common then they have at least one more point B in common.

I,8 There exist at least four points which do not lie in a plane.

A.2   Theorems

THEOREM 1. Two lines in a plane either have one point in common or none at all. Two planes have no point in common, or have one line and otherwise no other point in common. A plane and a line that does not lie in it either have one point in common or none at all.

THEOREM 2. Through a line and a point that does not lie on it, as well as through two distinct lines with one point in common, there always exists one and only one plane.

B   Summary of Group II

B.1   Axioms

II,1 If a point B lies between a point A and a point C then the points A, B, C are three distinct points of a line, and B then also lies between C and A.

II,2 For two points A and C, there always exists at least one point B on the line AC such that C lies between A and B.

II,3 Of any three points on a line there exists no more than one that lies between the other two.

II,4 Let A, B, C be three points that do not lie on a line and let a be a line in the plane ABC which does not meet any of the points A, B, C. If the line a passes through a point of the segment AB, it also passes through a point of the segment AC, or through a point of the segment BC.

B.2   Theorems

THEOREM 3. For two points A and C there always exists at least one point D on the line AC that lies between A and C.

THEOREM 4. Of any three points A, B, C on a line there always is one that lies between the other two.

THEOREM 5. Given any four points on a line, it is always possible to label them A, B, C, D in such a way that the point labeled B lies between A and C and also between A and D, and furthermore, that the point labeled C lies between A and D and also between B and D.

B.3   Some Formalisations

g22: ∀A C. ∃B. between A C B

exists_triangle: ∀A B. A ≠ B ⟹ ∃C. ¬COLLINEAR A, B, C

A Coherent Logic Based Geometry Theorem Prover Capable of Producing Formal and Readable Proofs*

Sana Stojanović, Vesna Pavlović, and Predrag Janičić

Faculty of Mathematics, University of Belgrade
Studentski trg 16, 11000 Belgrade, Serbia
{sana,vesnap,janicic}@matf.bg.ac.rs

Abstract. We present a theorem prover ArgoCLP based on coherent logic that can be used for generating both readable and formal (machine verifiable) proofs in various theories, primarily geometry. We applied the prover to various axiomatic systems and proved tens of theorems from standard university textbooks on geometry. The generated proofs can be used in different educational purposes and can contribute to the growing body of formalized mathematics. The system can be used, for instance, in showing that modifications of some axioms do not change the power of an axiom system. The system can also be used as an assistant for proving appropriately chosen subgoals of complex conjectures.

1 Introduction

Geometry has initiated a number of revolutions in mathematics. Also, it has always had a very important role in mathematical education because of paradigmatic reasoning that it requires. For a similar reason, for decades it has been a challenging domain for computer theorem proving, with most attention payed to Euclidean geometry. As early as from 1950's, there were interesting approaches to automated proving of geometry theorems, but real successes came in last decades of twentieth century. For example, theorem provers for Euclidean geometry based on Wu's method automatically proved hundreds of complex theorems [7] and this method is often considered the most efficient method for automated theorem proving overall. Today, there are two main directions in computer theorem proving in geometry:

- Interactive theorem proving using proof assistants such as Isabelle [29] or Coq [38]. The proofs in this context are made mainly manually, but are automatically verified by a computer. Interactive proving is often very demanding and time consuming, it requires an experienced user, and it is typically non-trivial to reuse pieces of existing proofs. Even more, since there is no automation (or it is of a very limited power), if one wants to formulate and prove the same theorem in just a slightly modified theory, that would often require doing the same amount of work all over again.

---

* This work has been partly supported by the grant 174021 of the Ministry of Science of Serbia and by the SNF SCOPES grant IZ73Z0_127979/1.

P. Schreck, J. Narboux, and J. Richter-Gebert (Eds.): ADG 2010, LNAI 6877, pp. 201–220, 2011.

– Automated theorem proving using algebraic methods (such as Wu's method [41] or Gröbner bases method [5,23]) or coordinates-free methods (such as the area method [10] or the full-angle method [11]). In this context, proofs are often generated very efficiently, but they are far from traditional, human-readable proofs.

The above two directions have somewhat different motivations: the former aims at building a corpus of verified mathematical knowledge, while the latter aims at applications in education (e.g., within dynamic geometry software) or in industry (when it is more important to know that a certain conjecture is valid than to have its proof). Nevertheless, there are also goals in the intersection of the above two directions. It would be beneficial (both for the growing body of formalized mathematics and for educational purposes) to have formal, machine verifiable geometry proofs automatically generated, if possible — efficiently and in the traditional geometry manner. In this paper we address these combined goals and describe our, coherent logic based, theorem prover ArgoCLP (*Automated Reasoning GrOup Coherent Logic Prover*) that automatically generates traditional, human readable, but in the same time formal proofs of geometry theorems (for various axiom systems). The generated step-by-step proofs are very similar to the proofs given in standard geometry textbooks. A suitable domain of the prover are foundational properties typically expressed in terms of applications of individual axioms. Hence, we do not aim at conjectures involving, for instance, metrical quantities, typically successfully proved by algebraic provers. Instead, we primarily aim at automatically proving that certain modifications of some axioms do not change the power of an axiom system. In addition, we believe that our theorem prover can serve as a machine assistant that can help mathematicians to prove complex theorems suitably broken apart into several smaller ones.

Organization of the paper. In Section 2 we give brief background information on some geometry axiomatizations, on formal mathematics, and on coherent logic. In Section 3 we present our algorithms for proving theorems in coherent logic, in Section 4 we briefly discuss the implementation of our theorem prover ArgoCLP, and in Section 5 we present applications of our prover to four axiom systems for Euclidean space geometry. In Section 6 we discuss the related work and in Section 7 we draw final conclusions and present some of the ideas for further work.

2   Background

Axiomatizations of Geometry. Euclid, with his book "Elements", is considered to be the first who systematically presented and used an axiomatic method in mathematics [19]. He succeeded to derive, using purely logical rules, many geometry properties that were known long time before him. This system, partly naive from today's point of view, was used for centuries.

In 1899, in his seminal book "Der Grundlagen der Geometrie", Hilbert proposed a new axiom system to elementary geometry that fixed many flaws and weaknesses of Euclid's system [20]. This Hilbert's work is one of the landmarks for XX century mathematics, but it is still not up to contemporary standards. The axiom system uses three sorts of primitive objects: points, lines and planes, while the set of axioms is divided into five groups (incidence axioms, axioms of order, axioms of congruence, axioms of parallels, and continuity axioms). Each group of axioms is accompanied with some fundamental theorems that can be proved using preceding axioms. One of more modern variants of Hilbert's system was given by Borsuk and Szmielev [4].

In mid-twenty century, Tarski presented a new, first-order axiomatisation (actually — several variants) for elementary geometry (with continuity features weaker compared to Hilbert's geometry), along with a decision procedure for that theory [37,34]. Tarski's axiom system is very simple: it is based only on one sort of primitive objects — points, it has only two predicates and eleven axioms.

Formal Mathematics. Over the last years, in all areas of mathematics and computer science, with a history of huge number of flawed published proofs, formal, machine verifiable proofs (given in object-level form — in terms of axioms and inference rules) have been gaining more and more importance. Formal proofs have important role in management of mathematical knowledge (e.g., in digitization of mathematical heritage), in education and e-learning, but also in industrial applications where correctness of some algorithms or calculations is critical. There are growing efforts in developing formal proofs, with many extremely complex theorems proved, with repositories of proved theorems, and also with many software tools for producing and checking formal proofs. Among the most popular theorem proving assistants (systems that implement formal logic and verify proofs) nowadays are Isabelle[1] [29], Coq[2] [38], Mizar[3] [39], and HOL-light[4] [18]. The level of automation in proof assistants is typically very limited.

Readable formal proofs and Isar. Most of the theorem proving assistants use proof scripts that explicitly list all axioms and inference rules used in every single proof step. Despite many results and successes in formalizing fragments of mathematics and computer science, they are still not used by a wide scientific community. The Intelligible semiautomated reasoning (Isar) approach to readable formal proof documents [40] aims to bridge the gap between internal notions of proof given by state-of-the-art interactive theorem proving systems and an appropriate level of abstraction for user-level work. Isar is an alternative to traditional proof tactic scripts, as it provides a proof language interface layer which is much more readable for the users. The Isabelle/Isar system provides an interpreter for the Isar formal proof document language, and readable Isar proof documents are converted and executed as series of low-level Isabelle inference steps. Therefore, Isar allows the user to express proofs in a somewhat human-friendly way, but they are still automatically verifiable by the underlying proof system.

---

[1] http://www.cl.cam.ac.uk/research/hvg/Isabelle/
[2] http://coq.inria.fr/
[3] http://www.mizar.org
[4] http://www.cl.cam.ac.uk/~jrh13/hol-light/

Coherent Logic. Coherent logic (CL) was initially defined by Skolem and in recent years it gained new attention [2,15,3]. CL allows certain existential quantifications so it can be considered as an extension of resolution logic. In contrast to the resolution method, the conjecture being proved is kept unchanged and directly proved (refutation, Skolemization and transformation to clausal form are not used). Proofs in CL are natural and intuitive and reasoning is constructive, so proof objects can be easily obtained [2]. Therefore, CL is a suitable framework for producing both readable and formal proofs. A number of theories and theorems can be formulated directly and simply in CL.

Formally, CL is a fragment of first-order logic (FOL) consisting of formulae of the following form:

$$A_1(\boldsymbol{x}) \wedge \ldots \wedge A_n(\boldsymbol{x}) \Rightarrow \exists \boldsymbol{y}_1\ B_1(\boldsymbol{x}, \boldsymbol{y}_1) \vee \ldots \vee \exists \boldsymbol{y}_m\ B_m(\boldsymbol{x}, \boldsymbol{y}_m)$$

which are implicitly universally quantified and where: $0 \leq n$, $0 \leq m$, $\boldsymbol{x}$ denotes a sequence of variables $x_1, x_2, \ldots, x_k$, A_i (for $1 \leq i \leq n$) denotes an atomic formula (involving some of the variables from $\boldsymbol{x}$), $\boldsymbol{y}_j$ denotes a sequence of variables $y_1^j, y_2^j, \ldots, y_{k_j}^j$, and B_j (for $1 \leq j \leq m$) denotes a conjunction of atomic formulae (involving some of the variables from $\boldsymbol{x}$ and $\boldsymbol{y}_j$). There are no function symbols with arity greater than 0. Function symbols of arity 0 are called *constants*. A *witness* is a new constant, not appearing in axioms used nor in the conjecture being proved. The name *constant* covers both constants that are parts of the signature and witnesses. A *term* is a constant or a variable. An *atomic formula* is either $\perp$ or $p(t_1, \ldots, t_n)$, where p is a predicate symbol of arity n and t_i $(1 \leq i \leq n)$ are terms. An atomic formula over constants is called a *fact*.

The only inference rules (in the style of natural deduction, a variant of the rules given in [3]) used in CL are as follows:

$$\frac{A_1(\boldsymbol{a}) \wedge \ldots \wedge A_n(\boldsymbol{a})}{A_i(\boldsymbol{a})} \wedge E \qquad \frac{A_1 \vee \ldots \vee A_n \quad \overset{[A_1]}{\underset{B}{\vdots}} \quad \ldots \quad \overset{[A_n]}{\underset{B}{\vdots}}}{B} \vee E \qquad \frac{\perp}{A}\ efq$$

$$\frac{A_1(\boldsymbol{a}) \quad \ldots \quad A_n(\boldsymbol{a}) \quad A_1(\boldsymbol{x}) \wedge \ldots \wedge A_n(\boldsymbol{x}) \Rightarrow \exists \boldsymbol{y}_1\ B_1(\boldsymbol{x}, \boldsymbol{y}_1) \vee \ldots \vee \exists \boldsymbol{y}_m\ B_m(\boldsymbol{x}, \boldsymbol{y}_m)}{B_1(\boldsymbol{a}, \boldsymbol{w}_1) \vee \ldots \vee B_m(\boldsymbol{a}, \boldsymbol{w}_m)}\ ax$$

where $\boldsymbol{a}$ is a vector of constants and $\boldsymbol{w}_j$ (for $1 \leq j \leq m$) are vectors of witnesses (i.e., fresh constants). When applied, the rule $\wedge E$ infers $A_i(\boldsymbol{a})$ for each i such that $1 \leq i \leq n$. The rule (ax) is applied only if there are no vectors $\boldsymbol{w}_j$ of constants such that $B_1(\boldsymbol{a}, \boldsymbol{w}_1) \vee \ldots \vee B_m(\boldsymbol{a}, \boldsymbol{w}_m)$ holds.

A formula

$$A_1(\boldsymbol{x}) \wedge \ldots \wedge A_n(\boldsymbol{x}) \Rightarrow \exists \boldsymbol{y}_1\ B_1(\boldsymbol{x}, \boldsymbol{y}_1) \vee \ldots \vee \exists \boldsymbol{y}_m\ B_m(\boldsymbol{x}, \boldsymbol{y}_m)$$

is a *CL-theorem*, if from premises $A_1(\boldsymbol{a}), \ldots, A_n(\boldsymbol{a})$ (where $\boldsymbol{a}$ denotes a sequence of fresh constants) all conjuncts of a formula $B_j(\boldsymbol{a}, \boldsymbol{w})$ can be derived for some j $(1 \leq j \leq m)$ and for some vector of constants $\boldsymbol{w}$.

There is a breadth-first proof procedure for coherent logic that is sound and complete: a coherent formula F can be proved if and only if F is true in all Tarskian models (with non-empty domains) of the set of the axioms and the facts $A_1(\boldsymbol{a}), \ldots, A_n(\boldsymbol{a})$ [2].

It can be proved that any first-order formula can be translated into a CL formula with preserved satisfiability [32]. This translation itself is not always constructive (i.e., it may rely on steps that involve classical logic).

Notice that the definition of CL does not involve negation. A single fact $\neg A$ can be represented in the form $A \Rightarrow \bot$, but this translation is not applicable in a general case. In order to reason about negated facts, for every predicate symbol R, typically a new symbol $\overline{R}$ is introduced that stand for $\neg R$ and the following axiom is used [32]: $R(\boldsymbol{x}) \wedge \overline{R}(\boldsymbol{x}) \Rightarrow \bot$.

3   ArgoCLP Proof Procedures

In this section we describe proof procedures that are used or can be used in our theorem prover ArgoCLP for CL (a description of the implemented procedures and techniques is given in Section 4). It is a generic theorem prover for coherent logic, so it can use any set of coherent axioms (not just geometrical). Sorts can be used (but, alternatively, corresponding unary predicates may be used).

Having in mind applications in geometry,[5] we will use axioms of the form $R(\boldsymbol{x}) \vee \overline{R}(\boldsymbol{x})$ (where R is a predicate symbol) that are special instances of the *tertium non datur* axiom schema. This addition still keeps the reasoning within the intuitionistic setting and do not compromise completeness of the breadth-first proof procedure or completeness of the proof procedures to be presented.

3.1   Basic Proof Procedure

An alternative to the breadth-first proof procedure is a simple proof procedure with forward chaining and with iterative deepening. Axioms are applied according to the inference rule (ax) given in Section 2. Definitions available are used as they were axioms. The axioms are applied in the waterfall manner: when one axiom has been successfully applied, then search for applicable axioms starts again from the first axiom. All constants are enumerated and there is a dedicated counter s that controls applications of axioms — an axiom can be applied only if all of its (universally quantified) variables are matched with constants whose order is less then s. Initially, s equals the number of constants appearing in the premises of the conjecture. The value s is increased once no axiom can be applied and the proof procedure continues. If no axiom can be applied anymore and the conjecture has not been proved, this means that the conjecture is not CL-theorem (however, for non-CL-theorems the proof procedure may not terminate). It can be proved (in a similar manner as it was proved for the breadth-first procedure in [2]) that this proof procedure is sound and complete: a coherent

---

[5] Proofs of many conjectures of Hilbert style geometry require instances of the *tertium non datur* axiom schema [13].

formula F can be proved if and only if F is true in all Tarskian models (with non-empty domains) of the set of the axioms and the facts $A_1(\boldsymbol{a}), \ldots, A_n(\boldsymbol{a})$ [2]. In addition, it can be proved that this proof procedure is sound and complete with respect to the inference system given in Section 2, i.e., a formula F can be proved if and only if F is CL-theorem.

Despite the completeness property, proving some conjectures in some theories is practically impossible with this basic proof procedure (i.e., impossible with reasonable memory and time resources).

3.2 Improved Proof Procedure

Efficiency of the basic proof procedure given above can be improved in a number of ways, while still preserving completeness. Here some possible improvements are listed, all of which aim at keeping control on the search space (i.e., on the number of introduced witnesses) and decreasing to some extent a combinatorial explosion (caused by derived facts that are irrelevant).

Ordering of axioms. The axioms are grouped into the following groups (it is assumed that $n \geq 0$ ($n > 0$ for the third and the fourth group), $m > 1$, and that one group of axioms excludes previous groups that are its special cases):

non-productive non-branching axioms: axioms of the form:
$$A_1(\boldsymbol{x}) \wedge \ldots \wedge A_n(\boldsymbol{x}) \Rightarrow B(\boldsymbol{x})$$
non-productive branching axioms: axioms of the form:
$$A_1(\boldsymbol{x}) \wedge \ldots \wedge A_n(\boldsymbol{x}) \Rightarrow B_1(\boldsymbol{x}) \vee \ldots \vee B_m(\boldsymbol{x})$$
productive non-branching axioms: axioms of the form:
$$A_1(\boldsymbol{x}) \wedge \ldots \wedge A_n(\boldsymbol{x}) \Rightarrow \exists \boldsymbol{y}\, B(\boldsymbol{x}, \boldsymbol{y})$$
productive branching axioms: axioms of the form:
$$A_1(\boldsymbol{x}) \wedge \ldots \wedge A_n(\boldsymbol{x}) \Rightarrow \exists \boldsymbol{y}_1\, B_1(\boldsymbol{x}, \boldsymbol{y}_1) \vee \ldots \vee \exists \boldsymbol{y}_m\, B_m(\boldsymbol{x}, \boldsymbol{y}_m)$$
strongly productive non-branching axioms: axioms of the form:
$$\exists \boldsymbol{y}\, B(\boldsymbol{y})$$
strongly productive branching axioms: axioms of the form:
$$\exists \boldsymbol{y}_1\, B_1(\boldsymbol{y}_1) \vee \ldots \vee \exists \boldsymbol{y}_m\, B_m(\boldsymbol{y}_m)$$
Axioms can be automatically assigned their types and are used in the proving process with priorities given to the groups as in the above ordering. There is no imposed ordering of axioms within a group (although their ordering within groups can also impact efficiency).

Early pruning. When testing an axiom for applicability, it is not necessary to instantiate all its variables and only then check if all relevant facts were already derived. Instead, a check for relevant facts can be performed as soon as possible, in order to enable early rejection of some axiom instances and pruning of the search space. For instance, when applying the following axiom:

$$\forall x : line\ \forall y : line\ \forall X : point\ \forall Y : point$$
$$(incident(X, x) \wedge incident(Y, x) \wedge incident(X, y) \wedge incident(Y, y) \wedge X \neq Y$$
$$\Rightarrow x = y)$$

instead of matching x, y, X, and Y with all admissible constants, x and X will be first unified with admissible constants, and the matching will backtrack immediately if the fact instantiated from $incident(X, x)$ has not been already derived. Generally, relevant facts are checked as soon as all involved arguments have been instantiated.

Breaking axioms that introduce several witnesses. As said in Section 2, an axiom like:

$$\forall x : line\ \exists X : point\ \exists Y : point\ (incident(X, x) \land incident(Y, x) \land X \neq Y)$$

will not be applied for a specific line a (instantiating x) if there are already constants $A : point$ and $B : point$ such that $incident(A, a)$, $incident(B, a)$, and $A \neq B$. However, for efficiency reasons, it is beneficial not to apply the above axiom even if there is a constant $A : point$ such that $incident(A, a)$ holds, and there is no constant $B : point$ such that $incident(B, a)$ and $A \neq B$ (because it would introduce two new points C and D). Instead, the following variant of the above axiom should be used:

$$\forall x : line\ \forall X : point\ (incident(X, x) \Rightarrow \exists Y : point\ (incident(Y, x) \land X \neq Y))$$

Therefore, instead of one axiom, two axioms will be used, with the general one having lower priority. The same mechanism can be applied for all axioms that involve more than one existential quantifier. Breaking such axioms into several versions is not always straightforward as in the above example. For example, the axiom:

$$\exists X : point\ \exists Y : point\ \exists Z : point\ \exists U : point\ noncoplanar(X, Y, Z, U)$$

should be broken into four axioms, with one of them:

$$\forall X : point\ \forall Y : point\ \forall Z : point\ \exists U : point\ noncoplanar(X, Y, Z, U)$$

However, this conjecture is invalid (in Euclidean geometry) and an additional premise ($noncolinear(X, Y, Z)$) is required. Because of this, if an axiom can be broken into several variants, each of them should be proved (again by the CL prover) before being used. If some variant cannot be proved (i.e., if it cannot be proved within some time limit), the user may be asked to modify it. Notice that additional axioms introduced in this way actually change the original axiom system, but since the new axiom system is equivalent to the original one (each of its axioms can be proved as a theorem by the other one and vice versa), this modification is legitimate (the new axioms can be considered only as lemmas).

Dealing with equality. For theories involving equality, the axioms of equality are not used explicitly. Instead, equivalence classes of equality of constants are maintained. Thanks to this, it suffices to work only with a canonical representative of a class instead of all objects that belong to that class. In the beginning of the proving process, every object represents its own class and the classes are maintained using Tarjan's *union-find* structures [36].

For example, if there are constants A : *point*, p : *line*, q : *line* and α : *plane* such that $incident(A, p)$, $incident(q, \alpha)$, and $p = q$ hold, the following axiom can be applied:

$$\forall X : point \quad \forall x : line \quad \forall \chi : plane \quad (incident(X, x) \wedge incident(x, \chi) \Rightarrow incident(X, \chi))$$

and, for $X = A$, $x = p$, $\chi = \alpha$, the fact $incident(A, \alpha)$ can be derived. Namely, for this instantiation of variables, when checking if the fact $incident(p, \alpha)$ hold, the representatives of p and α are first determined — say, q and α — and since $incident(q, \alpha)$ holds, the axiom can be applied. Although the axioms of equality are not used explicitly during the search process, they are used in building a proof trace from which a full (machine verifiable) proof object can be constructed.

Dealing with symmetrical predicate symbols. A predicate R is symmetrical (in argument positions i and j) if it holds (universal quantification is assumed):

$$R(x_1, \ldots, x_i, \ldots, x_j, \ldots, x_n) \Leftrightarrow R(x_1, \ldots, x_j, \ldots, x_i, \ldots, x_n)$$

For symmetrical predicates, only *representatives* of facts can be considered. For instance, instead of storing both $colinear(A, B, C)$ and $colinear(C, B, A)$, it suffices to store only $colinear(A, B, C)$. A representative of a class of facts can be determined in the following way: using the ordering of constants, sort arguments in symmetrical positions, and choose the minimum as the representative. This step is performed whenever a fact over a symmetrical predicate should be checked. This mechanism can be used in conjunction with the mechanism of equivalence classes w.r.t. equality to further reduce the number of facts stored. Like the equality axioms, statements ensuring that a predicate is symmetrical are not used during the search process, but they are used in building the proof trace from which a full (machine verifiable) proof object can be constructed.

For example, if there are constants A : *point*, B : *point*, C : *point* and D : *point* with the ordering $A < B < C < D$, and the facts $noncolinear(C, B, D)$ and $colinear(A, D, C)$ derived, if the fact $A = B$ is derived, the equivalence classes of these two objects will be merged and a contradiction can be detected. Namely, if A is the representative of a class containing A and B, by using the equivalence classes, the representative of $noncolinear(C, B, D)$ is $noncolinear(C, A, D)$, and, by symmetry properties, its representative is $noncolinear(A, C, D)$. By symmetry, the representative of $colinear(A, D, C)$ is $colinear(A, C, D)$, so, from $noncolinear(A, C, D)$ and $colinear(A, C, D)$, a contradiction can be derived.

Whether a predicate is symmetrical can be checked automatically (in the preprocessing phase): all relevant conjectures are generated and then tried to be proved. Instead of proving conjectures for all permutations of symmetrical arguments, it is sufficient to prove conjectures for permutation group generators. For instance, when trying to prove that the predicate *coplanar* is symmetrical on all four arguments, it is sufficient to prove conjectures only for permutation group generators (universal quantification is assumed):

$$coplanar(x_1, x_2, x_3, x_4) \Leftrightarrow coplanar(x_2, x_3, x_4, x_1)$$
$$coplanar(x_1, x_2, x_3, x_4) \Leftrightarrow coplanar(x_2, x_1, x_3, x_4)$$

Reuse of proved theorems. Proved conjectures that a predicate is symmetrical (along with their proofs) are used within wider proofs. However, this can be done also for other theorems of the theory proved by the system.

Even with all these techniques, many complex theorems cannot be proved in a reasonable time. Also, generated proofs contain many irrelevant derivations.

3.3   Techniques That Do Not Preserve Completeness

In order to improve efficiency of the prover, at least for some conjectures, some techniques that do not preserve completeness may be used:

Restriction on branching axioms. Branching axioms of the form $R(\boldsymbol{x}) \vee \overline{R}(\boldsymbol{x})$ are generated and used only for primitive (and not for defined) predicates. (Moreover, it can be proved that for some of defined predicates omitting axioms of the given form does not violate completeness.)

Restriction on axioms. In the proof procedure, only axioms that involve just predicates occurring in the conjecture are used. Another, relaxed variant of this restriction is: in the proof procedure, only axioms that involve at least one predicate occurring in the conjecture are used.

4   ArgoCLP Implementation

The prover ArgoCLP is implemented in C++. It consists of around 5000 lines of code, organized within 23 classes. Both the signature and the set of axioms are imported into the program through files, so the prover can be used for different CL theories. A conjecture is specified by:

Theory's signature: names of sorts are stated after the keyword *types*, for example:

```
types point line plane
```

followed by the list of predicate symbols given along with the list of types of each argument. For example, the *incidence* predicate over points and lines would be given as:

```
datatype inc_po_l point line
```

It is assumed that *eq_type* denotes equality over two objects of a type *type*.
Set of axioms: axioms are given in the following form:

```
point(1) point(2) ~eq_point(1,2)
=> line(3) inc_po_l(1,3) inc_po_l(2,3)
```

(variables are represented by natural numbers, universal quantification is assumed for variables appearing on the left hand side of the implication, existential quantification is assumed for variables appearing only on the right hand side of the implication).

Set of definitions: definitions are used for convenience and have the same form as axioms. For instance:

```
point(1) point(2) point(3) line(4)
inc_po_l(1,4) inc_po_l(2,4) inc_po_l(3,4)
=> colinear(1,2,3)
```

Conjecture: it is given in the same form as axioms. For example:
```
point(1) point(2) point(3) line(4)
inc_po_l(1,4) inc_po_l(2,4) bet(1,2,3)
=> inc_po_l(3,4)
```

Most of the techniques listed in Section 3.2 are already implemented within ArgoCLP: grouping and prioritizing axioms, early pruning, support for equality reasoning, support for symmetrical predicates. Lemmas obtained by breaking axioms that introduce several witnesses can be verified within the prover, but their automated generation (with possible assistance of the user) is still under development. Also, symmetrical predicates are used as explained, but automatic detection of symmetrical predicates and automatic generation of required properties are not fully implemented yet.

The user can state (through a configuration file) which of the techniques from Section 3.2 and Section 3.3 should be used in the proof search:

Equality flag indicates whether the built-in equality reasoning will be used.

Excluded middle flags indicate whether axioms of excluded middle are to be used and, more specifically, if only axioms of excluded middle for primitive predicate symbols (and not for defined ones) will be used.

Flags for non completeness-preserving techniques indicate whether only axioms that involve only predicates occurring in the conjecture should be used (this does not apply to equalities if the equality flag is set); whether only axioms that involve at least one predicate occurring in the conjecture should be used (this does not apply to equalities if the equality flag is set); whether the dedicated counter s is being incremented before trying to apply any of strongly productive axioms.

Along the proving process, ArgoCLP generates a proof trace with all relevant information. This proof trace can be exported to different output formats. Currently, ArgoCLP can generate (formally verifiable) proof objects in Isabelle/Isar form (that are accompanied by the axioms also exported from ArgoCLP), and to even more readable, natural language form (in English, in LaTeX format). In addition, there is a mechanism for eliminating all inference steps from a proof trace (including branching steps[6]) that were not relevant, yielding a „clean" (often significantly shorter) proof trace. Such clean proof traces, can be again exported to Isabelle/Isar or natural language form.

---

[6] A branching step is relevant only if both branches use the assumed case, otherwise, the branching can be eliminated and a branch that does not use the assumed case can be kept.

Example 1. Let us consider the following conjecture (of Hilbert-style Euclidean geometry): for three lines p, q, and r and a plane α which contains them all holds that if $p \neq q$ and $q \neq r$ and p and q do not intersect and q and r do not intersect and if there exists a point A which belongs to the plane α and to the lines p and r, then $p = r$.

The conjecture is specified in the following form:

```
premises
# TH_8
% for three lines and a plane which contains them all holds that
% if first and second are distinct and second and third are distinct
% and first and second do not intersect and second and third do not
% intersect and if there exists a point which belongs to the plane
% and to the first and third line, then first and third line are equal

line(1)
line(2)
line(3)
plane(4)
~eq_line(1,2)
~eq_line(2,3)
~int_l_l(1,2)
~int_l_l(2,3)
inc_l_pl(1,4)
inc_l_pl(2,4)
inc_l_pl(3,4)
point(5)
inc_po_pl(5,4)
inc_po_l(5,1)
inc_po_l(5,3)

conclusions

eq_line(1,3)
```

A key fragment of the generated („clean") Isabelle/Isar proof generated by the prover is given below.

```
...
lemma  TH_8:
assumes "LI1 ~= LI2"
and "LI2 ~= LI3"
and "\<not>int_l_l LI1 LI2"
and "\<not>int_l_l LI2 LI3"
and "inc_l_pl LI1 PL1"
and "inc_l_pl LI2 PL1"
and "inc_l_pl LI3 PL1"
and "inc_po_pl PO1 PL1"
and "inc_po_l PO1 LI1"
and "inc_po_l PO1 LI3"
shows "LI1 = LI3"
```

```
proof -

(*1*)
have  "LI1 = LI3 \<or> LI1 ~= LI3"
using ax_g_ex_mid_3 [of "LI1" "LI3"]
by auto
(*2*)  moreover
{ assume "LI1 = LI3"
(*3*)
from this
have ?thesis
by auto
} note note1 = this
(*4*)  moreover
{ assume "LI1 ~= LI3"
(*5*)  moreover
have  "inc_po_l PO1 LI2 \<or> \<not>inc_po_l PO1 LI2"
using ax_g_ex_mid_7 [of "PO1" "LI2"]
by auto
(*6*)  moreover
{ assume "inc_po_l PO1 LI2"
(*7*)  moreover
from 'LI1 ~= LI2' and 'inc_po_l PO1 LI1' and 'inc_po_l PO1 LI2'
have  "int_l_l LI1 LI2"
using ax_D5 [of "LI1" "LI2" "PO1"]
by auto
(*8*)  moreover
from 'int_l_l LI1 LI2' and '\<not>int_l_l LI1 LI2'
have False
by auto
(*9*)
ultimately
have False
by auto
} note note2 = this
(*10*)  moreover
{ assume "\<not>inc_po_l PO1 LI2"
(*11*)  moreover
from '\<not>int_l_l LI1 LI2'
have  "\<not>int_l_l LI2 LI1"
using ax_nint_l_l_21 [of "LI1" "LI2"]
by auto
(*12*)  moreover
from '\<not>inc_po_l PO1 LI2' and 'inc_po_pl PO1 PL1' and 'inc_l_pl LI2 PL1'
and 'inc_po_l PO1 LI1' and 'inc_l_pl LI1 PL1' and '\<not>int_l_l LI2 LI1'
and 'inc_po_l PO1 LI3' and 'inc_l_pl LI3 PL1' and '\<not>int_l_l LI2 LI3'
have  "LI1 = LI3"
using ax_E2 [of "PO1" "LI2" "PL1" "LI1" "LI3"]
by auto
(*13*)  moreover
```

```
from 'LI1 = LI3' and 'LI1 ~= LI3'
have False
by auto
(*14*)
ultimately
have False
by auto
} note note3 = this
(*15*)  from note2 and note3 and 'inc_po_l PO1 LI2 | \<not>inc_po_l PO1 LI2'
have False
by auto
(*16*)
ultimately
have False
by auto
} note note4 = this
(*17*)  from note1 and note4 and 'LI1 = LI3 | LI1 ~= LI3'
have ?thesis
by auto
ultimately
show ?thesis
by auto
qed
```

The („clean") proof generated in the natural language form (using the natural language description of the theory's signature), along with the natural language formulation generated from the conjecture specification, is given below (in order to have proofs that closely resemble proofs from mathematical textbooks, some additional transformations of the proof were automatically made, so it can be given in the *reductio ad absurdum* form).

Theorem TH_8:

Assuming that $p \neq q$, and $q \neq r$, and the line p is incident to the plane α, and the line q is incident to the plane α, and the line r is incident to the plane α, and the lines p and q do not intersect, and the lines q and r do not intersect, and the point A is incident to the plane α, and the point A is incident to the line p, and the point A is incident to the line r, show that $p = r$.

Proof:

Let us prove that $p = r$ by reductio ad absurdum.

1. Assume that $p \neq r$.

2. It holds that the point A is incident to the line q or the point A is not incident to the line q (by axiom of excluded middle).

3. Assume that the point A is incident to the line q.

4. From the facts that $p \neq q$, and the point A is incident to the line p, and the point A is incident to the line q, it holds that the lines p and q intersect (by axiom ax_D5).

5. From the facts that the lines p and q intersect, and the lines p and q do not intersect we get a contradiction.

Contradiction.

6. Assume that the point A is not incident to the line q.

7. From the facts that the lines p and q do not intersect, it holds that the lines q and p do not intersect (by axiom ax_nint_11_21).

8. From the facts that the point A is not incident to the line q, and the point A is incident to the plane α, and the line q is incident to the plane α, and the point A is incident to the line p, and the line p is incident to the plane α, and the lines q and p do not intersect, and the point A is incident to the line r, and the line r is incident to the plane α, and the lines q and r do not intersect, it holds that $p = r$ (by axiom ax_E2).

9. From the facts that $p = r$, and $p \neq r$ we get a contradiction.

Contradiction.

Therefore, it holds that $p = r$.

This proves the conjecture.

Theorem proved in 9 steps and in 0.02 s.

5   Applications

We applied ArgoCLP prover to four axiom systems for Euclidean (space) geometry in a uniform manner. These are Hilbert's system [20], Tarski's system [37,34], system given by Borsuk and Szmielev [4], and our system that is very close to Borsuk's one, but more suitable for CL-based proof procedure. We use the same signature for all the systems (so we could try to prove the same theorems within different systems), which is the union of all the sorts and the predicates used in each of these systems. Of course, if one system does not involve some predicates, it cannot be used for proving their properties (e.g., Tarski's system cannot be used for proving properties of incidence relations, since this system deals only with points). We encoded all axioms from these four systems, except axioms of continuity (for their complexity). Still, a large fragment of geometry can be built without them. We reformulated some axioms in order to avoid complex defined notions such as ray, half-plane, internal angle, etc, but we kept the original meaning of all axioms.

Encoding axioms in its own right is not trivial, because original formulations are often inaccurate, with some conditions only implicitly assumed. For instance, when Hilbert, in his axioms, uses the phrase "two points", he assumes that they are distinct (but does not explicitly state that). Meikle and Fleuriot also underlined this problem [26]. There is a number of problems of this sort and sometimes it is not trivial to show whether a modification would change the set of theorems of the system. Here we do not aim at a thorough comparison between these systems, but rather at illustrating the ArgoCLP prover and to make first steps in showing what fragments of one system can be built within some other system. The prover can also be used to show what modifications of certain axioms can be made.

We applied ArgoCLP on these axiom systems and on a number of theorems from standard geometry courses.[7] As expected, the results depended much on the set of the axioms used. As an illustration, we list 14 theorems (including some that were not proved by the prover within the time limit of 30 seconds) and the obtained results for the four systems (the intended meaning of sorts and predicates should be obvious from their names). All results were obtained with one fixed configuration of the prover (only axioms that involve just predicates occurring in the conjecture are used and only axioms of excluded middle for primitive predicate symbols are used).

Theorem 1. $\forall p : line \; \forall q : line \; (int(p, q) \Rightarrow \exists \alpha : plane \; (inc(p, \alpha) \wedge inc(q, \alpha)))$

Theorem 2. $\forall p : line \; \forall q : line \; \forall A : point \; \forall B : point \; (p \neq q \wedge inc(A, p) \wedge inc(A, q) \wedge inc(B, p) \wedge inc(B, q) \Rightarrow A = B)$

Theorem 3. $\forall p : line \; \forall \alpha : plane \; \forall A : point \; \forall B : point \; (\neg inc(p, \alpha) \wedge inc(A, p) \wedge inc(A, \alpha) \wedge inc(B, p) \wedge inc(B, \alpha) \Rightarrow A = B)$

Theorem 4. $\forall A : point \; \forall B : point \; \forall C : point \; (\neg col(A, B, C) \Rightarrow A \neq B \wedge A \neq C \wedge B \neq C)$

Theorem 5. $\forall A : point \; \forall B : point \; \forall C : point \; (\neg col(A, B, C) \Rightarrow \exists \alpha : plane(inc(A, \alpha) \wedge inc(B, \alpha) \wedge inc(C, \alpha)))$

Theorem 6. $\forall A : point \; \forall p : line \; (\neg inc(A, p) \Rightarrow \exists \alpha : plane \; (inc(A, \alpha) \wedge inc(p, \alpha)))$

Theorem 7. $\forall A : point \; \forall B : point \; \forall C : point \; \forall D : point \; \forall \alpha : plane \; (comp(A, B, C, D) \wedge \neg col(A, B, C) \wedge inc(A, \alpha) \wedge inc(B, \alpha) \wedge inc(C, \alpha) \Rightarrow inc(D, \alpha))$

Theorem 8. $\forall p : line \; \forall q : line \; \forall r : line \; \forall A : point \; \forall \alpha : plane \; (p \neq q \wedge q \neq r \wedge inc(p, \alpha) \wedge inc(q, \alpha) \wedge inc(r, \alpha) \wedge \neg int(p, q) \wedge \neg int(q, r) \wedge inc(A, \alpha) \wedge inc(A, p) \wedge inc(A, r) \Rightarrow p = r)$

Theorem 9. $\forall A : point \; \forall B : point \; \forall C : point \; \forall p : line \; (inc(A, p) \wedge inc(B, p) \wedge bet(A, B, C) \Rightarrow inc(C, p))$

Theorem 10. $\forall A : point \; \forall B : point \; \forall C : point \; (bet(A, B, C) \Rightarrow \neg bet(A, C, B))$

Theorem 11. $\forall A : point \; \forall B : point \; (A \neq B \Rightarrow \exists C : point \; bet(A, C, B))$

Theorem 12. $\forall A : point \; \forall B : point \; cong(A, B, A, B)$

Theorem 13. $\forall A : point \; \forall B : point \; \forall C : point \; \forall D : point \; (cong(A, B, C, D) \Rightarrow cong(C, D, A, B))$

Theorem 14. $\forall A : point \; \forall B : point \; \forall C : point \; \forall D : point \; (cong(A, B, C, D) \Rightarrow cong(B, A, D, C))$

---

[7] The prover ArgoCLP, along with descriptions of the used theories and conjectures, is available on-line from http://argo.matf.bg.ac.rs/downloads.html

Table 1. Performance of the prover; entries are given in the form time/n_1/n_2, where n_1 is the number of all axioms applied, and n_2 is the number of axioms applied in a „clean" proof (with eliminated all unnecessary steps) in the natural language form; '-' denotes timeout, NA denotes that the theorem does not belong to the language of the theory; experiments were ran on PC Core 2Quad 2.4GHz with 4GB RAM, running under Linux

#	ARGO system	Tarski's system	Borsuk's system	Hilbert's system
1	-	NA	-	-
2	0.01/5/3	NA	0.01/5/3	0.01/5/3
3	0.01/5/3	NA	0.01/5/3	0.01/5/3
4	-	NA	-	-
5	0.01/27/1	NA	0.03/28/1	-
6	-	NA	16.07/524/59	-
7	11.08/125/4	NA	8.09/119/4	-
8	0.01/12/9	NA	0.01/12/9	0.01/12/9
9	-	NA	-	-
10	0.01/2/1	-	0.01/2/1	-
11	-	-	0.07/71/8	-
12	0.01/5/2	0.01/6/2	0.01/6/2	-
13	0.25/13/3	0.16/24/3	0.22/24/3	-
14	1.26/26/7	0.52/30/7	0.57/30/7	-

6 Related Work

There is a number of axiom systems for Euclidean geometry. Most of them are variants of Euclid's, Hilbert's or Tarski's system and their comparison often require subtle analyses [31,26,27]. Developing new axiom systems is still an active research area, often motivated by machine formalizations. For instance, Avigad, Dean, and Mumma recently proposed an axiomatization [1] that rather faithfully captures basic ideas and methods of inference outlined in Euclid's "Elements", but in a rigorous manner.

A lot of efforts have been recently invested into formalization of geometry. Dehlinger, Dufourd and Shreck worked on formalization of first two groups of Hilbert's *Grundlagen* in Coq proof assistant following an intuitionistic approach [13]; they came to the conclusion that many theorems could not be proved this way. Meikle and Fleuriot [26] formalized the first three groups of Hilbert's axiomatics in Isabelle/Isar. They showed that some Hilbert's proofs relied on some implicit assumptions (most often based upon a graphical presentation of the problem) and in this way again emphasized the need of having formally verified proofs. Narboux formalized [28] in Coq the first eight chapters of Tarski's book [34] and demonstrated that geometry of Tarski is suitable for mechanization because of its simplicity and production of less degenerated cases. There are also other geometry related formalizations developed in Coq: Kahn's formalization of von Plato's constructive geometry [30,22], Guilhot's formalization of large portions of high school geometry [17], Duprat's formalisation of an axiom system for compass and ruler geometry [14], formalization of projective geometry by

Magaud, Narboux, and Schreck [24,25], etc. All of the mentioned formalizations were done completely manually, with no automation involved.

Automated theorem proving has a history more than fifty years long [8]. In 1959. Gelernter created a geometry theorem prover that could find solutions to a large number of problems taken from highschool textbooks in plane geometry [16]. The biggest successes in automated theorem proving in geometry were achieved (i.e., the most complex theorems were proved) by algebraic theorem provers based on Wu's method [41,7] and Gröbner bases method [5,23,6]. However, instead of readable, traditional geometry proofs, these methods produce only a yes/no answer with a corresponding algebraic argument. This is partly changed with coordinate-free methods, such as the area method [10], the full angle method [11,9], but for many conjectures these methods still deal with extremely complex expressions involving certain geometry quantities. An approach based on deductive database and forward chaining works over a suitably selected set of higher-order lemmas and can prove complex geometry theorems (yielding geometrical proofs), but still has a smaller scope than algebraic provers [12]. Quaife used a resolution theorem prover to prove theorems in Tarski's geometry [33]. Some challenging conjectures were proved, but no formal or readable proofs were produced.

Coherent logic may serve as a framework that enables automated generation of readable geometry proofs. It is well suited to foundational conjectures, close to the level of axioms. To our knowledge, the first automated theorem prover using CL was developed by Janičić and Kordić [21]. It used a fixed set of geometry axioms close to Borsuk's system [4] and was able to prove tens of foundational theorems from standard geometry textbooks. No formal proofs were generated. The system that we describe in this paper is related to this system but significantly extends it and improves it in several directions.

Over the last several years, CL was explored and popularized by Marc Bezem and his coauthors. Bezem and Coquand [2] developed in Prolog a CL prover that generates proof objects in Coq (some of the problems solved by this CL prover can be found on-line[8]). Berghofer and Bezem developed in ML an internal prover for CL in Isabelle. It has several advantages to "external" provers: it uses existing Isabelle's infrastructure and excludes the need for converting from/to "external" formats. Declarative programming languages such as Prolog and ML are well suited to this kind of problems but they can result in a slow executable code, so we believe that C++ implementation can tackle more realistic geometry theorems. As we are aware of, the above provers have not been used for dealing with fragments of geometry addressed in this paper.

7   Conclusions and Further Work

We presented a theorem prover ArgoCLP that uses coherent logic as its underlying logic and forward chaining and iterative deepening in its proof search. The prover can be used for any theory with coherent axioms and for conjectures in the coherent form. It can produce formal, machine verifiable proofs, but also readable

---

[8] http://www.ii.uib.no/~bezem/GL/

proofs given in a natural language form, consisting of steps typical for traditional geometry proofs, so they can be directly used in textbooks. We applied the prover to various axiomatic systems for Euclidean geometry and proved tens of theorems from standard university textbooks on geometry. Since the generated proofs are both formal and readable, they can be used in different educational purposes, and thanks to automation, the system can serve as a useful tool for building the body of formalized mathematics. The system, in its current version, still does not aim at proving all complex geometry theorems appearing in geometry textbooks, but rather at proving foundational theorems (close to the axiom level) of moderate hardness. For instance, a suitable problem for the prover would be checking if an axiom A could be replaced by another version A' (by proving A with A' and the rest of the system and by proving A' with A and the rest of the system). This is a very important issue for foundations of geometry — there are many axiom systems, sometimes with only slight modifications (following, for instance, different interpretations of the author's intention) and establishing their relationship could be very demanding (while cannot be dealt by algebraic theorem provers). Therefore, automation in this process is very much welcome. In addition, the system can be used as an assistant for proving appropriately chosen subgoals of complex conjectures, in a manner that was already applied in proving Hessenberg's theorem by a CL-based theorem prover [3].

We are planning to further develop our prover as there is still much space for improving efficiency. We have implemented a mechanism for cleaning up all irrelevant proof steps from a proof trace, but this cleaning is done only *post festum*, when the conjecture is already proved. We are planning to implement a similar mechanism that would be applied during the proving process itself since information about relevant/irrelevant facts can be useful in more efficient search guiding in the remaining process (i.e., in future branches). We are also planning to use techniques (e.g., backjumping and learning) used in other automated reasoning systems (e.g., SAT solvers) and we expect to obtain significant speed-ups and significant increase in number of theorems that can be proved within reasonable time limits. With a more efficient version of the prover, we are planning to formalize significant portions of different geometries, including the geometry developed by Avigad, Dean, and Mumma [1]. Instead of standard geometry axioms, we will also consider using higher-level lemmas, as in the deductive database method [12]. We are also planning to deeply explore different variants of the most significant axioms systems and their relationship by automatically proving axioms of one systems as theorems within another system. That work would answer a number of important questions about formulations of axioms. The domain of our prover is not limited to geometry, so we will apply it to other theories as well. In addition, we are planning to support input from the TPTP form[9] [35] and we are planning to add support for exporting proof objects to other proof assistants (e.g., Coq). Suport for TPTP would enable us to compare our prover with other coherent logic provers and also with other automated theorem provers, e.g., resolution-based provers.

---

[9] http://www.tptp.org

References

1. Avigad, J., Dean, E., Mumma, J.: A formal system for Euclid's Elements. The Review of Symbolic Logic (2009)
2. Bezem, M., Coquand, T.: Automating coherent logic. In: Sutcliffe, G., Voronkov, A. (eds.) LPAR 2005. LNCS (LNAI), vol. 3835, pp. 246–260. Springer, Heidelberg (2005)
3. Bezem, M., Hendriks, D.: On the Mechanization of the Proof of Hessenberg's Theorem in Coherent Logic. Journal of Automated Reasoning 40(1) (2008)
4. Borsuk, K., Szmielew, W.: Foundations of Geometry. Norht-Holland Publishing Company, Amsterdam (1960)
5. Buchberger, B.: An Algorithm for finding a basis for the residue class ring of a zero-dimensional polynomial ideal. PhD thesis. University of Innsbruck (1965)
6. Buchberger, B., et al.: Theorema: Towards computer-aided mathematical theory exploration. Journal of Applied Logic (2006)
7. Chou, S.-C.: Mechanical Geometry Theorem Proving. D. Reidel Publishing Company, Dordrecht (1988)
8. Chou, S.-C., Gao, X.-S.: Automated reasoning in geometry. In: Handbook of Automated Reasoning. Elsevier (2001)
9. Chou, S.-C., Gao, X.-S., Zhang, J.-Z.: Machine Proofs in Geometry. World Scientific, Singapore (1994)
10. Chou, S.-C., Gao, X.-S., Zhang, J.-Z.: Automated production of traditional proofs for constructive geometry theorems. In: IEEE Symposium on Logic in Computer Science LICS. IEEE Computer Society Press (1993)
11. Chou, S.-C., Gao, X.-S., Zhang, J.-Z.: Automated generation of readable proofs with geometric invariants, II. theorem proving with full-angles. Journal of Automated Reasoning 17 (1996)
12. Chou, S.-C., Gao, X.-S., Zhang, J.-Z.: A Deductive Database Approach to Automated Geometry Theorem Proving and Discovering. Journal Automated Reasoning 25(3) (2000)
13. Dehlinger, C., Dufourd, J.-F., Schreck, P.: Higher-order intuitionistic formalization and proofs in hilbert's elementary geometry. In: Richter-Gebert, J., Wang, D. (eds.) ADG 2000. LNCS (LNAI), vol. 2061, pp. 306–323. Springer, Heidelberg (2001)
14. Duprat, J.: Une axiomatique de la géométrie plane en coq. In: Actes des JFLA 2008, INRIA (2008)
15. Fisher, J., Bezem, M.: Skolem Machines and Geometric Logic. In: Jones, C.B., Liu, Z., Woodcock, J. (eds.) ICTAC 2007. LNCS, vol. 4711, pp. 201–215. Springer, Heidelberg (2007)
16. Gelernter, H., Hanson, J.R., Loveland, D.W.: Empirical explorations of the geometry-theorem proving machine. In: Computers and Thought. MIT Press (1995)
17. Guilhot, F.: Formalisation en coq d'un cours de géométrie pour le lycée. Journées Francophones des Langages Applicatifs (2004)
18. Harrison, J.: Hol light: A Tutorial Introduction. In: Srivas, M., Camilleri, A. (eds.) FMCAD 1996. LNCS, vol. 1166, pp. 265–269. Springer, Heidelberg (1996)
19. Heath, T.L.: The Thirteen Books of Euclid's Elements. Dover Publications, New-York (1956)
20. Hilbert, D.: Grundlagen der Geometrie, Leipzig (1899)
21. Janičić, P., Kordić, S.: EUCLID — the geometry theorem prover. FILOMAT 9(3) (1995)

22. Kahn, G.: Constructive geometry according to Jan von Plato. Coq contribution. Coq V5.10 (1995)
23. Kapur, D.: Using Gröbner bases to reason about geometry problems. Journal of Symbolic Computation 2(4) (1986)
24. Magaud, N., Narboux, J., Schreck, P.: Formalizing Projective Plane Geometry in Coq. In: Sturm, T., Zengler, C. (eds.) ADG 2008. LNCS, vol. 6301, pp. 141–162. Springer, Heidelberg (2011)
25. Magaud, N., Narboux, J., Schreck, P.: Formalizing Desargues' theorem in Coq using ranks. In: ACM Symposium on Applied Computing. ACM (2009)
26. Meikle, L.I., Fleuriot, J.D.: Formalizing Hilbert's Grundlagen in Isabelle/Isar. In: Basin, D., Wolff, B. (eds.) TPHOLs 2003. LNCS, vol. 2758, pp. 319–334. Springer, Heidelberg (2003)
27. Narboux, J.: Formalisation et automatisation du raisonnement géométrique en Coq. PhD thesis. Université Paris Sud (2006)
28. Narboux, J.: Mechanical theorem proving in tarski's geometry. In: Botana, F., Recio, T. (eds.) ADG 2006. LNCS (LNAI), vol. 4869, pp. 139–156. Springer, Heidelberg (2007)
29. Nipkow, T., Paulson, L.C., Wenzel, M.T.: Isabelle/HOL. LNCS, vol. 2283. Springer, Heidelberg (2002)
30. von Plato, J.: The axioms of constructive geometry. Annals of Pure and Applied Logic 76 (1995)
31. von Plato, J.: Formalization of Hilbert's geometry of incidence and parallelism. Synthese 110 (1997)
32. Polonsky, A.: Proofs, Types, and Lambda Calculus. PhD thesis. University of Bergen (2011)
33. Quaife, A.: Automated development of Tarski's geometry. Journal of Automated Reasoning 5(1) (1989)
34. Schwabhuser, W., Szmielew, W., Tarski, A.: Metamathematische Methoden in der Geometrie. Springer, Berlin (1983)
35. Sutcliffe, G.: The TPTP Problem Library and Associated Infrastructure: The FOF and CNF Parts, v3.5.0. Journal of Automated Reasoning 43(4) (2009)
36. Tarjan, R.E.: Efficiency of a good but not linear set union algorithm. Journal of ACM 22(2) (1975)
37. Tarski, A.: What is elementary geometry? In: The Axiomatic Method, with Special Reference to Geometry and Physics. North-Holland (1959)
38. The Coq development team. The Coq proof assistant reference manual, Version 8.2. TypiCal Project (2009)
39. Trybulec, A.: Mizar. In: Wiedijk, F. (ed.) The Seventeen Provers of the World. LNCS (LNAI), vol. 3600, pp. 20–23. Springer, Heidelberg (2006)
40. Wenzel, M.T.: Isar - a Generic Interpretative Approach to Readable Formal Proof Documents. In: Bertot, Y., Dowek, G., Hirschowitz, A., Paulin, C., Théry, L. (eds.) TPHOLs 1999. LNCS, vol. 1690, pp. 167–183. Springer, Heidelberg (1999)
41. Wu, W.-T.: On the decision problem and the mechanization of theorem proving in elementary geometry. Scientia Sinica 21 (1978)

Automated Generation of Readable Proofs for Constructive Geometry Statements with the Mass Point Method

Yu Zou[1] and Jingzhong Zhang[1,2]

[1] College of Computer Science and Educational Software,
Guangzhou University, Guangzhou, 510006, China
zouyu20082008@126.com
[2] Chengdu Institute for Computer Applications, Chinese Academy of Sciences,
Chengdu, 610041, China

Abstract. The existing readable machine proving methods deal with geometry problems using some geometric quantities. In this paper, we focus on the *mass point method* which directly deals with the *geometric points* rather than the geometric quantities. We propose two algorithms, *Mass Point Method* and *Complex Mass Point Method*, which can deal with the Hilbert intersection point statements in affine geometry and the linear constructive geometry statements in metric geometry respectively. The two algorithms are implemented in *Maple* as provers. The results of hundreds of non-trivial geometry statements run by our provers show that the mass point method is efficient and the machine proofs are human-readable.

1 Introduction

In the past over thirty years since Wu's method published in 1977, the research and practice for automated theorem proving in geometry has been considerably developed [1]. For statements in unordered geometry, the algebraic methods (such as the characteristic set method, also known as Wu's method [2,3], the elimination method [4], the Gröbner basis method [5,6], the Clifford algebra approach [7] and the numerical methods [8,9] can effectively judge "True or False", while the area method [10,11] and the search methods [12] can moreover generate readable proofs. Seeking for different automated proving methods will help us to enrich the knowledge of automated reasoning and deepen our understanding of geometry itself.

After the area method, some other methods to produce readable machine proofs have been proposed, such as the vector method [13], the full-angle method [14], the deductive database method [12], the geometric algebra method [15,16,17] and the advanced invariants method proposed recently [18]. There are a few implementations for these methods, and researches in these directions are going on in depth. As known that all these methods deal with geometry problems using the geometric quantities, but, to our knowledge, only the area method

P. Schreck, J. Narboux, and J. Richter-Gebert (Eds.): ADG 2010, LNAI 6877, pp. 221–258, 2011.
© Springer-Verlag Berlin Heidelberg 2011

has been developed as a complete machine proving algorithm, and moreover, implementing the area method is usually a very challenging task.

We focus on the *mass point method* which directly deals with the *geometric points* other than the geometric quantities. The main idea of the mass point method is to express the hypotheses of a theorem using two or three starting points and a set of constructive statements each of them introducing a new point (similar to the area method), and to check whether or not the geometric points (rather than the geometric quantities) appearing in the conclusion satisfy a certain relationship (different from the area method). The proof is generated by keeping finding, and recording the relationship of the newly introduced point and the previously introduced points, and at last finding the relationship of all points involved in the conclusion, along with the process of eliminating points.

The mass point method is also a coordinates-free and diagram-independent method. It can be implemented as provers capable of proving many difficult geometry theorems, including Pappus' Theorem, Pascal's Theorem, Feuerbach's Theorem and Morley's Triangle Theorem. The generated proofs are indeed human-readable and easily understood by a mathematician, and moreover, each expression in a proof has clear and intuitive geometrical meaning, although some of them may involve seemingly huge expressions. Since we can apply arithmetic operations directly to geometric points, the algorithms and implementations for the mass point method are much easier and more concise than that of the area method.

We propose two algorithms, *Mass Point Method* (**MPM**) and *Complex Mass Point Method* (**CMPM**), which can deal with the Hilbert intersection point statements in affine geometry and the linear constructive geometry statements in metric geometry respectively. The results of hundreds of non-trivial statements run by our provers show that the mass point method is not only efficient, but also most proofs are human-readable. The algorithm **CMPM** is effective against most constructive statements in unordered geometry, thus is another complete algorithm which is different from the area method.

The paper is organized as follows: first, in Section 2 and Section 3, we provide a detailed description for the mass point method, propose the algorithm **MPM** and implement it in *Maple* as a prover which can generate the machine proofs entirely automatically for the Hilbert intersection point statements in affine geometry. To deal with constructive statements in metric geometry, in Section 4 and Section 5, we then develop the complex mass point geometry, provide a detailed description for the complex mass point method and propose another algorithm **CMPM** which is also implemented in *Maple* as a prover for proving the linear constructive geometry statements in metric geometry. In Section 6, we summarize some data of 350 running examples and draw conclusions.

2   Mass Point Geometry

Mass point geometry, colloquially known as mass points, is usually taken as a geometry problem-solving technique which applies the physical principle of the center of mass to geometry problems involving triangles and intersecting cevians

[19]. Though modern mass point geometry was developed in the 1960s, the concept has been found to have been used as early as 1827 by August Ferdinand Möbius in his theory of homogenous coordinates [20]. The history and sources for mass point geometry have already been described in [21,22].

2.1 Mass Point Geometry Preliminaries

The existing theory of mass points is usually defined according to the following definitions [21,22,23]:

Mass Point: A mass point is a pair (m, P), also written as mP, including a mass, m, a positive real number, and an ordinary point, P on a plane.

Coincidence: Two points mP and nQ coincide if and only if (iff) $m = n$ and $P = Q$.

Addition: The sum of two mass points mP and nQ has mass $m+n$ and point R where R is the point on PQ such that $PR : RQ = n : m$ (also expressed by $mP + nQ = (m + n)R$). In other words, R is the fulcrum point that perfectly balances the points P and Q. Mass point addition is closed, idempotent, commutative, and associative.

Scalar Multiplication: Given a mass point mP and a positive real scalar k, scalar multiplication is defined to be $k(m, P) = (km, P)$. Mass point scalar multiplication is distributive over mass point addition.

Substraction: If $m > n$ then $mP = nQ + xX$ may be solved for unknown mass point xX. Namely, $xX = (m - n)R$ where P on RQ such that $RP : PQ = n : (m - n)$.

Mass point geometry involves systematically assigning "weights" to points, which can then be used to deduce lengths, using the fact that the lengths must be inversely proportional to their weight (just like a balanced lever). Additionally, the point dividing the line has a mass equal to the sum of the weights on either end of the line (like the fulcrum of a lever). Using mass points can greatly simplify the proofs of many theorems which can also be solved using similar triangles, vectors, or area ratios. There are many materials with plenty of examples solved using mass points and many webpages at the WWW[21,22].

In essence, mass points involves using a local coordinate system to identify points by the ratios into which they divide line segments, thus mass points are closely related with barycentric coordinates or area coordinates.

In the following, we extend the previous mass point method and summarize some useful propositions for mass point geometry.

2.2 Basic Propositions of Mass Point Geometry

Let $\mathbb{R}$ be the field of the real numbers. If there is no special instruction below, we always use capital English letters with or without subscripts to denote points in the plane. We denote by $\overline{AB}$ a vector from A to B, by $\triangle ABC$ the triangle ABC, and by S_{ABC} the signed area of oriented triangle ABC.

The propositions below are based on the following new definition for mass point.

Definition 1. aP is called a mass point in mass point geometry, and $xM - xN$ is called a vector, where $a, x \in \mathbb{R}$, $a \neq 0$, P, M and N are points in Euclidean plane. If $x = 0$ or $M = N$, then $xM - xN$ is called a zero vector.

As shown in the previous definitions for mass point, the mass of a mass point is required to be a positive real number and there is actually no real substraction, i.e., if $mP = nQ + xX$ for unknown mass point xX, then xX is not denoted by $mP - nQ$, because $mP - nQ$ is not always significative, e.g., when $m = n$. If we take substraction as the inverse operation of addition, then the concept *vector* defined in Definition 1 ensures that the set of all mass points and vectors with the addition operator "+" is an additive (Abelian) group, which enables us to carry the addition operation among mass points freely, just like the way we carry the arithmetic polynomial operations.

The following six propositions are the basic propositions required by our mass point method.

Proposition 2.1. Let $m + n \neq 0$, then $(m + n)C = nA + mB$ iff C is a point on line AB and $\overline{AC} : \overline{CB} = m : n$.

Notes. a) In fact, we have $n(A - C) = m(C - B)$ from $\overline{AC} : \overline{CB} = m : n$, which implies that $(m + n)C = nA + mB$; b) Let $\frac{m}{m+n} = x$, then $C = (1 - x)A + xB$, which implies that any point on line AB could be denoted as the linear combination of A, B uniquely depending on the value of x.

Proposition 2.2. There exist x, y, m, n satisfying $x + y = m + n = t(\neq 0)$ such that $xA + yB = tO = mP + nQ$ iff O is the intersection of two lines AB and PQ.

Proposition 2.3. $x\overline{AB} = y\overline{PQ}$ iff $xA - xB = yP - yQ$.

Proposition 2.4. Any point P on plane ABC can be expressed as the linear combination of A, B, C uniquely in the form $P = aA + bB + (1 - a - b)C$.

Notes. In fact, $(a, b, 1-a-b)$ is the area coordinates or barycentric coordinates of P with respect to (wrt) $\triangle ABC$. In the following text, $P = aA + bB + (1-a-b)C$ is called the *standard form* of P wrt A, B, C.

Proposition 2.5. Let P, Q and R be points on plane ABC, and $P = aA + bB + (1 - a - b)C, Q = xA + yB + (1 - x - y)C, R = sA + tB + (1 - s - t)C$, then

$$\frac{S_{PQR}}{S_{ABC}} = \begin{vmatrix} a & b & 1 - a - b \\ x & y & 1 - x - y \\ s & t & 1 - s - t \end{vmatrix} .$$

Proposition 2.6. P, Q and R are collinear iff $S_{PQR} = 0$.

A direct derivation (omitted here) for the above propositions is not difficult. Based on these propositions, we can effectively solve many complicated problems which are usually difficult by using the previous techniques related to mass points.

3   Mass Point Method

The mass point method is also a decision procedure for a fragment of constructive geometry statements stated by sequences of specific geometric constructions. We begin introducing the method by way of examples.

3.1   Introductory Examples

We here use two examples to illustrate how to use the basic propositions given above to prove geometry theorems.

Example 1 (Theorem of Centroid). Let D, E and F be the midpoints of the sides BC, CA and AB of triangle ABC respectively, G the intersection of AD and BE. Show that C, F and G are collinear and $\frac{\overline{AG}}{\overline{GD}} = \frac{\overline{BG}}{\overline{GE}} = \frac{\overline{CG}}{\overline{GF}} = 2$ (Fig. 1).

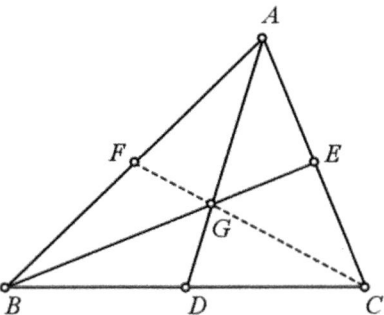

Fig. 1. Theorem of Centroid

The construction steps and proof steps:

1. Take three non-collinear points A, B and C arbitrarily;
2. Take the midpoint D of BC , by Proposition 2.1, $2D = B + C$;
3. Take the midpoint E of CA , by Proposition 2.1, $2E = A + C$;
4. Take the midpoint F of AB , by Proposition 2.1, $2F = A + B$;
5. Take the intersection point G of AD and BE, by Proposition 2.2, step1 and step2, we have $A + 2D = B + 2E = 3G$, and further, $3G = A + B + C$.
6. The conclusions can be written as $A + 2D = B + 2E = 3G = C + 2F$. Here we still need to find the relationship of G, F, C. By step4 and step5, noticing that $3G = A + B + C = 2F + C$, by Proposition 2.1, the conclusions hold.

Example 2 (Desargues' Theorem). Given two triangles ABC and $A_1B_1C_1$, if the three lines AA_1, BB_1, CC_1 meet in a point S, show that the three points $P = BC \cap B_1C_1$, $Q = CA \cap C_1A_1$ and $R = AB \cap A_1B_1$ are collinear (Fig. 2).

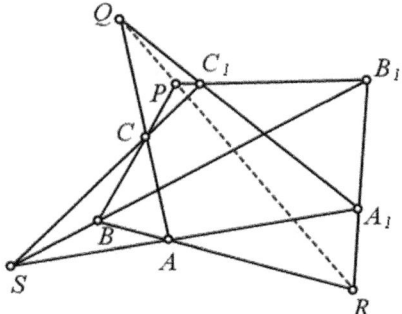

Fig. 2. Desargues' Theorem

The construction steps and proof steps:

1. Take three non-collinear points A, B and C arbitrarily;
2. Take an arbitrary point $S(\neq A, B, C)$ on plane ABC, by Proposition 2.4, S can be expressed as $S = aA + bB + (1 - a - b)C$;
3. Take a point $A_1(\neq S, A)$ on line SA arbitrarily, by Proposition 2.1, A_1 can be expressed as $(1 + x)A_1 = S + xA$;
4. Take a point $B_1(\neq S, B)$ on line SB arbitrarily, by Proposition 2.1, B_1 can be expressed as $(1 + y)B_1 = S + yB$;
5. Take a point $C_1(\neq S, C)$ on line SC arbitrarily, by Proposition 2.1, C_1 can be expressed as $(1 + z)C_1 = S + zC$;
6. Take the intersection point P of BC and B_1C_1, by Proposition 2.2, step4 and step5, $yB - zC = (1 + y)B_1 - (1 + z)C_1 = (y - z)P$;
7. Take the intersection point Q of AC and A_1C_1, by Proposition 2.2, step3 and step5, $xA - zC = (1 + x)A_1 - (1 + z)C_1 = (x - z)Q$;
8. Take the intersection point R of AB and A_1B_1, by Proposition 2.2, step3 and step4, $xA - yB = (1 + x)A_1 - (1 + y)B_1 = (x - y)R$;
9. To prove the conclusion, we need to find the relationship of P, Q, R. By step6, step7 and step8, we have $(y - z)P + (x - y)R = (x - z)Q$, by Proposition 2.1, P, Q, R are collinear.

The two examples above briefly illustrate some key features of the mass point method. Usually, when proving a geometry theorem using the mass point method, we follow three steps:

Firstly, formulate the geometry theorem in a constructive way. Initially, three free points are introduced, and each of the rest construction steps introduces a new point.

Secondly, based on the basic propositions in Sect. 2.2, interpret the constructions into mass point expressions; beginning from the second construction, try to express the newly introduced points as the linear combination of the previously introduced points, keep gaining new mass point relations until the last construction.

Finally, check whether the points involved in the conclusion satisfy certain mass point relations.

3.2   The Hilbert Intersection Point Statements

The mass point method here is restricted to prove a special class of the constructive geometry conjectures defined below, the Hilbert intersection point statements.

Before developing the mass point method into the general machine proving method for the Hilbert intersection point statements, we need to first describe the set of available constructions and then the set of conjectures can be expressed.

The Constructions. Given below is the list of constructions covered by the mass point method, along with the equivalent forms. The ndg-conditions are omitted.

C1 $Point(X)$: Take an arbitrary point X on a plane. Point X is a free point.

Taking the first three different free points A, B, C is usually denoted by $Points(A, B, C)$.

C2 $Aratio(X, A, B, C, a, b)$: Having taken three free points A, B, C, take another point X such that $X = aA + bB + (1 - a - b)$, where a, b are rational numbers, independent variables or rational expressions.

C3 $Lratio(X, A, B, r)$: Take a point X on line AB such that $\overline{AX} = r\overline{AB}$ (or $X = (1 - r)A + rB$), where r is a rational number, a rational expression or a variable.

Taking a point X on line AB such that $\overline{AX} = r\overline{XB}$ (or $(1+r)X = A+rB$) is denoted by $Mratio(X, A, B, r)$, where r is the same as $Lratio(X, A, B, r)$; especially, $Midpoint1(X, A, B)$ means taking the midpoint of AB (or $2X = A + B$), while $Symmetry(X, A, B)$ means taking the symmetric point X of A wrt B (or $X = 2B - A$).

C4 $Pratio(X, U, A, B, r)$: Take a point X on the line passing from U and parallel to line AB such that $\overline{UX} = r\overline{AB}$ (or $X = U + r(B - A)$) where r is the same as **C3**.

A special case of **C4** is $Parallel(X, U, A, B)$, which means taking a point X such that $XUAB$ is a parallelogram, i.e., $\overline{UX} = \overline{AB}$ (or $X = U + B - A$).

C5 $Inter(X, U, V, A, B)$: Take the intersection point X of line UV and line AB.

C6 $Pinter(X, A, B, W, U, V, Y)$: Take the intersection point X of line AB and the line passing from W and parallel to line UV.

This construction is equivalent to $Parallel(Y, W, U, V) + Inter(X, W, Y, A, B)$, where Y is a newly introduced point.

C7 $PPinter(X, C, A, B, W, U, V, Y, Z)$: Take the intersection point X of the line passing from C and parallel to line AB and the line passing from W and parallel to line UV.

This construction is equivalent to $Parallel(Y, C, A, B) + Parallel(Z, W, U, V) + Inter(X, C, Y, W, Z)$, where Y, Z are newly introduced points.

The point X in each of the above constructions is said to be introduced by that construction. **C1** is called the initial construction. Thus the constructions in Example 1 can be represented in terms of the given constructions above as follows: $Points(A, B, C)(\mathbf{C1})$; $Midpoint1(D, B, C)(\mathbf{C3})$; $Midpoint1(E, C, A)(\mathbf{C3})$; $Midpoint1(F, A, B)(\mathbf{C3})$; $Inter(G, A, D, B, E)(\mathbf{C5})$.

The Form of the Conclusions. Before we can describe a geometry statement completely, it is necessary to explain how to express the conclusions.

Using the mass point method, all the constructions will be converted to some mass point relations, and at last, most of the conclusions are also expressed as mass point relations. For all kinds of conclusions, there are usually two kinds of things we need to do: checking or calculating.

For instance, checking whether the two lines are parallel, whether three points are collinear, whether a point is the midpoint of two other points as well as whether two points coincide when using the identity method; moreover, calculating the ratio of two oriented segments on one line or on two parallel lines, the ratio of two signed area of oriented triangles or quadrilaterals, and the rational expressions of the ratios.

In fact, checking is to find the Boolean value and calculating is to find the numerical value. The basic propositions in Sect. 2.2 provide us a method to check the conclusions or to do the corresponding calculations.

Given below is the list of main predicates expressing geometry conclusions covered by the mass point method.

- $ifparallel(X, Y, A, B)$: Check whether XY is parallel to AB. If it holds, then find the ratio t of them and output $X - Y = t(A - B)$, else output the corresponding non-parallel mark.
- $ifmidpoint(C, A, B)$: Check whether C is the midpoint of AB. If it holds, then output $2C = A + B$, else output $2C \neq A + B$.
- $ifcollinear(A, B, C)$: Check whether A, B and C are collinear and output the corresponding mark.
- $ifequal(X, Y)$: Check whether X coincides with Y. If it holds, then output $X = Y$, else output $X \neq Y$.
- $ratio(A, B, X, Y)$: Find the signed ratio of two parallel segments AB, XY.
- $arearatio2(A, B, C, X, Y, Z)$: Find the ratio of two signed area of oriented triangles ABC and XYZ.

The Hilbert Intersection Point Statements. In the mass point method, geometry statements have the following specific form.

Definition 2 (Hilbert Intersection Point Statement). *A Hilbert intersection point statement, is a list $S = (C_1, C_2, \ldots, C_k, G)$ where*

a) *C_i, for $i = 1, \ldots, k$, are constructions of this section such that the point introduced by each C_i must be different from the points introduced by the previous constructions and other points occurring in C_i must be introduced by $C_j (j = 1, \ldots, j - 1)$; and*

b) *$G = (E_1, E_2, \ldots)$, is the conclusion of the statement, where E_i is one of the predicates of conclusions.*

The class of all Hilbert intersection point statements is denoted by **CH**.

For a statement $S = (C_1, C_2, \ldots, C_k, G)$ from **CH**, the ndg-conditions, which are omitted here for the limitation of pages, must be satisfied.

Given the definition of statement of class **CH**, we can now describe Example 2 in this way: $Points(A, B, C)$; $Aratio(S, A, B, C, a, b)$; $Mratio(A_1, S, A, x)$; $Mratio(B_1, S, B, y)$; $Mratio(C_1, S, C, z)$; $Iner(P, B, C, B_1, C_1)$; $Iner(Q, A, C, A_1, C_1)$; $Iner(R, A, B, A_1, B_1)$; $ifcollinear(P, Q, R)$.

3.3 The Algorithm and Its Features

It can be seen from the examples in Sect. 3.1 that the crucial steps of the mass point method are to eliminate some points from the existing mass point relations to get a new mass point relation. Although the whole process of eliminating points can proceed in an orderly manner, each time when we eliminate a point, especially when we deal with an intersection of two lines, it seems that only by resorting to human observation can we know which point(s) should be eliminated and how the point(s) are eliminated. Therefore, the key to the design of algorithm is how to implement all these steps automatically.

In this section, we present the mass point method's algorithm.

Three Parts of the Algorithm. The mass point method can be developed as an automated theorem proving method, whose algorithm is called *Mass Point Method* (**MPM**) in this paper.

The algorithm **MPM** consists of three parts: the construction function set (**CFS**), the aim function set (**AFS**) and the controller (**MMC**).

For each construction in Sect. 3.2, there is a corresponding homonymous construction function in **CFS**.

For each predicate of conclusion in Sect. 3.2, there is also a corresponding homonymous function in **AFS**.

Given a statement $S = (C_1, C_2, \ldots, C_k, G)$ in **CH**, the controller **MMC** reads the constructions one by one; calls the corresponding function in **CFS** repeatedly to deal with the constructions; and finally calls the corresponding function in **AFS** to deal with the conclusion G.

Normalization. Generally, any non-trivial geometry statement would involve at least three non-collinear points. In the mass point method, the three points introduced by the initial construction, usually but not necessarily are A, B and C, are called *basis points*. In the following, we prove that all points introduced by the rest constructions in Sect. 3.2 could be expressed by the linear combination of the basis points in the *standard form*.

Actually, the new point introduced by **C2** is of the standard form; the point introduced by **C3** or **C4** has been denoted as the linear combination of the previously introduced points, and can further be expressed as the standard form, given that all the previously introduced points are of the standard form; **C6** and **C7** can be represented as a sequence of **C4** and **C5**. Thus, it is left to discuss **C5**.

The point X introduced by **C5**: $Inter(X, U, V, P, Q)$ is the intersection of line UV and line PQ. Let U, V, P, Q be of the standard form. Because each standard form is corresponding to a ternary array (usually called the area coordinates), and the four ternary arrays must be linearly dependent, thus there must be four, not all zero, numbers or rational expressions m, n, r, s such that $mU + nV = rP + sQ$. We only consider the non-degenerate case, i.e., $m + n = r + s = t \neq 0$ and $m \cdot n \cdot r \cdot s \neq 0$, by Proposition B.2, $tX = mU + nV = rP + sQ$, then it is easy to get the standard form of X wrt three basis points by representing X as $\frac{mU+nV}{t}$ or $\frac{rP+sQ}{t}$.

So far, we have proven that all the introduced points can be expressed as the standard form wrt three basis points. For each point X, there is a corresponding ternary array (a, b, c) with $a + b + c = 1$ and a, b, c being rational numbers, rational expressions or independent variables. Obviously, three basis points are corresponding to $(1, 0, 0), (0, 1, 0)$ and $(0, 0, 1)$ respectively.

For $mU + nV = rP + sQ$, there are two degenerate cases:

a) If $m + n = r + s = t \neq 0$ and $m \cdot n \cdot r \cdot s = 0$, then X must coincide with one of U, V, P, Q, thus X is not a new point;

b) If $m + n = r + s = t = 0$, then it implies that line UV is parallel to line PQ, thus it is impossible to produce a new point.

In the algorithm, case a) and case b) should be taken into account, i.e., when one of them appears, the algorithm stops and reminds the user of checking the validity of input.

Three Global Variables. If we have expressed each point as the standard form, then we can label each point with its area coordinates. In this way, the operations on the points such as addition, subtraction and scalar multiplication, even the process of eliminating points can be converted into calculations of the points' area coordinates, which are applicable, and much easier for computer.

Now, what we need to consider is, firstly, how to store the points and their area coordinates, and secondly, how to let the computer call the stored points and their area coordinates accurately to carry calculation when needed. Additionally, we also need to let the computer output the stored information in the form of mass point relations.

Given a geometry statement, the number of points involved is finite. We can use a list, named *varlist*, to store all points in the order in which the points are introduced by constructions, and use another list, named *coordlist*, to store the corresponding points' area coordinates.

For example, in Example 1, $varlist = \{A, B, C, D, E, F, G\}$, $coordlist =$

$$\{(1,0,0),(0,1,0),(0,0,1),(0,\tfrac{1}{2},\tfrac{1}{2}),(\tfrac{1}{2},0,\tfrac{1}{2}),(\tfrac{1}{2},\tfrac{1}{2},0),(\tfrac{1}{3},\tfrac{1}{3},\tfrac{1}{3})\}.$$

Notice that the location of a point in *varlist* is consistent with that of its area coordinates in *coordlist*, thus the ith elements of *varlist* and *coordlist* can be denoted by $varlist[i]$ and $coordlist[i]$ respectively. Knowing the location of a point in *varlist*, we can call the point's area coordinates of the corresponding location in *coordlist* to do some desired computation.

Using this technique, for Example 1, when checking whether G, F, C are collinear, we first search the locations of G, F, C respectively. Since G, F, C are the seventh, sixth and third element in *varlist* above, then we call $coordlist[7]$, $coordlist[6]$ and $coordlist[3]$ to check if the determinant is equal to zero.

When storing the points into *varlist* one by one, we need a counter, named *pnts*, to record the number of points in *varlist*. Each time when a construction introduces a new point X: $pnts \rightarrow pnts + 1$, $varlist[pnts + 1] = X$ and $coordlist[pnts+1] = v_X$, where v_X is the area coordinates of X. The counter *pnts* keeps increasing until the last construction. All these steps are done by functions in **CFS**.

In algorithm **MPM**, *pnts*, *varlist* and *coordlist* are three global variables which enable the controller **MMC** to implement all involved steps automatically.

Dealing with the Area Coordinates. In the mass point method, almost each step involves calculation of the area coordinates. Given a geometry statement, after all points are introduced, the functions in **CFS** have expressed all points as the standard forms, while checking whether the conclusions hold or not is the main work of functions in **AFS**.

Let v_P be the area coordinates of P. Given follow are the main rulers, which are embedded into the corresponding functions in **AFS**, to check conclusions by dealing with the area coordinates.

- XY is parallel to AB iff $v_X - v_Y = t(v_A - v_B)$;
- C is the midpoint of AB iff $2v_C = v_A + v_B$;
- A, B, C are collinear iff the determinant of the matrix (v_A, v_B, v_C) formed by v_A, v_B, v_C equals zero or $v_A - v_B = r(v_A - v_C)$;
- X coincides with Y iff $v_X = v_Y$;
- $\frac{\overline{AB}}{\overline{XY}} = \frac{v_A - v_B}{v_X - v_Y}$;
- $\frac{S_{ABC}}{S_{XYZ}} = \frac{|(v_A,v_B,v_C)|}{|(v_X,v_Y,v_Z)|}$ where $|(v_A,v_B,v_C)|$ and $|(v_X,v_Y,v_Z)|$ are determinants.

The Algorithm MPM. Given a statement $S = (C_1, C_2, \ldots, C_k, G)$ in **CH**, the mass point method checks whether it is a theorem or not, i.e., it checks whether the points involved in the conclusion satisfy a certain mass point relation by eliminating unrelated points. The points are introduced one by one, and each point belongs to more than one mass point relations which are together taken as the proof for the statement.

Algorithm: Mass Point Method
Input: $S = (C_1, C_2, \ldots, C_k, G)$ is a statement in **CH**.
Output: The algorithm produces mass point relations and checks whether S is a theorem or not.

1. Initially, *pnts* $= 0$, *varlist* is a null list and *coordlist* is also a null list;
2. After the initial construction $Points(P_1, P_2, P_3)$, *pnts* $= 3$, *varlist* $= \{P_1, P_2, P_3\}$, *coordlist* $= \{(1, 0, 0), (0, 1, 0), (0, 0, 1)\}$;
3. For $i = 2, \ldots, k$, call the corresponding functions in **CFS** repeatedly to process the construction C_i:
    (a) Store the points into *varlist*;
    (b) Search and record the locations of previously introduced points involved in C_i from *varlist* and call their area coordinates from *coordlist*;
    (c) Express the newly introduced point as the linear combination of related points and further as the standard form wrt the basis points;
    (d) Store the newly introduced point's area coordinates into *coordilist*;
    (e) Output all related mass point relations;
    (f) If there is no warning, then continue to run the next step;
4. Call the corresponding functions in **AFS** to process the conclusion G:
    (a) Search and record the locations of all points involved in G from *varlist*;
    (b) Call the involved point's area coordinates of the corresponding locations in *coordlist* and check if the conclusion holds or not by dealing with the area coordinates;
    (c) Output related mass point relations or other related results.

3.4   Implementing the Algorithm MPM in *Maple*

In this section we briefly describe our implementation for the algorithm **MPM**. We implement the algorithm **MPM** in *Maple*. *Maple* is a famous technical computing software which is good at rapid computation and contains rich commands which can be used directly.

All steps involved in the algorithm **MPM** can be easily implemented in *Maple*. Two crucial steps, one is calling the stored points and their area coordinates accurately to carry calculation and another one is dealing with an intersection of two lines, are implemented as follows:

– the command *Search* in package *ListTools* can easily find the location of a point in *varlist*, which makes it easy to call the point's area coordinates of the corresponding location in *corrdlist*;

– the command *LinearSolve* in package *LinearAlgebra* can rapidly find the re-lation of points U, V, A, B appearing in the construction $Inter(X, U, V, A, B)$ according to their area coordinates.

For each function in **CFS** and **AFS**, there are codes corresponding to it. In order to satisfy all kinds of possible cases when checking conclusions, we write as many checking functions as possible. All of the codes are integrated into a big program which is called the **MPM**prover. After loading the program, just input $S = (C_1, C_2, \ldots, C_k, G)$ directly in *Maple* and run, the machine proof can be produced automatically.

To improve the readability of the machine proof, we ask the prover to insert some signs, such as $P = (AB) \cap (UV)$, to tell what is going on in the proof. It is also possible to add other kinds of signs. Each mass point relation is corre-sponding to a certain function.

3.5 Running Examples

Here are two examples whose proofs are generated by the **MPM**prover entirely mechanically.

Example 3 (Pappus' Theorem). Let ABC and $A_1B_1C_1$ be two lines, and $P = A_1B \cap AB_1$, $Q = A_1C \cap AC_1$, $R = B_1C \cap BC_1$. Then P, Q and R are collinear (Fig. 3).

Example 4 (Butterfly Theorem for Quadrilaterals). Let $ABCD$ be a quadrilat-eral such that the intersection M of its diagonals is the midpoint of AC. Passing through M two lines are drawn which meet the sides of $ABCD$ at P, Q, R, S. Let $G = PR \cap AC$, $H = SQ \cap AC$. Show that $GM = MH$ (Fig. 4).

Notes. When reading the input to the **MPM**prover, it is not difficult to distin-guish the functions in **CFS** and **AFS**. The obvious difference is that the initial letters of functions in **CFS** are capital, while the initial letters of functions in **AFS** are lowercase.

Table 1 below lists the time used to prove the examples in Section 3 by our **MPM**prover in *Maple* 12 in an *Intel Pentium(R) Dual-Core Machine* (*CPU@E5200 2.5GHz, 2GB RAM, Windows Vista*).

Table 1. Proving time of examples in Section 3

Examples	Example1	Example2	Example3	Example4
Time used(s)	0.031	0.062	0.078	0.078

It is worth mentioning that the time used by the **MPM**prover to prove a statement depends on two main factors: the number of free points and the num-ber of times of two-line intersecting.

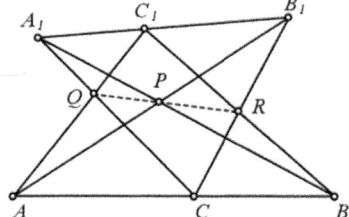

```
Points(A, B, A1);
Aratio(B1, A, B, A1, a, b);
Mratio(C, A, B, x);
Mratio(C1, A1, B1, y);
Inter(P, A1, B, A, B1);
Inter(Q, A1, C, A, C1);
Inter(R, B, C1, B1, C);
ifcollinear(P, Q, R);
```

(a) Diagram for Pappus' Theorem

(b) Input to the **MPM**prover

$$A, B, A1$$

$$B1 = a\ A + b\ B + (1 - a - b)\ A1$$

$$(1 + x)\ C = A + x\ B$$

$$(1 + y)\ C1 = A1 + y\ B1$$

$$P = (A\ B1) \cap (A1\ B)$$

$$A1 - \frac{b\ B}{-1 + a + b} = \frac{a\ A}{-1 + a + b} - \frac{B1}{-1 + a + b}$$

$$P = -\frac{b\ B}{-1 + a} + \frac{(-1 + a + b)\ A1}{-1 + a}$$

$$Q = (A\ C1) \cap (A1\ C)$$

$$A1 - \frac{y\ b\ (1 + x)\ C}{x\ (-1 - y + y\ a + y\ b)} = \frac{(a\ x - b)\ y\ A}{x\ (-1 - y + y\ a + y\ b)} - \frac{(1 + y)\ C1}{-1 - y + y\ a + y\ b}$$

$$Q = -\frac{y\ b\ A}{y\ a\ x - x\ y - x - y\ b} - \frac{y\ b\ x\ B}{y\ a\ x - x\ y - x - y\ b} + \frac{x\ (-1 - y + y\ a + y\ b)\ A1}{y\ a\ x - x\ y - x - y\ b}$$

$$R = (B\ C1) \cap (B1\ C)$$

$$B + \frac{(-1 - y + a + y\ a + b + y\ b)\ C1}{a\ x - b} = \frac{(-1 - y + y\ a + y\ b)\ B1}{a\ x - b} + \frac{(1 + x)\ a\ C}{a\ x - b}$$

$$R = \frac{y\ (-1 + a + b)\ a\ A}{a\ x - 1 - y + a + y\ a + y\ b} + \frac{(-y\ b + b\ y\ a + y\ b^2 + a\ x - b)\ B}{a\ x - 1 - y + a + y\ a + y\ b} - \frac{(-1 + a + b)\ (-1 - y + y\ a + y\ b)\ A1}{a\ x - 1 - y + a + y\ a + y\ b}$$

$$P, Q, R, \text{"are collinear!"}$$

(c) Machine proof for Pappus' Theorem

Fig. 3. Pappus' Theorem

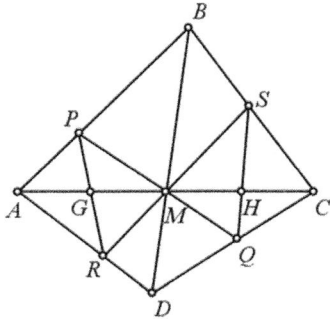

```
Points(A, C, B);
Midpoint1(M, A, C);
Lratio(D, M, B, a);
Lratio(P, A, B, b);
Lratio(R, A, D, c);
Inter(Q, P, M, C, D);
Inter(S, R, M, B, C);
Inter(G, A, M, P, R);
Inter(H, M, C, S, Q);
ifmidpoint(M, G, H);
```

(a) Diagram for Butterfly Theorem

(b) Input to the **MPM-**prover

$$A, C, B$$

$$2\ M = A + C$$

$$D = a\ M + (1 - a)\ B$$

$$P = b\ A + (1 - b)\ B$$

$$R = c\ A + (1 - c)\ D$$

$$Q = (C\ D) \cap (P\ M)$$

$$P + \frac{(-2\ a + b\ a + b)\ M}{a} = (-1 + b)\ C + \frac{b\ D}{a}$$

$$Q = -\frac{1}{2}\ \frac{b\ (-1 + a)\ A}{-a + b\ a + b} + \frac{1}{2}\ \frac{(-2\ a + b\ a + b)\ C}{-a + b\ a + b} + \frac{b\ a\ B}{-a + b\ a + b}$$

$$S = (B\ C) \cap (R\ M)$$

$$R - (2 - c - c\ a)\ M = c\ a\ B + (-1 + c)\ C$$

$$S = \frac{(-1 + c)\ C}{-1 + c + c\ a} + \frac{c\ a\ B}{-1 + c + c\ a}$$

$$G = (A\ M) \cap (P\ R)$$

$$A - \frac{c\ (-1 + a)\ b\ M}{-b\ c + b\ c\ a + b - c\ a} = -\frac{c\ a\ P}{-b\ c + b\ c\ a + b - c\ a} + \frac{b\ R}{-b\ c + b\ c\ a + b - c\ a}$$

$$G = -\frac{1}{2}\ \frac{(-b\ c + b\ c\ a + 2\ b - 2\ c\ a)\ A}{-b + c\ a} + \frac{1}{2}\ \frac{c\ (-1 + a)\ b\ C}{-b + c\ a}$$

$$H = (M\ C) \cap (S\ Q)$$

$$M - \frac{(-b\ c + b\ c\ a + b - c\ a)\ C}{c\ (-1 + a)\ b} = \frac{(-1 + c + c\ a)\ S}{c\ (-1 + a)} - \frac{(-a + b\ a + b)\ Q}{b\ (-1 + a)}$$

$$H = \frac{1}{2}\ \frac{c\ (-1 + a)\ b\ A}{-b + c\ a} - \frac{1}{2}\ \frac{(-b\ c + b\ c\ a + 2\ b - 2\ c\ a)\ C}{-b + c\ a}$$

$$2\ M = G + H$$

(c) Machine proof for Butterfly Theorem

Fig. 4. Butterfly Theorem for Quadrilaterals

3.6   Further Discussion

So far we have presented a new readable machine proving method for geometry theorems, *the mass point method*, and proposed an algorithm, *Mass Point Method*, for dealing with the Hilbert intersection point statements in affine geometry.

As stated above, we can handle constructive statements in affine geometry successfully based on the basic propositions 2.1-2.6 of mass point geometry in Sect.2.2, while it is not enough for us to use only these propositions for dealing with all kinds of constructive statements in metric geometry.

For example, when proving the Orthocenter Theorem, i.e., given a triangle ABC, letting H be the intersection of two heights CD and BE, it is not applicable for the **MPM**prover to prove $AH \perp BC$, since the propositions 2.1-2.6 neither contain the method to express the foot from a point to a certain line, nor involve the method for checking whether two lines are perpendicular.

As known for proving statements in metric geometry, the area method needs to introduce another geometric quantity – Pythagoras difference [10,11]. In Section 4 and Section 5, we will develop the *complex mass point geometry* and extend the mass point method to the *complex mass point method* such that it is suitable for dealing with constructive statements in metric geometry.

4   The Complex Mass Point Geometry

In order to handle more geometry problems by the mass point method, in this section, we develop the complex mass point geometry. We first introduce the basic idea of the complex mass point geometry and then give the related properties.

4.1   The Basic Idea

Let $\mathbb{R}$ be the field of the real numbers. If there is no special instruction below, we always use capital English letters with or without subscripts to denote points in the plane. We denote by $\triangle ABC$ the triangle ABC, and by S_{ABC} the signed area of oriented triangle ABC, by $\overline{AB}$ the vector from A to B, by $|\overline{AB}|$ the distance of A and B and by $\angle(AB, MN)$ the directed angle from line AB to line MN.

The basic idea of the complex mass point geometry is based on the following two facts:

1. As known to us, a complex number can be viewed as a point or position vector in the complex plane, and given two complex numbers $z_1(\neq 0)$ and z_2, there must be a complex number z such that $z = \frac{z_2}{z_1} = r \cdot e^{i\theta}$, i.e., $z_2 = r \cdot e^{i\theta} \cdot z_1$ where $i^2 = -1$, θ is the rotation angle from z_1 to z_2 and $r = \frac{|z_2|}{|z_1|}$ is a nonnegative real number. Noticing the relationship of complex numbers and vectors, for any two vectors $\overline{AB}(A \neq B)$ and $\overline{MN}$ in Euclidean

plane, we can also depict their geometric relationship as $\overline{MN} = \overline{AB} \cdot r \cdot e^{i\theta}$, where $\theta = \angle(AB, MN)$ is the directed angle formed by $\overline{AB}$ and $\overline{MN}$ and $r = \frac{|\overline{MN}|}{|\overline{AB}|}$.

2. As defined by Definition 1 in Section 2, the difference of two mass points with the same mass is regarded as a vector. Thus, $\overline{AB}$ and $\overline{MN}$ can also be denoted by the difference of two points respectively.

Given the two above facts, the relationship of $\overline{AB}$ and $\overline{MN}$ can be expressed as $M - N = r \cdot e^{i\theta}(A - B)$. Let $r \cdot e^{i\theta} = a + bi(a, b \in \mathbb{R})$, then $M - N = (a + bi)(A - B)$.

Let X be an arbitrary point different from A and B in Euclidean plane, then for the two vectors $\overline{AX}$ and $\overline{AB}$, there should be two real numbers u, v such that $A - X = (u + vi)(A - B)$, i.e., $X = (1 - u - vi)A + (u + vi)B$.

For $X = (1 - u - vi)A + (u + vi)B$, there are two cases:

- if $v = 0$, then $X = (1 - u)A + uB$, by Propositions 2.1, X lies on line AB, and $A - X = u(A - B)$ or $\overline{AX} = u\overline{AB}$;
- if $v \neq 0$, then it is obvious that the coefficients of A and B are complex numbers. If we write $X = (1 - u - vi)A + (u + vi)B$ as $X = ((1 - u)A + uB) + vi(B - A)$, letting $D = (1 - u)A + uB$, then D lies on AB, $\overline{AD} = u\overline{AB}$ and $X = D + vi(B - A)$. Further, if writing $X = D + vi(B - A)$ in the form of $X - D = vi(B - A)$, then we have:
  - if $v > 0$, then $X - D = v \cdot e^{\frac{\pi}{2}i} \cdot (B - A)$, thus $XD \perp AB$, i.e., D is the foot from X to line AB, $|\overline{XD}| = v \cdot |\overline{AB}|$ and $\angle(AB, XD) = \frac{\pi}{2}$;
  - if $v < 0$, then $X - D = |v| \cdot e^{-\frac{\pi}{2}i} \cdot (B - A)$, thus $XD \perp AB$, $|\overline{XD}| = |v| \cdot |\overline{AB}|$, but $\angle(AB, XD) = -\frac{\pi}{2}$.

The discussion above shows that the relative location of X with respect to (wrt) A, B depends on the value of u, v in $X = (1 - u - vi)A + (u + vi)B$, i.e., X lies on the line passing through point D and perpendicular to line AB such that $|\overline{XD}| = |v| \cdot |\overline{AB}|$, where D is a point on line AB such that $\overline{AD} : \overline{AB} = u$. Moreover, when $v > 0$, the orientation of $A - B - X$ is counterclockwise, which is consistent with the sign of the signed area S_{ABC}; when $v = 0$, X coincides with D; when $v < 0$, the orientation of $A - B - X$ is clockwise.

In this way, the geometric relationship of any three different points in Euclidean plane could be depicted by a certain complex coefficient mass point relation. The following definition is based on the above discussion.

Definition 3. *Let D be a point on line AB, $\overline{AD} : \overline{AB} = u$, C be a point such that $CD \perp AB$, $|\overline{CD}| = |\overline{AB}|$ and $S_{ABC} > 0$. If X is a point on line CD such that $\overline{XD} = v\overline{CD}(v \in \mathbb{R})$, then $X = (1 - u - vi)A + (u + vi)B$ or $X - A = (u + vi)(B - A)$* (Fig.5).

Obviously, by Definition 3, point C in Fig. 5 can be denoted as $C = (1 - u - i)A + (u + i)B$. Since $X = (1 - u - vi)A + (u + vi)B$, then $C - X = (1 - v)i(B - A)$ which depicts the metric relationship of two vectors $\overline{CX}$ and $\overline{AB}$. Therefore, Definition 3 provides us a method to describe the metric relationship of any three points

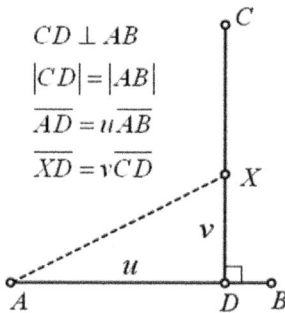

$$CD \perp AB$$
$$|CD| = |AB|$$
$$\overline{AD} = u\overline{AB}$$
$$\overline{XD} = v\overline{CD}$$

Fig. 5. The basic idea for the complex mass point geometry

(not necessarily collinear) on the plane, and moreover, the metric relationship of any two vectors (not necessarily parallel).

In Section 2, the coefficients of a mass point relation in mass point geometry are real numbers, while in the above definition, the coefficients of A and B in $X = (1 - u - vi)A + (u + vi)B$ are complex numbers. Therefore, to make a distinction between mass point geometry and the above idea, we propose a new concept, *complex mass point geometry*.

For the concept of complex mass point geometry, it does not mean we have complex mass point, so far as we are aware of, a single complex mass point does not make any sense. But it is worth mentioning that, the coefficients of A, B, X in $X = (1 - u - vi)A + (u + vi)B$, i.e., $1 - u - vi, u + vi, 1$, can form a triangle being similar to triangle ABX in the complex plane.

4.2   The Properties of the Complex Mass Point Geometry

Beginning from Definition 3, we can deduce many properties of the complex mass point geometry which are necessary for us to deal with statements in metric geometry.

Definition 4. *Let $P - Q = (x + yi)(B - A)$, then $\theta = \arctan(|\frac{y}{x}|)$ is called the included angle formed by $\overrightarrow{PQ}$ and $\overrightarrow{AB}$.*

Proposition 4.1. *Let A, B be two distinct points in Euclidean plane, then for any point P of the same plane, there exists an unique real pairs (a, b) such that $P = (1 - a - bi)A + (a + bi)B$.*

Remark. In the following text, $P = (1 - a - bi)A + (a + bi)B$ is called the *standard form* of P wrt A, B.

Proposition 4.2. *Let $C = (1 - u - vi)A + (u + vi)B$.*
(1) If $v = 0$, then C, A and B are collinear;
(2) If $v > 0$, then the orientation of $A - B - C$ is counterclockwise; If $v < 0$, then the orientation of $A - B - C$ is clockwise;

(3) If $v \neq 0$, when $u = 0$, $\angle CAB = \frac{\pi}{2}$; when $u > 0$, $\angle CAB$ is an acute angle; when $u < 0$, $\angle CAB$ is an obtuse angle.

Proposition 4.3. *Let $C = (1 - u - vi)A + (u + vi)B$, then X is the foot from C to line AB iff $X = (1 - u)A + uB$.*

Proposition 4.4. *Let $P - Q = (x + yi)(M - N)$.*
(1) If $y = 0, x \neq 0$, then $\overline{PQ} \parallel \overline{MN}$;
(2) If $x = 0, y \neq 0$, then $\overline{PQ} \perp \overline{MN}$;
(3) If $|x+yi| = t$, then $|\overline{PQ}| = t \cdot |\overline{MN}|$; especially, when $t = 1$, $|\overline{PQ}| = |\overline{MN}|$.

Proposition 4.5. *Let $P = (1 - m - ni)A + (m + ni)B$, $Q = (1 - a - bi)A + (a + bi)B$. If $PX \perp AB, QY \perp AB$, X, Y are the feet, then*
(1) $\overline{PX} : \overline{QY} = n : b$;
(2) If $n \neq b$, then PQ and AB intersect at a point; let $O = PQ \cap AB$, then $(n - b)O = nQ - bP$.

Proposition 4.6. *Let $C = (1-a-bi)A+(a+bi)B$, $P = (1-x-yi)Q+(x+yi)R$.*
(1) If $\frac{a}{|b|} = \frac{x}{|y|}$, then $\angle CAB = \angle PQR$;
(2) If $\frac{a}{|b|} = -\frac{x}{|y|}$, then $\angle CAB + \angle PQR = \pi$;
(3) If $a = x, b = \pm y \neq 0$, then $\triangle CAB$ is similar to $\triangle PQR$. Especially, when $|\overline{AB}| = |\overline{QR}|$, $\triangle CAB$ is congruent to $\triangle PQR$.

Proposition 4.7. *Let $P = (1 - a - bi)S + (a + bi)R$ and $P = (1 - x - yi)Q + (x + yi)R$. If $a = x = 0$ or $|\frac{b}{a}| = |\frac{y}{x}|$, then P, Q, R and S are concyclic.*

Definition 5. *Let $\triangle CAB$ be similar to $\triangle PQR$. If the orientation of $A - B - C$ is the same with that of $P - Q - R$, then $\triangle CAB$ is said to be **normal similar** to $\triangle PQR$, otherwise, $\triangle CAB$ is said to be **inverse similar** to $\triangle PQR$.*

Proposition 4.8. *Let $\triangle CAB$ be similar to $\triangle PQR$ and $C = (1 - a - bi)A + (a + bi)B$.*
(1) If $\triangle PQR$ is normal similar to $\triangle ABC$, then $P = (1-a-bi)Q+(a+bi)R$;
(2) If $\triangle PQR$ is inverse similar to $\triangle ABC$, then $P = (1-a+bi)Q+(a-bi)R$.

Proposition 4.9. *Let $P = (1 - u - vi)Q + (u + vi)R$, then the signed area of oriented triangle PQR is $S_{PQR} = \frac{1}{2}v \cdot |\overline{QR}|^2$.*

A direct derivation for the above propositions is not difficult. The definitions and propositions above allow expressing all kinds of geometry properties in Euclidean geometry (in form of mass point relations) such as collinearity of three points, parallelism of two lines, perpendicularity of two lines, etc. Based on the propositions in this section and those propositions in Section 2, we can effectively deal with statements in metric geometry.

5   Complex Mass Point Method

Similar to the mass point method, we begin introducing the complex mass point method by way of examples.

5.1   Introductory Examples

We here use two examples to illustrate how to use the propositions given in Section 2 and Section 4 to prove geometry theorems.

Example 5 (Nine-point Circle). In triangle ABC, let D, E and F be the midpoints of sides AB, BC and CA respectively, G be the foot from C to side AB. Show that D, E, F and G are concyclic (Fig.6).

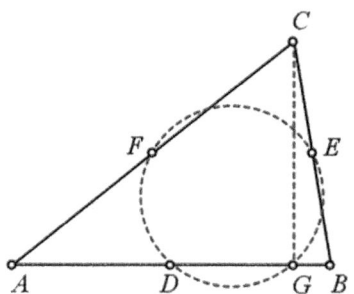

Fig. 6. Nine-point Circle

The construction steps and proof steps:

1. Take two distinct points A, B arbitrarily;
2. Take another point C (C does not belong to line AB) arbitrarily, by Definition 3 in Sect. 4.1, we can let $C = (1 - a - bi)A + (a + bi)B$;
3. Take the midpoint D of AB , by Proposition 2.1, $D = \frac{1}{2}A + \frac{1}{2}B$;
4. Take the midpoint E of BC , by Proposition 2.1, $E = \frac{1}{2}B + \frac{1}{2}C$, and further, $E = \frac{1-a-bi}{2}A + \frac{1+a+bi}{2}B$;
5. Take the midpoint F of AC , by Proposition 2.1, $F = \frac{1}{2}A + \frac{1}{2}C$, and further, $F = (1 - \frac{a+bi}{2})A + \frac{a+bi}{2}B$;
6. Take the foot G from C to side AB, by Proposition 4.3, $G = (1 - a)A + aB$;
7. To prove the conclusion, we need to find the specific relationship of the four points D, E, F and G. Firstly, by Step3, Step4 and Step5, it is not difficult to get

$$E = \left(1 - \frac{1 - a + bi}{(1 - a)^2 + b^2}\right) F + \frac{1 - a + bi}{(1 - a)^2 + b^2} D \ . \tag{1}$$

Secondly, by Step3, Step4 and Step6, it is not difficult to get

$$E = \frac{a + bi}{2a - 1} G - \frac{1 - a + bi}{2a - 1} D \ . \tag{2}$$

By Proposition 4.7, D, E, F and G are concyclic.

Notes. In Step3, Step4 and Step5 above, D, E and F are expressed as the standard forms wrt A, B, which is convenient for eliminating points. For instance, in Step7, to get (1), we actually eliminate points A, B from the three related mass point relations, i.e.,

$$\begin{pmatrix} E = \frac{1-a-bi}{2}A + \frac{1+a+bi}{2}B \\ F = (1 - \frac{a+bi}{2})A + \frac{a+bi}{2}B \\ D = \frac{1}{2}A + \frac{1}{2}B \end{pmatrix} \xrightarrow[A,B]{eliminate}$$

$$E = \left(1 - \frac{1-a+bi}{(1-a)^2 + b^2}\right)F + \frac{1-a+bi}{(1-a)^2 + b^2}D .$$

When checking the conclusion at the last step, by Proposition 4.7, we only pay attention to the coefficients of D in (1) and (2), and actually compare the absolute value of the real part and the imaginary part of the coefficients as well as the ratio of the two absolute values.

Example 6 (Brahmagupta's Theorem). Let A, B, C and D be four points on a circle such that the two diagonals AB and CD are perpendicular. Let E be the intersection of AB and CD, F the midpoint of AC, then $EF \perp BD$ (Fig. 7).

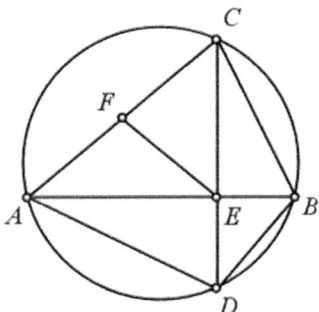

Fig. 7. Brahmagupta's Theorem

The construction steps and proof steps:

1. Take two distinct points A, B arbitrarily;
2. Take another point C (C does not belong to line AB) arbitrarily, by Definition 3 in Sect. 4.1, we can let $C = (1 - a - bi)A + (a + bi)B$;
3. Take the foot E from C to side AB, by Proposition 4.3, $E = (1-a)A + aB$;
4. Take the midpoint F of AC , by Proposition 2.1, $F = \frac{1}{2}A + \frac{1}{2}C$, and further, $F = (1 - \frac{a+bi}{2})A + \frac{a+bi}{2}B$;
5. Take the intersection point $D(\neq C)$ of line CE and the circumcircle of triangle ABC. By Definition 5, $\triangle DEB$ is inverse similar to $\triangle AEC$. By Step2 and Step3, it is easy to get $A = (1 - \frac{a}{b}i)E + \frac{a}{b}iC$. By Proposition 4.8, $D = (1 + \frac{a}{b}i)E - \frac{a}{b}iB$, and further, $D = (1 - a + \frac{a-a^2}{b}i)A + (a - \frac{a-a^2}{b}i)B$.
6. To prove the conclusion, by Step3, Step4 and Step5, we have $F - E = \frac{a-bi}{2}(A - B), D - B = (1 - a + \frac{a-a^2}{b}i)(A - B)$, thus $D - B = \frac{2(1-a)i}{b}(F - E)$. By Proposition 4.4, $EF \perp BD$.

Notes. By Proposition 4.4, when checking whether $EF \perp BD$ or not, we need to find x, y such that $D - B = (x + yi)(F - E)$, which is divided into three steps:

- firstly, find the relationship of $F - E$ and $A - B$ (where A, B are the two initially introduced points);
- secondly, find the relationship of $D - B$ and $A - B$; and
- finally, find the relationship of $F - E$ and $D - B$.

It is easy to find the relationship of $D - B$(or $F - E$) and $A - B$ since all points have been expressed as the standard form wrt A, B after all constructions.

5.2   Some Features

The two examples above show that the complex mass point method is also a process of eliminating points.

The main steps of the complex mass point method are similar to those of the mass point method discussed in Section 3, i.e., firstly, formulate the geometry theorem in a constructive way; secondly, interpret the constructions into mass point expressions; and finally, check whether the conclusion holds or not. But there are some details which need to be explained:

1. The initial construction introduces only two points rather than three ones, these two points are also called *basis points*, and the rest points introduced by other constructions are usually expressed as the linear combination of two basis points in the standard form.
2. When dealing with a two-line intersecting, though the method described in Section 3 is sometimes applicable, we usually follow the method described by Proposition 4.5, for instance, if $O = PQ \cap AB$, four procedures are involved:
    (a) Express P as the linear combination of A and B in the standard form by eliminating points;
    (b) Express Q as the linear combination of A and B in the standard form by eliminating points;
    (c) Compare the imaginary part of the coefficients of B (or A) in the above two mass point relations;
    (d) Express O as the linear combination of P and Q, and further, express O as the linear combination of the basis points in standard form.

There are also other features, for example, we can check whether or not two triangles are similar directly using the complex mass point method.

5.3   The Linear Constructive Geometry Statements

The complex mass point method is also used for proving constructive geometry conjectures: statements about properties of objects constructed by some fixed set of constructions. In this section, we first describe the set of available constructions and then the set of conjectures which can be expressed in the complex mass point method.

The Constructions. The constructive geometry statements are assertions about configurations that can be drawn only by ruler and compass. Using ruler and compass, we can construct an arbitrary point, an arbitrary line, a line such that two given points belong to it, a circle such that its center is one given point and such that the second point belongs to it, the intersection of two lines (if such point exists), the intersection of a given line and a given circle (if such points exist) and the intersection of two given circles (if such points exist). However, when a line intersecting a circle or two circles intersecting, there are possibly two intersection points.

The complex mass point method cannot deal with all geometry statements involving constructions by ruler and compass. Since the complex mass point method always deals with points (only finding the relationship of points), it does not support construction of an arbitrary line, and it supports intersections of a line and a circle or two circles only in a limited way, i.e., when a line intersects a circle or two circles intersect, we always assume that one of the intersection points has been introduced at first, and we want to introduce the second intersection point by the new construction (if such point exists).

In the complex mass point method, only lines and circles determined by specific points can be used. In the following text,

- $Line(P,Q)$ will denote the line passing through two given points P and Q;
- $Pline(W,U,V)$ will denote the line passing through a given point W and parallel to $Line(U,V)$;
- $Tline(W,U,V)$ will denote the line passing through a given point W and perpendicular to $Line(U,V)$
- $Bline(U,V)$ will denote the perpendicular bisector of segment with endpoints U and V; and
- $Circle(O,U)$ will denote the circle with a given point O being its center and passing a given point U.

Given below is the list of basic constructions covered by the complex mass point method, along with some equivalent forms. The ndg-conditions should be satisfied, which are omitted here.

C1 $Point(X)$: Take an arbitrary point X on a plane. Point X is a free point.

Taking the first two different free points A, B is usually denoted by $Points(A, B)$.

C2 $Point1(X, A, B, a, b)$: Having taken two points A, B, take another point X such that $X = (1 - a - bi)A + (a + bi)B$, where a, b are real numbers, rational expressions or variables.

C3 $Aratio(X, A, B, C, a, b)$: Having taken three points A, B, C, take another point X such that $X = aA + bB + (1 - a - b)C$, where a, b are real numbers, rational expressions or variables.

C4 $Lratio(X, A, B, r)$: Take a point X on a line $Line(A, B)$ such that $\overline{AX} = r\overline{AB}$ (or $X = (1 - r)A + rB$), where r is a real number, a rational expression or a variable.

There are three special cases: $Mratio(X, A, B, r)$ $((1 + r)X = A + rB)$, $Midpoint1(X, A, B)$ $(2X = A + B)$ and $Symmetry(X, A, B)$ $(X = 2B - A))$.

C5 $Pratio(X, U, A, B, r)$: Take a point X on a line $Pline(U, A, B)$ such that $\overline{UX} = r\overline{AB}$ (or $X = U + r(B - A)$) where r is the same as **C4**.

A special case of **C5** is $Parallel(X, U, A, B)$ $(X = U + B - A)$.

C6 $Tratio(X, A, B, r)$: Take a point X on a line $Tline(A, A, B)$ such that $\overline{AX} = r\overline{AC}$ where C is introduced by $Point1(C, A, B, 0, 1)$ $(C = (1 - i)A + iB)$.

$Tratio(X, A, B, r)$ is equivalent to $Point1(C, A, B, 0, r)$. An extended case of **C6** is $Bratio(X, A, B, r)$ which means taking a point X on a line $Bline(A, B)$ such that $\overline{YX} = r\overline{YC}$ where C is introduced by $Point1(C, A, B, \frac{1}{2}, 1)$ and Y is introduced by $Midpoint1(Y, A, B)$. Actually, $Bratio(X, A, B, r)$ is equivalent to $Point1(C, A, B, \frac{1}{2}, r)$.

C7 $Inter(X, U, V, A, B)$: Take the intersection point X of a line $Line(U, V)$ and a line $Line(A, B)$.

C8 $Foot1(X, C, A, B)$: Take the foot X from a given point C to a line $Line(A, B)$.

C9 $Cinter(X, O, A, P)$: Take the intersection X (other than A) of a line $Line(A, P)$ and a circle $Circle(O, A)$.

$Cinter(X, O, A, P)$ can be represented as a sequence of $Foot1(Y, O, A, P)$ and $Symmetry(X, A, Y)$.

C10 $CCinter(X, P, O_1, O_2)$: Take the intersection X (other than P) of a circle $Circle(O_1, P)$ and a circle $Circle(O_2, P)$.

$CCinter(X, P, O_1, O_2)$ can be represented as a sequence of $Foot1(Y, P, O_1, O_2)$ and $Symmetry(X, P, Y)$.

C11 $Similartrianglepoint(X, P, Q, A, B, C)$: Given a triangle ABC and two points P, Q, take a point X such that $\triangle XPQ$ is *normal similar* to $\triangle ABC$.

C12 $Similartrianglepoint1(X, P, Q, A, B, C)$: Given a triangle ABC and two points P, Q, take a point X such that $\triangle XPQ$ is *inverse similar* to $\triangle ABC$.

The point X in each of the above basic constructions is said to be introduced by that construction. **C1** is called the initial construction.

To make the information of a geometry statement more concise, much clearer or more readable, we package some of the basic constructions to be one combined construction.

Firstly, since there are four kinds of lines, thus there are still other kinds of two-line intersecting:

- $Pinter(X, A, B, W, U, V)$: $X = Line(A, B) \cap Pline(W, U, V)$;
- $PPinter(X, C, A, B, W, U, V)$: $X = PLine(C, A, B) \cap Pline(W, U, V)$;
- $Tinter(X, A, B, W, U, V)$: $X = Line(A, B) \cap Tline(W, U, V)$;
- $TTinter(X, C, A, B, W, U, V)$: $X = Tline(C, A, B) \cap Tline(W, U, V)$;
- $Binter(X, A, B, U, V)$: $X = Line(A, B) \cap Bline(U, V)$;
- $BBinter(X, A, B, U, V)$: $X = Bline(A, B) \cap Bline(U, V)$;
- $PTinter(X, C, A, B, W, U, V)$: $X = Pline(C, A, B) \cap Tline(W, U, V)$;
- $TBinter(X, W, U, V, A, B)$: $X = Tline(W, U, V) \cap Bline(A, B)$;
- $PBinter(X, W, U, V, A, B)$: $X = Pline(W, U, V) \cap Bline(A, B)$.

All these kinds of two-line intersecting can be represented as a sequence of some constructions of **C2**-**C7**, since $Pline(W, U, V), Tline(W, U, V)$ and $Bline(U, V)$ can be converted into lines each passing through two certain points.

Many other combined constructions are list below:

- $Centroid(X, A, B, C)$: Take the centroid X of $\triangle ABC$;
- $Circumcenter(X, A, B, C)$: Take the circumcenter X of $\triangle ABC$;
- $Orthocenter(X, A, B, C)$: Take the orthocenter X of $\triangle ABC$;
- $Incenter(X, B, C, I)$: Given three points I, B, C, take a point X such that I is the incenter of $\triangle XBC$;
- $Symmetricpoint(X, P, A, B)$: Take the symmetric point X of a given point P wrt $Line(A, B)$;
- $Conjugatepoint(X, A, B, C, P)$: Take the isogonal conjugate point X of a given point P wrt $\triangle ABC$;
- $Pointoncircle(X, A, B, C, P)$: Given $\triangle ABC$ and a point P on $Line(A, B)$, take the intersection point X (other than C) of $Line(C, P)$ and the circumcircle of $\triangle ABC$.
- $ASApoint(X, A, B, \alpha, \beta)$: Given two points A, B and two angles α, β, take a point X such that $\angle XAB = \alpha, \angle XBA = \beta$, where $0 \leq \alpha, \beta, \alpha + \beta < \pi$ and the orientation of $A - B - C$ is counterclockwise.

Given the above constructions, it is possible to construct an arbitrary point on a plane by $Point(X)$, $Point1(X, A, B, a, b)$ or $Aratio(X, A, B, C, a, b)$ where a and b are indeterminate and A, B, C are free points, or on a line $Line(U, V)$ by $Lratio(X, U, V, r)$ where r is indeterminate. For constructing an arbitrary point on a circle, there are three ways:

- take an arbitrary point X on a given circle $Circle(O, A)$ by $Cinter(X, O, A, P)$ where P is an arbitrary point on a plane;
- take an arbitrary point X on a given circle $Circle(O, A)$ by $Point1(X, O, A, \sin\theta, \cos\theta)$ where θ is an arbitrary angle with $0 \leq \theta < 2\pi$;
- take an arbitrary point X on the circumcircle of $\triangle ABC$ by $Pointoncircle(X, A, B, C, P)$ where P is an arbitrary point on $Line(A, B)$.

Now the constructions given in Example 6 in Sect. 5.1 can be represented in terms of the given constructions above as follows: $Points(A, B)$; $Point1(C, A, B, a, b)$; $Foot1(E, C, A, B)$; $Midpoint1(F, A, C)$; $Pointoncircle(D, A, B, C, E)$.

The Form of the Conclusions. Before we can describe a geometry statement completely, it is still necessary to explain how we express the conclusions with the complex mass point method. For all kinds of conclusions of constructive geometry statements, there are still two kinds of things for us: checking or calculating. Checking is to find the Boolean value and calculating is to find the numerical value.

Given below is the list of main predicates expressing geometry conclusions covered by the complex mass point method.

- $relation3(P, Q, R)$: Check the linear relationship of three points P, Q, R, i.e., find a, b such that $P = (1 - a - bi)Q + (a + bi)R$, where a, b are specific values or expressions. We can judge the geometry relationship of P, Q, R according to a, b:
  - If $b = 0$, then $P = (1 - a)Q + aR$, thus P, Q, R are collinear and $\overline{QP} : \overline{PR} = a : (1 - a)$;
  - If $b = 0$ and $a = \frac{1}{2}$, then P is the midpoint of QR;
  - If $a = 0$, then $\angle PQR = \frac{\pi}{2}$; If $a = 1$, then $\angle PRQ = \frac{\pi}{2}$;
  - If $a = b$, then $\angle PQR = \frac{\pi}{4}$; If $a = -b$, then $\angle PQR = \frac{3\pi}{4}$;
  - If $b \neq 0$ and $a = \frac{1}{2}$, then $P = (\frac{1}{2} - bi)Q + (\frac{1}{2} + bi)R$, thus P lies on the perpendicular bisector of QR;
  - If $a = \frac{1}{2}$ and $b = \pm\frac{1}{2}$, then $\triangle PQR$ is an isosceles triangle with $\angle RPQ = \frac{\pi}{2}$ and $|\overline{QP}| = |\overline{PR}|$;
  - If $a = \frac{1}{2}$ and $b = \pm\frac{\sqrt{3}}{2}$, then $\triangle PQR$ is an equilateral triangle.
- $relation4(P, Q, M, N)$: Check the relationship of two vectors $\overline{PQ}, \overline{MN}$, i.e., find x, y such that $P - Q = (x + yi)(M - N)$ where x, y are specific values or expressions, and compute the value of $t = |x + yi|$ or $t^2 = x^2 + y^2$. We can judge the geometry relationship of $\overline{PQ}, \overline{MN}$ according to x, y:
  - If $y = 0, x \neq 0$, then $\overline{PQ} \parallel \overline{MN}$ and $\overline{PQ} = x\overline{MN}$;
  - If $x = 0, y \neq 0$, then $\overline{PQ} \perp \overline{MN}$ and $|\overline{PQ}| = |y| \cdot |\overline{MN}|$;
  - If $x \neq 0, y \neq 0$, then $|\overline{PQ}| = t \cdot |\overline{MN}|$;
  - If $t = 1$, then $|\overline{PQ}| = |\overline{MN}|$.
- $equalangle(P, Q, R, A, B, C)$: Check whether $\angle PQR$ is equal to $\angle ABC$ by the method given in Proposition 4.6. If it holds, then output $\angle PQR = \angle ABC$, else output $\angle PQR \neq \angle ABC$.
- $cocircle(P, Q, R, S)$: Check whether P, Q, R, S are concyclic by the method given in Proposition 4.7, and then output the corresponding mark.
- $similartriangle(P, Q, R, A, B, C)$: Check whether $\triangle PQR$ is similar to $\triangle ABC$ by the method given in Proposition 4.6 and then output the corresponding mark.
- $congruenttriangle(P, Q, R, A, B, C)$: Check whether $\triangle PQR$ is congruent to $\triangle ABC$ by the method given in Proposition 4.6 and then output the corresponding mark.
- $cctangent(O_1, P, O_2, Q)$: Check whether two circles $Circle(O_1, P)$ and $Circle(O_2, Q)$ are tangent, i.e., whether $|\overline{O_1O_2}|^2 = (|\overline{O_1P}| \pm |\overline{O_2Q}|)^2$ or not, and then output the corresponding mark.

- *onradical*(X, O_1, P, O_2, Q): Check whether X lies on the axis of two circles $Circle(O_1, P)$ and $Circle\ (O_2, Q)$, i.e., whether $|\overline{O_1X}|^2 - |\overline{O_2X}|^2 = |\overline{O_1P}|^2 - |\overline{O_2Q}|^2$ or not, and then output the corresponding mark.
- *inversion*(P, Q, O, A): Check whether point P is the inversion of point Q wrt the circle $Circle(O, A)$, i.e., whether $|\overline{OA}|^2 = |\overline{OP}| \cdot |\overline{OQ}|$ or not, and then output the corresponding mark.
- *equalproduct*(P, Q, M, N, U, V, S, T): Check whether $|\overline{PQ}| \cdot |\overline{MN}|$ is equal to $|\overline{UV}| \cdot |\overline{ST}|$.
- *area3*(P, Q, R): Find the relative signed area of $\triangle ABC$ by the method given in Proposition 4.9.
- *arearatio2*(A, B, C, X, Y, Z): Find the ratio of two signed area of oriented $\triangle ABC$ and $\triangle XYZ$.
- *lengthsquare1*(P, Q): Find the relative distance of two points P and Q wrt two basis points, i.e., find t such that $|\overline{PQ}|^2 = t \cdot |\overline{P_1P_2}|^2$ where P_1, P_2 are the two basis points.
- *squaresum3*(P, Q, M, N, U, V): Find the value of t such that $t \cdot |\overline{PQ}|^2 = |\overline{MN}|^2 + |\overline{UV}|^2$.
- *ratiosum3*$(A, B, M, N, P, Q, X, Y, S, T, E, F)$: Find the value of t such that $\frac{\overline{AB}}{\overline{MN}} + \frac{\overline{PQ}}{\overline{XY}} + \frac{\overline{ST}}{\overline{EF}} = t$.
- *ratioproduct2*(A, B, M, N, P, Q, X, Y): Find the value of t such that $\frac{\overline{AB}}{\overline{MN}} \cdot \frac{\overline{PQ}}{\overline{XY}} = t$.

In order to satisfy all kinds of need when checking the conclusions, we can add many other predicates. For instance, we can add more predicates for the sum or product of ratios of segments, such as *squaresum4*, *squaresum5*, *ratiosum4*, *ratiosum5*, *ratioproduct3* and *ratioproduct4* in the same way.

The Linear Constructive Geometry Statements. In the complex mass point method, geometry statements have the following specific form.

Definition 6 (Linear Constructive Geometry Statement). *A Linear Constructive Geometry Statements, is a list $S = (C_1, C_2, \ldots, C_k, G)$ where*

a) *C_i, for $i = 1, \ldots, k$, are constructions given in this section such that the point introduced by each C_i must be different from the points introduced by the previous constructions and other points occurring in C_i must be introduced by $C_j (j = 1, \ldots, i - 1)$; and*
b) *$G = (E_1, E_2, \ldots)$, is the conclusion of the statement, where E_i is one of the predicates of conclusions.*

The class of all linear constructive geometry statements is denoted by **CL**.

For a statement $S = (C_1, C_2, \ldots, C_k, G)$ of class **CL**, the ndg-conditions, which are omitted here for the limitation of pages, must be satisfied.

Given the definition of statement of class **CL**, we can now describe Example 5 in this way: $Points(A, B)$; $Point1(C, A, B, a, b)$; $Midpoint1(D, A, B)$; $Midpoint1(E, B, C)$; $Midpoint1(F, A, C)$; $Foot1(G, C, A, B)$; $cocircle(D, E, F, G)$.

5.4   The Algorithm Complex Mass Point Method

The complex mass point method can also be developed as an automated theorem proving method, whose algorithm is called *Complex Mass Point Method* (**CMPM**) in this paper.

In the complex mass point method, the two points introduced by the initial construction are called *basis points*. It is not difficult to prove that all points introduced by the rest constructions could be expressed as the linear combination of the basis points in the *standard form*. Let the basis points be P_1 and P_2, then the newly introduced point, said X, can be expressed as $X = (1 - x - yi)P_1 + (x + yi)P_2$ by some computations. In this way, each point X will be in a one-to-one correspondence to a binary array $(1 - x - yi, x + yi)$ with x, y are real numbers, expressions or independent variables. Obviously, P_1 and P_2 are corresponding to $(1, 0)$ and $(0, 1)$ respectively.

Since every three binary arrays must be linearly dependent, thus we can always find the linear relationship of three points and can express one of them as linear combination of the other two ones in the *standard form*. By the propositions of the complex mass point geometry, it is convenient for us to check all kinds of conclusions.

Similar to the algorithm **MPM**, the algorithm **CMPM** also consists of three parts: the construction function set (**CCFS**), the aim function set (**CAFS**) and the controller (**CMMC**).

In algorithm **CMPM**, three global variables *pnts, varlist* and *coordlist* are still needed to enable the controller **CMMC** to implement all involved steps automatically, where *varlist* is used to store all points in the order in which the points are introduced by constructions, *coordlist* is used to store all points' corresponding binary arrays, and *pnts* is used to record the number of points in *varlist*.

For each construction in Sect. 5.3, there is a corresponding homonymous construction function in **CCFS**.

For each predicate of conclusion in Sect. 5.3, there is a corresponding homonymous function in **CAFS**.

Given a statement $S = (C_1, C_2, \ldots, C_k, G)$ in **CL**, the controller **CMMC** reads the constructions C_i one by one; calls the corresponding function in **CCFS** repeatedly to deal with the constructions; and finally calls the corresponding function in **CAFS** to deal with the conclusion G.

For a given geometry statement, the functions in **CCFS** keep storing points into *varlist* one by one, expressing the newly introduced points as the standard form wrt the two basis points, and then storing all points' corresponding binary arrays into *coordlist*, where the location of a point in *varlist* is consistent with that of its binary array in *coordlist*. Functions in **CAFS** check whether the conclusions hold or not by dealing with the involved points' corresponding binary arrays.

Algorithm: Complex Mass Point Method
Input: $S = (C_1, C_2, \ldots, C_k, G)$ is a statement in **CL**.
Output: The algorithm produces mass point relations and checks whether S is a theorem or not.

1. Initially, *pnts* = 0 , *varlist* is a null list and *coordlist* is a null list;
2. After the initial construction, $Points(P_1, P_2)$, *pnts* = 2 , *varlist* = $\{P_1, P_2\}$, *coordlist* = $\{(1, 0), (0, 1)\}$;
3. For $i = 2, \ldots, k$, call the corresponding functions in **CCFS** repeatedly to process the construction C_i:
    (a) Store the points into *varlist*;
    (b) Search and record the locations of previously introduced points involved in C_i from *varlist* and call their binary arrays from *coordilist*;
    (c) Express the newly introduced point as the linear combination of related points and further as the standard form wrt the basis points;
    (d) Store the newly introduced point's binary array into *coordilist*;
    (e) Output all related mass point relations;
    (f) If there is no warning, then continue to run the next step;
4. Call the corresponding functions in **CAFS** to process the conclusion G:
    (a) Search and record the locations of all points involved in G from *varlist*;
    (b) Call the involved points' binary arrays of the corresponding locations in *coordlist* and check if the conclusion holds or not by dealing with the binary arrays;
    (c) Output related mass point relations or other related results.

5.5   Implementing the Algorithm CMPM in *Maple*

The algorithm **CMPM** is also implemented in *Maple*. All steps involved in the algorithm **CMPM** can be easily implemented in *Maple*, which we won't go into details.

Compared to the algorithm **MPM**, there are much more functions in **CCFS** and **CAFS** to be dealt with. All the codes written for the functions are integrated into a big program which is called the CMPMprover. After loading the program, just input $S = (C_1, C_2, \ldots, C_k, G)$ directly in *Maple* and run, the machine proof can be produced automatically.

To improve the readability of the machine proof, we ask the prover to insert some signs, such as $P = (AB) \cap (UV)$ or $P = (CP) \perp (AB)$, to tell what is going on in the proof. It is also possible to add other kinds of signs. Actually, much geometry information is implied in the mass point relation, such as $2M = A + B$ implies that M is the midpoint of A and B.

To shorten the whole proof, only the most important and most necessary mass point relations are generated. For instance, when dealing with a two-line intersecting, by the method given in Sect. 5.2, four mass point relations would be generated, but in the machine proof, only two mass point relations are generated: one expresses the newly introduced point as the linear combination of the two points which determine one of the intersecting line, another one expresses the intersection point as the standard form wrt the two basis points. In other cases, the newly introduced point is always expressed as the linear combination of directly related points other than the standard form wrt the two basis points.

The techniques to shorten the proof are not always necessary. If we want to check the proof more carefully, it is also convenient to let the prover generate more detailed information.

5.6   Running Examples

Here are two examples whose proofs are generated by the **CMPM**prover entirely mechanically.

Example 7 (Pascal's Theorem on a Circle). Given six points A, B, C, D, E and F on a circle, let $AB \cap DE = P$, $BC \cap EF = Q$, $CD \cap FA = R$. Show that P, Q and R are collinear (Fig. 8).

Notes. (1) For the input to the **CMPM**prover, the initial letters of functions in **CCFS** are capital, while the initial letters of functions in **CAFS** are lowercase.

(2) The sign "$\sim$" at the top right corner of each variable is generated by the software *Maple* automatically, which means the variable is a real number.

(3) As shown in the machine proof, the imaginary unit "i" is usually denoted by "I" in *Maple* automatically.

Example 8 (Morley's Theorem). The three points of intersections of the adjacent trisectors of angles of any triangle form an equilateral triangle (Fig. 9, Fig. 10 and Fig.11).

Table 2 below lists the time used to prove the examples in Section 5 by the **CMPM**prover in *Maple* 12 in a *Intel Pentium(R) Dual-Core Machine (CPU @ E5200 2.5GHz, 2GB RAM, Windows Vista)*.

Table 2. Proving time of examples in Section 5

Examples	Example5	Example6	Example7	Example8
Time used(s)	0.093	0.046	1.544	1.778

It is worth mentioning that the time cost by the **CMPM**prover to prove a geometry statement depends on two main factors to a great extent:

- the number of free points, especially the number of free points on the circle, since each time when we introduce one more free point, we add one or two more variables, which increase the amount of computation for the prover;
- the number of times of two-line intersecting or line-circle intersecting.

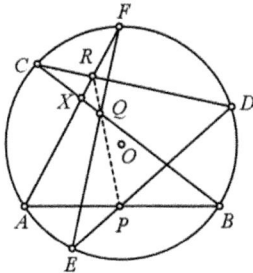

(a) Diagram for Pascal's Theorem on a Circle

```
Points(A, B);
Point1(C, A, B, a, b);
Circumcenter(O, A, B, C);
Mratio(X, B, C, r);
Cinter(F, O, A, X);
Mratio(Q, B, C, x);
Cinter(E, O, F, Q);
Mratio(P, A, B, y);
Cinter(D, O, E, P);
Inter(R, C, D, A, X);
relation3(Q, P, R);
```

(b) Input to the **CMPM**-prover

$$A, B$$

$$C = (1 - a'' - I\ b'')\ A + (a'' + I\ b'')\ B$$

$$O = -\frac{1}{2}\ \frac{(I\ b'^2 - b'' + I\ a'^2 - I\ a'')\ A}{b''} + \frac{1}{2}\ \frac{(b'' + I\ b'^2 - I\ a'' + I\ a'^2)\ B}{b''}$$

$$(1 + r'')\ X = B + r''\ C$$

$$F = -\frac{1}{1 + 2\ a''\ r'' + a'^2\ r'^2 + b'^2\ r'^2}(r''\ (-b'^2\ r'' + b'^2\ a''\ r'' - a'^2\ r'' + a'^3\ r'' + b'^2$$
$$+ a'^2 - a'' + I\ b''\ a'^2\ r'' + I\ b'^3\ r'' + I\ b'')\ A)$$
$$+ \frac{(1 + a''\ r'' + I\ b''\ r'')\ (a'^2\ r'' + b'^2\ r'' + 1)\ B}{1 + 2\ a''\ r'' + a'^2\ r'^2 + b'^2\ r'^2}$$

$$(1 + x'')\ Q = B + x''\ C$$

$$E = \frac{(a'^2\ r'' - a''\ r'' - a''\ x'' + I\ b''\ r'' + b'^2\ r'' - I\ b''\ x'' + x'')\ x''\ A}{a'^2\ r'^2 - 2\ a''\ r''\ x'' + b'^2\ r'^2 + x'^2}$$
$$- \frac{1}{a'^2\ r'^2 - 2\ a''\ r''\ x'' + b'^2\ r'^2 + x'^2}((-a'^2\ r'^2 + a'^2\ r''\ x'' + a''\ r''\ x'' - a''\ x'^2$$
$$- b'^2\ r'^2 + b'^2\ r''\ x'' + I\ b''\ r''\ x'' - I\ b''\ x'^2)\ B)$$

$$(1 + y'')\ P = A + y''\ B$$

$$D = -((-a'^2\ r'^2 + a'^2\ r''\ x''\ y'' + 2\ a'^2\ r''\ x'' - x'^2\ a'^2 - x'^2\ y''\ a'^2 - a''\ r''\ x''\ y''$$
$$+ a''\ x'^2\ y'' - b'^2\ r'^2 + I\ b''\ r''\ x''\ y'' + b'^2\ r''\ x''\ y'' + 2\ b'^2\ r''\ x'' - b'^2\ x'^2$$
$$- I\ b''\ x'^2\ y'' - b'^2\ x'^2\ y'')\ A)/(a'^2\ r'^2 - 2\ a'^2\ r''\ x''\ y'' - 2\ a'^2\ r''\ x''$$
$$+ 2\ x'^2\ y''\ a'^2 + x'^2\ a'^2 + a'^2\ x'^2\ y'^2 + 2\ a''\ r''\ x''\ y'' - 2\ a''\ x'^2\ y'^2 - 2\ a''\ x'^2\ y''$$
$$+ b'^2\ r'^2 - 2\ b'^2\ r''\ x'' - 2\ b'^2\ r''\ x''\ y'' + 2\ b'^2\ x'^2\ y'' + b'^2\ x'^2 + x'^2\ y'^2$$
$$+ y'^2\ b'^2\ x'^2) + (x''\ y''\ (-a'^2\ r'' + a'^2\ x'' + y''\ a'^2\ x'' + a''\ r'' - a''\ x''$$
$$- 2\ y''\ a''\ x'' + I\ b''\ r'' - b'^2\ r'' + b'^2\ x'' + b'^2\ y''\ x'' - I\ b''\ x'' + y''\ x'')\ B)/(a'^2\ r'^2$$
$$- 2\ a'^2\ r''\ x''\ y'' - 2\ a'^2\ r''\ x'' + 2\ x'^2\ y''\ a'^2 + x'^2\ a'^2 + a'^2\ x'^2\ y'^2 + 2\ a''\ r''\ x''\ y''$$
$$- 2\ a''\ x'^2\ y'^2 - 2\ a''\ x'^2\ y'' + b'^2\ r'^2 - 2\ b'^2\ r''\ x'' - 2\ b'^2\ r''\ x''\ y'' + 2\ b'^2\ x'^2\ y''$$
$$+ b'^2\ x'^2 + x'^2\ y'^2 + y'^2\ b'^2\ x'^2)$$

$$R = (A\ X) \cap (C\ D)$$

$$\frac{(-r'' + x'' + y''\ x'' + r''\ y''\ x'')\ R}{r'' - x''} = A - \frac{y''\ x''\ (1 + r'')\ X}{r'' - x''}$$

$$R = -\frac{(r'' - x'' - r''\ y''\ x'' + a''\ r''\ x''\ y'' + I\ r''\ y''\ x''\ b'')\ A}{-r'' + x'' + y''\ x'' + r''\ y''\ x''}$$
$$+ \frac{(1 + a''\ r'' + I\ b''\ r'')\ y''\ x''\ B}{-r'' + x'' + y''\ x'' + r''\ y''\ x''}$$

$$Q = \frac{(-x'' - y''\ x'' + r'' + r''\ y'')\ P}{y''\ r''\ (1 + x'')} + \frac{(-r'' + x'' + y''\ x'' + r''\ y''\ x'')\ R}{y''\ r''\ (1 + x'')}$$

(c) Machine proof for Pascal's Theorem on a Circle

Fig. 8. Pascal's Theorem on a Circle

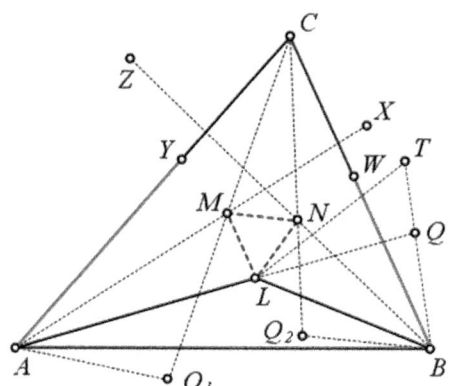

(a) Diagram for Morley's Theorem

```
Points(A, B);
Point1(L, A, B, a, b);
Symmetricpoint(X, B, A, L);
Symmetricpoint(Y, L, A, X);
Symmetricpoint(Z, A, B, L);
Symmetricpoint(W, L, B, Z);
Inter(C, A, Y, B, W);
Bratio(T, L, B, √3/2);
Inter(Q, A, L, B, T);
Similartrianglepoint1(Q1, C, A, Q, L, T);
Similartrianglepoint(Q2, C, B, Q, L, T);
Inter(M, A, X, C, Q1);
Inter(N, B, Z, C, Q2);
relation4(M, N, L, N);
```

(b) Input to the **CMPM**prover

Fig. 9. Morley's Theorem

$$A,\ B$$

$$L = (1 - \tilde{a} - I\ \tilde{b})\, A + (\tilde{a} + I\ \tilde{b})\, B$$

$$B = \frac{(-\tilde{a} + a^2 + b^2 + I\ \tilde{b})\, A}{a^2 + b^2} - \frac{(-\tilde{a} + I\ \tilde{b})\, L}{a^2 + b^2}$$

$$X = -\frac{(-a^2 - b^2 + \tilde{a} + I\ \tilde{b})\, A}{a^2 + b^2} + \frac{(\tilde{a} + I\ \tilde{b})\, L}{a^2 + b^2}$$

$$L = (-\tilde{a} + I\ \tilde{b} + 1)\, A + (\tilde{a} - I\ \tilde{b})\, X$$

$$Y = (1 - \tilde{a} - I\ \tilde{b})\, A + (\tilde{a} + I\ \tilde{b})\, X$$

$$A = -\frac{(-a^2 - b^2 + \tilde{a} + I\ \tilde{b})\, B}{1 - 2\ \tilde{a} + a^2 + b^2} + \frac{(-\tilde{a} + I\ \tilde{b} + 1)\, L}{1 - 2\ \tilde{a} + a^2 + b^2}$$

$$Z = \frac{(-\tilde{a} + a^2 + b^2 + I\ \tilde{b})\, B}{1 - 2\ \tilde{a} + a^2 + b^2} - \frac{(-1 + \tilde{a} + I\ \tilde{b})\, L}{1 - 2\ \tilde{a} + a^2 + b^2}$$

$$L = (\tilde{a} - I\ \tilde{b})\, B + (-\tilde{a} + I\ \tilde{b} + 1)\, Z$$

$$W = (\tilde{a} + I\ \tilde{b})\, B + (1 - \tilde{a} - I\ \tilde{b})\, Z$$

$$C = (A\ Y) \cap (B\ W)$$

$$\frac{1}{4}\ \frac{(6\ a^2\ b^2 - 6\ \tilde{a}\ b^2 + 3\ b^4 + 3\ a^2 - 6\ a^3 + 3\ a^4 - b^2)\, C}{b^2\ (a^2 - 2\ \tilde{a} + b^2)} = B$$

$$+\ \frac{1}{4}\ \frac{(1 - 2\ \tilde{a} + a^2 + b^2)\ (3\ a^2 - b^2)\, W}{b^2\ (a^2 - 2\ \tilde{a} + b^2)}$$

$$C = -\Big(\big((-1 + 3\ \tilde{a} - 3\ a^2 + 3\ b^2 + a^3 - 3\ \tilde{a}\ b^2 + 3\ I\ b^2 - 6\ I\ \tilde{a}\ \tilde{b} + 3\ I\ b\ a^2$$
$$-\ I\ b^3)\ (3\ a^2 - b^2)\, A\big)\big/\big(6\ a^2\ b^2 - 6\ \tilde{a}\ b^2 + 3\ b^4 + 3\ a^2 - 6\ a^3 + 3\ a^4$$
$$-\ b^2\big) + \big((18\ a^2\ b^2 - 9\ \tilde{a}\ b^2 + 3\ a^3 - 6\ a^4 + 3\ a^5 - 10\ a^3\ b^2 + 9\ I\ b\ a^2$$
$$-\ 18\ I\ a^3\ \tilde{b} + 9\ I\ \tilde{a}\ \tilde{b} - 6\ I\ a^2\ b^3 + 3\ b^4\ \tilde{a} - 3\ I\ b^3 + 6\ I\ b^3\ \tilde{a} + I\ b^5)\, B\big)$$
$$\big/\big(6\ a^2\ b^2 - 6\ \tilde{a}\ b^2 + 3\ b^4 + 3\ a^2 - 6\ a^3 + 3\ a^4 - b^2\big)$$

$$T = \left(\frac{1}{2} - \frac{1}{2}\ I\ \sqrt{3}\right) L + \left(\frac{1}{2} + \frac{1}{2}\ I\ \sqrt{3}\right) B$$

$$Q = (A\ L) \cap (B\ T)$$

$$\frac{(-\sqrt{3}\ \tilde{a} + \sqrt{3}\ a^2 - \tilde{b} + \sqrt{3}\ b^2)\, Q}{-\sqrt{3}\ \tilde{a} + \sqrt{3}\ a^2 + \tilde{b} + \sqrt{3}\ b^2} = B - \frac{2\ \tilde{b}\ T}{-\sqrt{3}\ \tilde{a} + \sqrt{3}\ a^2 + \tilde{b} + \sqrt{3}\ b^2}$$

$$Q = -\frac{(1 - \tilde{a} - \sqrt{3}\ \tilde{b} - I\ \tilde{b} - I\ \sqrt{3} + I\ \sqrt{3}\ \tilde{a})\ \tilde{b}\ A}{-\sqrt{3}\ \tilde{a} + \sqrt{3}\ a^2 - \tilde{b} + \sqrt{3}\ b^2}$$

$$+\ \frac{(-\tilde{b} + \sqrt{3}\ \tilde{a} - \sqrt{3})\ (\tilde{a} + I\ \tilde{b})\, B}{-\sqrt{3}\ \tilde{a} + \sqrt{3}\ a^2 - \tilde{b} + \sqrt{3}\ b^2}$$

$$Q = -\frac{1}{2}\ \big((3\ a^4 + 3\ I\ \sqrt{3}\ a^2 + 6\ I\ \sqrt{3}\ a^2\ b^2 - 6\ a^3 + 3\ a^2 - 6\ I\ \tilde{a}\ \sqrt{3}\ b^2$$
$$+\ 3\ I\ b^4\ \sqrt{3} + 6\ a^2\ b^2 - 6\ \tilde{a}\ b^2 + 3\ I\ \sqrt{3}\ a^4 - 6\ I\ \sqrt{3}\ a^3 + 3\ b^4 - b^2$$
$$-\ I\ \sqrt{3}\ b^2)\, L\big)\big/\big(-3\ a^4 + 6\ a^3 - 6\ a^2\ b^2 - 3\ a^2 + 2\ b^2\ \sqrt{3}\ a^2 - 2\ b^2\ \sqrt{3}\ \tilde{a}$$
$$+\ 6\ a^2\ b^2 + 2\ b^3\ \sqrt{3} - 3\ b^4 - b^2\big) + \frac{1}{2}\ \big((-a^4\ \sqrt{3} + 2\ \sqrt{3}\ a^3 - 4\ \tilde{a}\ \tilde{b}$$
$$-\ 2\ b^2\ \sqrt{3}\ a^2 - \sqrt{3}\ a^2 + 4\ a^2\ \tilde{b} + 2\ \tilde{a}\ \sqrt{3}\ b^2 - b^4\ \sqrt{3} - \sqrt{3}\ b^2 + 4\ b^3$$
$$-\ I\ b^2 + 3\ I\ b^4 - 6\ I\ \tilde{a}\ b^2 - 6\ I\ a^3 + 6\ I\ a^2\ b^2 + 3\ I\ a^2 + 3\ I\ a^4)\ \sqrt{3}\ T\big)\big/\big($$
$$-3\ a^4 + 6\ a^3 - 6\ a^2\ b^2 - 3\ a^2 + 2\ b^2\ \sqrt{3}\ a^2 - 2\ b^2\ \sqrt{3}\ \tilde{a} + 6\ a^2\ b^2$$
$$+\ 2\ b^3\ \sqrt{3} - 3\ b^4 - b^2\big)$$

$$Q_1 = \frac{1}{2}\ \big((3\ I\ \sqrt{3}\ a^2 + 6\ I\ \sqrt{3}\ a^2\ b^2 - 6\ I\ \tilde{a}\ \sqrt{3}\ b^2 + b^2 - 6\ a^2\ b^2 + 3\ I\ b^4\ \sqrt{3}$$
$$+\ 6\ \tilde{a}\ b^2 - 3\ a^4 - 3\ b^4 - 3\ a^2 + 6\ a^3 + 3\ I\ \sqrt{3}\ a^4 - 6\ I\ \sqrt{3}\ a^3 - I\ \sqrt{3}\ b^2)$$
$$C\big)\big/\big(-3\ a^4 + 6\ a^3 - 6\ a^2\ b^2 - 3\ a^2 + 2\ b^2\ \sqrt{3}\ a^2 - 2\ b^2\ \sqrt{3}\ \tilde{a} + 6\ a^2\ b^2$$
$$+\ 2\ b^3\ \sqrt{3} - 3\ b^4 - b^2\big) - \frac{1}{2}\ \big(\sqrt{3}\ (a^4\ \sqrt{3} - 2\ \sqrt{3}\ a^3 + 4\ \tilde{a}\ \tilde{b}$$
$$+\ 2\ b^2\ \sqrt{3}\ a^2 + \sqrt{3}\ a^2 - 4\ a^2\ \tilde{b} - 2\ \tilde{a}\ \sqrt{3}\ b^2 + b^4\ \sqrt{3} + \sqrt{3}\ b^2 - 4\ b^3$$
$$-\ I\ b^2 + 3\ I\ b^4 - 6\ I\ \tilde{a}\ b^2 - 6\ I\ a^3 + 6\ I\ a^2\ b^2 + 3\ I\ a^2 + 3\ I\ a^4)\, A\big)\big/\big($$
$$-3\ a^4 + 6\ a^3 - 6\ a^2\ b^2 - 3\ a^2 + 2\ b^2\ \sqrt{3}\ a^2 - 2\ b^2\ \sqrt{3}\ \tilde{a} + 6\ a^2\ b^2$$
$$+\ 2\ b^3\ \sqrt{3} - 3\ b^4 - b^2\big)$$

Fig. 10. Part 1 of the machine proof for Morley's Theorem

$$Q = -\frac{1}{2}\left(\left(3\,a^{\sim4} + 3\,I\,\sqrt{3}\,a^{\prime2} + 6\,I\,\sqrt{3}\,a^{\prime2}\,b^{\prime2} - 6\,a^{\sim3} + 3\,a^{\prime2} - 6\,I\,a^{\sim}\,\sqrt{3}\,b^{\prime2}\right.\right.$$
$$+ 3\,I\,b^{\sim4}\,\sqrt{3} + 6\,a^{\prime2}\,b^{\prime2} - 6\,a^{\sim}\,b^{\prime2} + 3\,I\,\sqrt{3}\,a^{\sim4} - 6\,I\,\sqrt{3}\,a^{\sim3} + 3\,b^{\sim4} - b^{\prime2}$$
$$\left.- I\,\sqrt{3}\,b^{\prime2}\right)\,L\right)/\left(-3\,a^{\sim4} + 6\,a^{\sim3} - 6\,a^{\prime2}\,b^{\prime2} - 3\,a^{\prime2} + 2\,b^{\sim}\,\sqrt{3}\,a^{\prime2} - 2\,b^{\sim}\,\sqrt{3}\,a^{\sim}\right.$$
$$\left.+ 6\,a^{\sim}\,b^{\prime2} + 2\,b^{\sim3}\,\sqrt{3} - 3\,b^{\sim4} - b^{\prime2}\right) + \frac{1}{2}\left(\left(-a^{\sim4}\,\sqrt{3} + 2\,\sqrt{3}\,a^{\sim3} - 4\,a^{\sim}\,b^{\sim}\right.\right.$$
$$- 2\,b^{\prime2}\,\sqrt{3}\,a^{\prime2} - \sqrt{3}\,a^{\prime2} + 4\,a^{\prime2}\,b^{\sim} + 2\,a^{\sim}\,\sqrt{3}\,b^{\prime2} - b^{\sim4}\,\sqrt{3} - \sqrt{3}\,b^{\prime2} + 4\,b^{\sim3}$$
$$\left.- I\,b^{\prime2} + 3\,I\,b^{\sim4} - 6\,I\,a^{\sim}\,b^{\prime2} - 6\,I\,a^{\sim3} + 6\,I\,a^{\prime2}\,b^{\prime2} + 3\,I\,a^{\prime2} + 3\,I\,a^{\sim4}\right)\,\sqrt{3}\,T\right)/\left(\right.$$
$$-3\,a^{\sim4} + 6\,a^{\sim3} - 6\,a^{\prime2}\,b^{\prime2} - 3\,a^{\prime2} + 2\,b^{\sim}\,\sqrt{3}\,a^{\prime2} - 2\,b^{\sim}\,\sqrt{3}\,a^{\sim} + 6\,a^{\sim}\,b^{\prime2}$$
$$\left.+ 2\,b^{\sim3}\,\sqrt{3} - 3\,b^{\sim4} - b^{\prime2}\right)$$

$$Q2 = -\frac{1}{2}\left(\left(3\,a^{\sim4} + 3\,I\,\sqrt{3}\,a^{\prime2} + 6\,I\,\sqrt{3}\,a^{\prime2}\,b^{\prime2} - 6\,a^{\sim3} + 3\,a^{\prime2} - 6\,I\,a^{\sim}\,\sqrt{3}\,b^{\prime2}\right.\right.$$
$$+ 3\,I\,b^{\sim4}\,\sqrt{3} + 6\,a^{\prime2}\,b^{\prime2} - 6\,a^{\sim}\,b^{\prime2} + 3\,I\,\sqrt{3}\,a^{\sim4} - 6\,I\,\sqrt{3}\,a^{\sim3} + 3\,b^{\sim4} - b^{\prime2}$$
$$\left.- I\,\sqrt{3}\,b^{\prime2}\right)\,c\right)/\left(-3\,a^{\sim4} + 6\,a^{\sim3} - 6\,a^{\prime2}\,b^{\prime2} - 3\,a^{\prime2} + 2\,b^{\sim}\,\sqrt{3}\,a^{\prime2} - 2\,b^{\sim}\,\sqrt{3}\,a^{\sim}\right.$$
$$\left.+ 6\,a^{\sim}\,b^{\prime2} + 2\,b^{\sim3}\,\sqrt{3} - 3\,b^{\sim4} - b^{\prime2}\right) + \frac{1}{2}\left(\left(-a^{\sim4}\,\sqrt{3} + 2\,\sqrt{3}\,a^{\sim3} - 4\,a^{\sim}\,b^{\sim}\right.\right.$$
$$- 2\,b^{\prime2}\,\sqrt{3}\,a^{\prime2} - \sqrt{3}\,a^{\prime2} + 4\,a^{\prime2}\,b^{\sim} + 2\,a^{\sim}\,\sqrt{3}\,b^{\prime2} - b^{\sim4}\,\sqrt{3} - \sqrt{3}\,b^{\prime2} + 4\,b^{\sim3}$$
$$\left.- I\,b^{\prime2} + 3\,I\,b^{\sim4} - 6\,I\,a^{\sim}\,b^{\prime2} - 6\,I\,a^{\sim3} + 6\,I\,a^{\prime2}\,b^{\prime2} + 3\,I\,a^{\prime2} + 3\,I\,a^{\sim4}\right)\,\sqrt{3}\,B\right)/\left(\right.$$
$$-3\,a^{\sim4} + 6\,a^{\sim3} - 6\,a^{\prime2}\,b^{\prime2} - 3\,a^{\prime2} + 2\,b^{\sim}\,\sqrt{3}\,a^{\prime2} - 2\,b^{\sim}\,\sqrt{3}\,a^{\sim} + 6\,a^{\sim}\,b^{\prime2}$$
$$\left.+ 2\,b^{\sim3}\,\sqrt{3} - 3\,b^{\sim4} - b^{\prime2}\right)$$

$$M = (A\ X) \cap (C\ Q1)$$

$$\left(\left(6\,a^{\sim3}\,\sqrt{3}\,b^{\prime2} - 10\,b^{\prime2}\,\sqrt{3}\,a^{\prime2} + 3\,a^{\sim}\,\sqrt{3}\,b^{\sim4} + 3\,\sqrt{3}\,a^{\sim3} - 6\,a^{\sim4}\,\sqrt{3} + 3\,a^{\sim5}\,\sqrt{3}\right.\right.$$
$$+ 3\,a^{\sim}\,\sqrt{3}\,b^{\prime2} + 6\,a^{\prime2}\,b^{\sim3} - 6\,b^{\sim3}\,a^{\sim} + 3\,b^{\sim5} + 3\,a^{\prime2}\,b^{\sim} - 6\,b^{\sim}\,a^{\sim3} + 3\,a^{\sim4}\,b^{\sim}$$
$$\left.+ 3\,b^{\sim3} - 4\,b^{\sim4}\,\sqrt{3}\right)\,M\right)/\left(\left(\sqrt{3}\,a^{\sim} - b^{\sim}\right)\left(6\,a^{\prime2}\,b^{\prime2} - 6\,a^{\sim}\,b^{\prime2} + 3\,b^{\sim4} + 3\,a^{\prime2} - 6\,a^{\sim3}\right.\right.$$
$$\left.\left.+ 3\,a^{\sim4} - b^{\prime2}\right)\right) = C - \left(2\left(-3\,a^{\sim4} + 6\,a^{\sim3} - 6\,a^{\prime2}\,b^{\prime2} - 3\,a^{\prime2} + 2\,b^{\sim}\,\sqrt{3}\,a^{\prime2}\right.\right.$$
$$- 2\,b^{\sim}\,\sqrt{3}\,a^{\sim} + 6\,a^{\sim}\,b^{\prime2} + 2\,b^{\sim3}\,\sqrt{3} - 3\,b^{\sim4} - b^{\prime2}\right)\,b^{\sim}\,Q1\Big)/\left(\left(\sqrt{3}\,a^{\sim}\right.\right.$$
$$\left.\left.- b^{\sim}\right)\left(6\,a^{\prime2}\,b^{\prime2} - 6\,a^{\sim}\,b^{\prime2} + 3\,b^{\sim4} + 3\,a^{\prime2} - 6\,a^{\sim3} + 3\,a^{\sim4} - b^{\prime2}\right)\right)$$

$$M = -\left(\left(3\,a^{\sim4}\,\sqrt{3} - 12\,I\,a^{\prime2}\,\sqrt{3}\,b^{\sim} - 9\,\sqrt{3}\,a^{\sim3} - 4\,b^{\prime2}\,\sqrt{3}\,a^{\prime2} + 9\,\sqrt{3}\,a^{\prime2}\right.\right.$$
$$- 2\,I\,a^{\sim}\,\sqrt{3}\,b^{\sim3} - 3\,a^{\prime2}\,b^{\sim} - 3\,\sqrt{3}\,a^{\sim} + 6\,I\,a^{\sim3}\,\sqrt{3}\,b^{\sim} + 6\,I\,a^{\sim}\,\sqrt{3}\,b^{\sim} + 6\,a^{\sim}\,b^{\sim}$$
$$+ 3\,a^{\sim}\,\sqrt{3}\,b^{\prime2} + b^{\sim4}\,\sqrt{3} - 3\,b^{\sim3} - 3\,b^{\sim} + \sqrt{3}\,b^{\prime2}\right)\,A\Big)/\left(3\,\sqrt{3}\,a^{\sim3} - 6\,\sqrt{3}\,a^{\prime2}\right.$$
$$\left.+ 3\,a^{\prime2}\,b^{\sim} + 3\,\sqrt{3}\,a^{\sim} + 3\,a^{\sim}\,\sqrt{3}\,b^{\prime2} - 6\,a^{\sim}\,b^{\sim} + 3\,b^{\sim3} + 3\,b^{\sim} - 4\,\sqrt{3}\,b^{\prime2}\right)$$
$$^{\prime2}\,b^{\sim} + 3\,\sqrt{3}\,a^{\sim} + 3\,a^{\sim}\,\sqrt{3}\,b^{\prime2} - 6\,a^{\sim}\,b^{\sim} + 3\,b^{\sim3} + 3\,b^{\sim} - 4\,\sqrt{3}\,b^{\prime2}\right)$$

$$N = (B\ Z) \cap (C\ Q2)$$

$$\left(\left(\sqrt{3}\,b^{\prime2} + a^{\sim}\,\sqrt{3}\,b^{\prime2} - 8\,b^{\prime2}\,\sqrt{3}\,a^{\prime2} + 3\,a^{\sim}\,\sqrt{3}\,b^{\sim4} + 6\,a^{\sim3}\,\sqrt{3}\,b^{\prime2} + b^{\sim4}\,\sqrt{3}\right.\right.$$
$$+ 6\,b^{\sim3}\,a^{\sim} + 9\,\sqrt{3}\,a^{\sim3} + 3\,a^{\sim5}\,\sqrt{3} - 3\,\sqrt{3}\,a^{\prime2} - 9\,a^{\sim4}\,\sqrt{3} - 3\,b^{\sim3} - 3\,b^{\sim5}$$
$$\left.- 6\,a^{\prime2}\,b^{\sim3} - 3\,a^{\sim4}\,b^{\sim} - 3\,a^{\prime2}\,b^{\sim} + 6\,b^{\sim}\,a^{\sim3}\right)\,N\right)/\left(\left(\sqrt{3}\,a^{\sim} + b^{\sim} - \sqrt{3}\right)\left(6\,a^{\prime2}\,b^{\prime2}\right.\right.$$
$$\left.\left.- 6\,a^{\sim}\,b^{\prime2} + 3\,b^{\sim4} + 3\,a^{\prime2} - 6\,a^{\sim3} + 3\,a^{\sim4} - b^{\prime2}\right)\right) = C + \left(2\left(-3\,a^{\sim4} + 6\,a^{\sim3}\right.\right.$$
$$- 6\,a^{\prime2}\,b^{\prime2} - 3\,a^{\prime2} + 2\,b^{\sim}\,\sqrt{3}\,a^{\prime2} - 2\,b^{\sim}\,\sqrt{3}\,a^{\sim} + 6\,a^{\sim}\,b^{\prime2} + 2\,b^{\sim3}\,\sqrt{3} - 3\,b^{\sim4}$$
$$\left.- b^{\prime2}\right)\,b^{\sim}\,Q2\Big)/\left(\left(\sqrt{3}\,a^{\sim} + b^{\sim} - \sqrt{3}\right)\left(6\,a^{\prime2}\,b^{\prime2} - 6\,a^{\sim}\,b^{\prime2} + 3\,b^{\sim4} + 3\,a^{\prime2} - 6\,a^{\sim3}\right.\right.$$
$$\left.\left.+ 3\,a^{\sim4} - b^{\prime2}\right)\right)$$

$$N = -\frac{\left(1 - 2\,a^{\sim} + a^{\prime2} - b^{\prime2} - 2\,I\,b^{\sim} + 2\,I\,b^{\sim}\,a^{\sim}\right)\sqrt{3}\,\left(3\,a^{\prime2} - b^{\prime2}\right)\,A}{3\,\sqrt{3}\,a^{\sim3} - 3\,\sqrt{3}\,a^{\prime2} - 3\,a^{\prime2}\,b^{\sim} + 3\,a^{\sim}\,\sqrt{3}\,b^{\prime2} + \sqrt{3}\,b^{\prime2} - 3\,b^{\sim3}} + \left(\left(3\,a^{\sim4}\,\sqrt{3}\right.\right.$$
$$+ 6\,I\,a^{\sim3}\,\sqrt{3}\,b^{\sim} - 3\,\sqrt{3}\,a^{\sim3} - 4\,b^{\prime2}\,\sqrt{3}\,a^{\prime2} - 3\,a^{\prime2}\,b^{\sim} - 6\,I\,\sqrt{3}\,a^{\prime2}\,b^{\sim}$$
$$\left.- 2\,I\,a^{\sim}\,\sqrt{3}\,b^{\sim3} + 5\,a^{\sim}\,\sqrt{3}\,b^{\prime2} + b^{\sim4}\,\sqrt{3} - 3\,b^{\sim3} + 2\,I\,\sqrt{3}\,b^{\sim3}\right)\,B\right)/\left(3\,\sqrt{3}\,a^{\sim3}\right.$$
$$\left.- 3\,\sqrt{3}\,a^{\prime2} - 3\,a^{\prime2}\,b^{\sim} + 3\,a^{\sim}\,\sqrt{3}\,b^{\prime2} + \sqrt{3}\,b^{\prime2} - 3\,b^{\sim3}\right)$$

$$L - N = \left(\frac{1}{2} + \frac{1}{2}\,I\,\sqrt{3}\right)(M - N)$$

$$L\ N^2 = M\ N^2$$

Fig. 11. Part 2 of the machine proof for Morley's Theorem

6   Experimental Results and Conclusions

In this section we summarize our experimental results and state conclusions.

6.1   Experimental Results

Statistics for the Proving Time. We have run 110 geometry theorems of class **CH** by our **MPM**prover and 240 nontrivial geometry theorems of class **CL** (not including statements of class **CH**) by our **CMPM**prover in *Maple* 12 on an *Intel Pentium Dual-Core E5200 Machine.* Table 3 and Table 4 contain some information about the proving time of the 350 theorems.

Table 3. Statistics for the proving time of 110 theorems of class **CH**

Proving time(s)	Number of theorems	Percentage(%)
<0.1	88	80.0
<0.2	101	91.8
<0.5	109	99.1
<1.1	110	100.0

Table 4. Statistics for the proving time of 240 theorems of class **CL**

Proving time(s)	Number of theorems	Percentage(%)
≤ 0.1	49	20.4
≤ 0.2	103	42.9
≤ 0.3	144	60.0
≤ 0.5	200	83.3
≤ 1.0	227	94.6
≤ 2.0	236	98.3
<100.0	240	100.0

It can be seen from Table 3 and Table 4 that our provers are efficient.

For statements of class **CH**, it is obviously applicable for the **CMPM**prover to generate the machine proofs, since **CH** is a subset of **CL**. But, for a certain statement of class **CH**, the machine proof produced by the **CMPM**prover is different from that produced by the **MPM**prover in some aspects which we won't go into details for the limitation of pages. Usually, if a statement is of class **CH**, we often use the **MPM**prover to prove it, because the **MPM**prover is made for class **CH**.

Features of the Machine Proof. It can also be seen from the running examples in the paper that the machine proofs produced by our provers are indeed human-readable and easily understood by a mathematician. The readability is reflected in the following aspects:

- *Clear and concise*: The mass point relations are generated one by one according to the order in which points are introduced. For each mass point relation, the capital letters denote geometric points, while the coefficients of points are numbers, independent variables or rational expressions in lowercase letters. There are also signs telling what is going on, and at last the conclusions are checked and output.
- *Easy to understand*: All the mass point relations can be understood as long as one understands the basic propositions of mass point geometry, although the proofs generated by our provers are obviously different from the traditional proof. To read the proof, we never need to remember any complex formulae.
- *With clear geometric meaning*: Each expression in a proof has clear and intuitive geometrical meaning, although some of them may involve seemingly huge expressions. Because the statements proved by the mass point method are of the unordered geometry problems, it is not necessary to resort to diagrams when reading the proof.
- *Rich information included*: In addition to the final conclusion, the proof contains rich non-trivial information related to the geometry statement. The quantitative relationship implied in each non-trivial mass point relation is also a certain geometry statement.

6.2   Conclusions and Future Work

Conclusions. In this paper we first gave a detailed description of the mass point method, a new method for automated theorem proving in geometry. The method can efficiently prove many non-trivial theorems of affine geometry.

The algorithm of the mass point method is much more concise and simpler than that of the area method. The algorithm of area method includes a lot of eliminating point formulae because it can only deal with geometric quantities other than the geometric point itself, so when eliminating the same point from different quantities one has to use different formulae, while using the mass point method, one eliminates points more freely, i.e., only eliminates those undesired points at each step other than eliminates points one by one in reverse order.

We made an implementation of the algorithm in *Maple*, called **MPM**prover, which can generate machine proof entirely mechanically for statements of affine gemoentry. Theoretically, it is applicable for the mass point method described in this paper to prove other kinds of geometry statements, not restricted to affine geometry statements, given that one knows all points' area coordinates of a statement. How to implement this idea automatically is still under development.

To deal with metric geometry problems, we then developed the complex mass point geometry, and gave a detailed description of the complex mass point method which can efficiently prove many difficult statements of metric geometry. The algorithm **CMPM**, which is based on the complex mass point method, is very similar to, but still different from the algorithm **MPM**. We also made an implementation of the algorithm **CMPM** in *Maple*, called **CMPM**prover. Many difficult theorems, which would cost the area method much time or could not be proved within a reasonable time, can be proved by the **CMPM**prover efficiently, for instance, the time cost for proving Morley's Triangle Theorem is less than 2 seconds.

The algebraic methods in automated reasoning in geometry root in the work of Descartes and in the translation of geometric problems to algebraic problems, thus deal with polynomials that are often extremely complex for a human to understand, and also with no direct link to the geometric contents. However, the (complex) mass point method takes points as geometric objects and only pays attention to the quantitative relationship of geometric points. In the machine proof generated by the **CMPM**prover or **MPM**prover, though there may be large expressions, each mass point relation has clear geometrical meaning, since all the large expressions are just the coefficients of certain points of certain mass point relations.

Future Work. The two mass point methods described in the two parts of this series provide us a new tool for automated theorem proving in geometry. Ongoing work and future work include:

- Improving the **MPM**prover's and the **CMPM**prover's ability to solve problems such that they can deal with more geometry statements which are not of class **CH** or **CL**;
- Continuing to develop the (complex) mass point geometry such that the statements of solid geometry or higher-dimensional geometry could be covered;
- Continuing to develop or revisiting the existing machine proving methods based on the mass point method, such as the area method, the vector method, the complex number method, the forward search method, etc.

References

1. Zhang, J.Z., Li, Y.B.: Automaic theorem proving for three decades. Journal of Systems Science and Mathematical Sciences (J. Sys. Sci. & Math. Scis.) 29, 1155–1168 (2009)
2. Wu, W.T.: On the Decision Problem and The Mechanization of Theorem Proving in Elemengtary Geometry. In: Automated Theorem Proving: After 25 years, vol. 29, pp. 213–214. American Mathematical Society (1984)
3. Chou, S.C.: Proving and dicoversing geometry theorems using Wu's method. Ph.D. thsis. The University of Texas, Austin (1985)
4. Wang, D.: Reasoning about geometric problems using an elimination method. In: Pfalzgraf, J., Wang, D. (eds.) Automated Practical Reasoning, pp. 147–185. Springer, New York (1995)
5. Kapur, D.: Using Gröbner bases to reason about geometry problems. Journal of Symbolic Computation 2, 399–408 (1986)
6. Chou, S.C.: Automated reasoning in geometries using the characteristic set method and Gröbner basis method. In: Proc. ISSAC 1990, Tokyo, pp. 255–260 (1990)
7. Li, H.: Clifford algebra approaches to mechanical geometry theorem proving. In: Gao, X.S., Wang, D. (eds.) Mathematics Mechanization and Applications, pp. 205–299. Academic Press, San Diego (2000)

8. Hong, J.: Can we prove geometry theorems by computing an example? Sci. Sinica. 29, 824–834 (1986)
9. Yang, L., Zhang, J.Z., Li, C.Z.: A prover for parallel numerical verification of a class of constructive geometry theorems. In: Proc. IWMM 1992, pp. 244–255. Inter. Academic Publishers, Beijing (1992)
10. Chou, S.C., Gao, X.S., Zhang, J.Z.: Automated production of traditional proofs for constructive geometry theorems. In: Vardi, M. (ed.) Proceedings of the 8th Annual IEEE Symposium on Logic in Computer Science LICS, Montreal, pp. 48–56 (1993)
11. Chou, S.C., Gao, X.S., Zhang, J.Z.: Machine proofs in geometry: Automated production of readable proofs for geometry theorems. World Scientific, Singapore (1994)
12. Chou, S.C., Gao, X.S., Zhang, J.Z.: A deductive database approach to automated geometry theorem proving and discovering. Journal of Automated Reasoning 25, 219–246 (1996)
13. Chou, S.C., Gao, X.S., Zhang, J.Z.: Mechanical theorem proving by vector calculation. In: Proc ISSAC 1993, Keiv, pp. 284–291 (1993)
14. Chou, S.C., Gao, X.S., Zhang, J.Z.: Automated production of readable proofs with geometric invariants, theorem proving with full-angles. Journal of Automated Reasoning 17, 349–370 (1996)
15. Li, H., Hestenes, D.: Rockwood A. Generalized homogeneous coordinates for computational geometry. In: Sommer, G. (ed.) Geometric Computing with Clifford Algebra, pp. 27–60. Springer, Heidelberg (2000)
16. Li, H.: Automated Geometric Theorem Proving, Clifford bracket algebra and Clifford expansions. In: Trends in Mathematics: Advances in Analysis and Geometry, pp. 345–363. Birkhäuser, Basel (2004)
17. Li, H.: Symbolic computation in the homogeneous geometric model with Clifford algebra. In: Gutierrez, J. (ed.) Proc. ISSAC 2004, pp. 221–228. ACM Press (2004)
18. Li, H.: Invariant algebras and geometric reasoning. World Scientific, Singapore (2008)
19. Rhoad, R., Milauskas, G., Whipple, R.: Geometry for Enjoyment and Challenge. McDougal, Littell & Company (1991)
20. Pedoe, D.: Notes on the History of Geometrical Ideas I: Homogeneous Coordinates. Math. Magazine, 215–217 (1975)
21. Tom, R.: Mass point geometry,
    http://mathcircle.berkeley.edu/archivedocs/2007_2008/lectures/0708lecturesps/MassPointsBMC07.ps
22. Tom, R.: Mass point geometry (Barycentirc Coordinates),
    http://mathcircle.berkeley.edu/archivedocs/1999_2000/lectures/9900lecturespdf/mpgeo.pdf
23. Coxeter, H.S.M.: Introduction to Geometry, pp. 216–221. John Wiley & Sons, Inc. (1969)

Author Index